Field Theories in
Condensed Matter Physics

Field Theories in Condensed Matter Physics

Edited by

Sumathi Rao

*Harish-Chandra Research Institute
Allahabad*

Institute of Physics Publishing
Bristol and Philadelphia

© 2001 Hindustan Book Agency (India)

No part of the material protected by this copyright notice may be reproduced or utilized in any form or by any means, electronic or mechanical, including photocopying, recording or by any information storage and retrieval system, without written permission from the copyright owner, who has also the sole right to grant licences for translation into other languages and publication thereof.

British Library Cataloguing-in-Publication Data

A catalogue record for this book is available from the British Library.

ISBN 0 7503 0876 1

Library of Congress Cataloging-in-Publication Data are available

Authorised edition by IOP Publishing Ltd for exclusive distribution outside India

This edition published 2002

Published by Institute of Physics Publishing, wholly owned by The Institute of Physics, London

Institute of Physics Publishing, Dirac House, Temple Back, Bristol BS1 6BE, UK

US Office: Institute of Physics Publishing, The Public Ledger Building, Suite 1035, 150 South Independence Mall West, Philadelphia, PA 19106, USA

Printed in India

Related titles from Institute of Physics Publishing

Path Integrals in Physics
Volume I: Stochastic Processes and Quantum Mechanics
Volume II: Quantum Field Theory, Statistical Physics and other Modern Applications
M Chaichian and A Demichev

Geometry, Topology and Physics
M Nakahara

Symmetries in Quantum Mechanics: from Angular Momentum to Supersymmetry
M Chaichian and R Hagedorn

Problems on Statistical Mechanics
D A R Dalvit, J Frastai and I D Lawrie

Soft and Fragile Matter: Nonequilibrium Dynamics, Metastability and Flow
Edited by M E Cates and M R Evans

Statistical Mechanics: Fundamentals and Model Solutions
T Dorlas

Quantum Dynamics of Simple Systems
Edited by G-L Oppo, S M Barnett, E Riis and M Wilkinson

Quantum Mechanics: An Introduction for Device Physicists and Electrical Engineers, Second Edition
D Ferry

Contents

Preface   xiii

Introduction   1

1 Quantum Many Particle Physics   7
Pinaki Majumdar

     1.1 Preamble 8
     1.2 Introduction 8
     1.3 Introduction to many particle physics 10
         1.3.1 Phases of many particle systems 10
         1.3.2 Quantities of physical interest 12
         1.3.3 Fermi and Bose liquids 14
     1.4 Phase transitions and broken symmetry 22
         1.4.1 Phase transitions and symmetry breaking . 22
         1.4.2 Symmetry breaking and interactions in BEC   25
     1.5 Normal Fermi systems: model problems 31
         1.5.1 Neutral fermions: dilute hardcore Fermi gas   34
         1.5.2 Charged fermions: the electron gas 39
     1.6 Electrons and phonons: Migdal-Eliashberg theory . 46
         1.6.1 Weak coupling theory: BCS 48
         1.6.2 The normal state: Migdal theory 52
         1.6.3 BCS theory: Greens function approach ... 56
         1.6.4 Superconductivity: Eliashberg theory ... 58
     1.7 Conclusion: 'field theory' and many particle physics   63

2 Critical Phenomena   69
Somendra M. Bhattacharjee

- 2.1 Preamble 70
  - 2.1.1 Large system: Thermodynamic limit 72
- 2.2 Where is the problem? 72
- 2.3 Recapitulation - A few formal stuff 74
  - 2.3.1 Extensivity 74
  - 2.3.2 Convexity: Stability 76
- 2.4 Consequences of divergence 78
- 2.5 Generalized scaling 81
  - 2.5.1 One variable: Temperature 81
  - 2.5.2 Solidarity with thermodynamics 87
  - 2.5.3 More variables: Temperature and field ... 88
  - 2.5.4 On exponent relations 92
- 2.6 Relevance, irrelevance and universality 93
- 2.7 Digression 95
  - 2.7.1 A first-order transition: $\alpha=1$ 95
  - 2.7.2 Example: Polymers : no "ordering" 97
- 2.8 Exponents and correlations 99
  - 2.8.1 Correlation function 99
  - 2.8.2 Relations among the exponents 101
  - 2.8.3 Length-scale dependent parameters 103
- 2.9 Models as examples: Gaussian and ϕ^4 105
  - 2.9.1 Specific heat for the Gaussian model 106
  - 2.9.2 Cut-off and anomalous dimensions 107
  - 2.9.3 Through correlations 110
- 2.10 Epilogue 112

3 Phase Transitions and Critical Phenomena   119
Deepak Kumar

- 3.1 Introduction 120
- 3.2 Thermodynamic stability 121
- 3.3 Lattice gas : mean field approximation 126
- 3.4 Landau theory 134

3.5	Spatial correlations	138
3.6	Breakdown of mean field theory	141
3.7	Ginzburg-Landau free energy functional	143
3.8	Renormalisation group (RG)	144
3.9	RG for a one dimensional Ising chain	146
3.10	RG for a two-dimensional Ising model	150
3.11	General features of RG	158
	3.11.1 Irrelevant variables	163
3.12	RG scaling for correlation functions	164
3.13	RG for Ginzburg-Landau model	167
	3.13.1 Tree-level approximation	170
	3.13.2 Critical exponents for $d > 4$	172
	3.13.3 Anomalous dimensions	175
3.14	Perturbation series for $d < 4$	176
3.15	Generalisation to a n-component model	183

4 Topological Defects — 189
Ajit M. Srivastava

4.1	The subject of topological defect	191
4.2	What is a topological defect?	193
	4.2.1 Meaning of order parameter	194
	4.2.2 Spontaneous symmetry breakdown(SSB)	195
	4.2.3 SSB in particle physics	197
	4.2.4 Order parameter space	197
4.3	The domain wall	198
	4.3.1 Why defect?	200
	4.3.2 Why topological?	201
	4.3.3 Energy considerations	202
4.4	Examples of topological defects	203
4.5	Condensed matter versus particle physics	209
4.6	Detailed understanding of a topological defect	213
	4.6.1 Free homotopy of maps	216
	4.6.2 Based homotopy and the fundamental group	217
4.7	Classification of defects using homotopy groups	219
4.8	Defect structure in liquid crystals	227

	4.8.1 Defects in nematics	228
	4.8.2 Non abelian π_1 - biaxial nematics	230
4.9	Formation of topological defects	231

5 Introduction to Bosonization   239
Sumathi Rao and Diptiman Sen

5.1	Fermi and Luttinger liquids	240
5.2	Bosonization	247
	5.2.1 Bosonization of a fermion with one chirality	248
	5.2.2 Bosonisation with two chiralities	257
	5.2.3 Field theory near the Fermi momenta . . .	265
5.3	Correlation functions and dimensions of operators	268
5.4	RG analysis of perturbed models	272
5.5	Applications of bosonization	281
5.6	Quantum antiferromagnetic spin 1/2 chain	282
5.7	Hubbard model	300
5.8	Transport in a Luttinger liquid - clean wire	309
5.9	Transport in the presence of isolated impurities . .	319
5.10	Concluding remarks	328

6 Quantum Hall Effect   335
R. Rajaraman

6.1	Classical Hall effect	336
6.2	Quantized Hall effect	337
6.3	Landau problem	338
6.4	Degeneracy counting	340
6.5	Laughlin wavefunction	341
6.6	Plasma analogy	342
6.7	Quasi-holes and their Laughlin wavefunction . . .	344
6.8	Localization physics and the QH plateaux	345
6.9	Chern-Simons theory	348
6.10	Vortices in the CS field and quasiholes	354
6.11	Jain's theory of composite fermions	355

7 Low-dimensional Quantum Spin Systems 359
Indrani Bose

7.1 Introduction 360
7.2 Ground and excited states 365
7.3 Theorems and rigorous results for antiferromagnets 369
    7.3.1 Lieb-Mattis theorem 369
    7.3.2 Marshall's sign rule 370
    7.3.3 Lieb, Schultz and Mattis theorem 372
    7.3.4 Mermin-Wagner theorem 376
7.4 Possible ground states and excitation spectra . . . 376
7.5 The Bethe Ansatz 387

Preface

The idea of holding an SERC school on 'Field theories in condensed matter physics' came through discussions with Diptiman Sen, from the Centre for Theoretical Studies in Bangalore, Somendra Bhattacharjee from the Institute of Physics in Bhubaneswar and R. Shankar from the Institute of Mathematical Sciences in Chennai. All of us felt that field theoretic methods are widely used in studying a variety of phenomena in condensed matter physics; so it would be a good idea to expose graduate students from different parts of the country to these methods. After a few rounds of discussions, we converged on a set of topics and lecturers, and then I wrote up a proposal to the Department of Science and Technology based on this, which got approved.

The school was held at the Harish-Chandra Research Institute (formerly Mehta Research Institute of Mathematics and Mathematical Sciences) from February 13 to March 4, 2000. All the lecturers we contacted proved enthusiastic about the idea and came for the school fully prepared for their courses. With over forty participants and twelve lecturers from all over the country, it turned out to be a great experience, where the lecturers as well as the students benefited from the interaction. The success of such a school depends mainly on the students; I must thank them for their whole-hearted participation during the school, which was very encouraging for the lecturers and for their warm and genuine feedback. Most of the lecturers agreed to write up their lecture notes for publication and this book is the result of their cooperation. My thanks to all of them. I also thank Peali Majumdar, Dileep Jatkar and Radhika Vathsan for their help with the preparation of the lecture notes for publication.

It goes without saying that the organisation of such a school is impossible without the efforts and cooperation of several people. Thanks are definitely due to my two condensed matter colleagues, Dattu Gaitonde and Pinaki Majumdar, who were with me right from the beginning. I also received a lot of practical organisational help from our students and post-doctoral fellows, Pradeep Mohanty, Saurabh Bannerjee, Sourin Das, Tribikram Gupta and Sanjeev Kumar. Special thanks are also due to our highly efficient administrative staff, who ensured the smooth running of the school. I would also like to thank the rest of the academic members of the institute for participating in the school; in particular, I would like to thank Ashoke Sen for constant encouragement and for attending all the lectures with great interest. Finally, I must thank Prof. H.S. Mani, the director of the institute who was supportive of the school, right from its inception to its culmination now in the form of this book.

List of Lecturers

1. Prof. G. Baskaran, *Institute of Mathematical Sciences, Chennai*

2. Prof. Somendra M. Bhattacharjee, *Institute of Physics, Bhubaneswar*

3. Prof. Indrani Bose, *J. C. Bose Institute, Kolkata*

4. Prof. P. Durganandini, *Dept. of Physics, Pune University, Pune*

5. Prof. Deepak Kumar, *Jawaharlal Nehru University, New Delhi*

6. Prof. Pinaki Majumdar, *Harish-Chandra Research Institute, Allahabad*

7. Prof. R. Rajaraman, *Jawaharlal Nehru University, New Delhi*

8. Prof. T. V. Ramakrishnan, *Indian Institute of Science, Bangalore*

9. Prof. Sumathi Rao, *Harish-Chandra Research Institute, Allahabad*

10. Prof. R. Shankar, *Institute of Mathematical Sciences, Chennai*

11. Prof. Diptiman Sen, *Indian Institute of Science, Bangalore*

12. Prof. Ajit Srivastava, *Institute of Physics, Bhubaneswar*

List of Participants

1. Mr. Atul Bharadwaj, *Delhi University, New Delhi*

2. Ms. Sarmishtha Bose, *Indian Statistical Institute, Kolkata*

3. Mr. Saugata Bhattacharyya, *Indian Association for the Cultivation of Science, Kolkata*

4. Dr. Sourabh Banerjee, *Harish-Chandra Research Institute, Allahabad*

5. Mr. Subhashish Banerjee, *Jawaharlal Nehru University, New Delhi*

6. Mr. Anirban Chakraborti, *Saha Institute of Nuclear Physics, Kolkata*

7. Mr. Indrajit Chakravarty, *Harish-Chandra Research Institute, Allahabad*

8. Ms. Emily Chattopadhyay, *J. C. Bose Institute, Kolkata*

9. Ms. Jayita Chatterjee, *Saha Institute of Nuclear Physics, Kolkata*

10. Mr. Amit Dutta, *Saha Institute of Nuclear Physics, Kolkata*

11. Mr. E. Krishna Das, *MG University, Kottayam*

12. Mr. Sourin Das, *Harish-Chandra Research Institute, Allahabad*

13. Mr. Sreedhar Dutta, *Institute of Mathematical Sciences, Chennai*

14. Mr. Sankalpa Ghosh, *Jawaharlal Nehru University, New Delhi*

15. Mr. Tarun Kanti Ghosh, *Institute of Mathematical Sciences, Chennai*

16. Dr. Dattu Gaitonde, *Harish-Chandra Research Institute, Allahabad*

17. Mr.Tribikram Gupta, *Harish-Chandra Research Institute, Allahabad*

18. Mr. Sandeep K. Joshi, *Institute of Physics, Bhubaneswar*

19. Mr. R. S. Johal, *Punjab University, Chandigarh*

20. Ms. Kavita Jain, *Tata Institute of Fundamental Research, Mumbai*

21. Mr. Brijesh Kumar, *Indian Institute of Science, Bangalore*

22. Mr. Rajesh Karan, *Indian Institute of Science, Bangalore*

23. Ms. Raishma Krishnan, *Hyderabad University, Hyderabad*

24. Mr. Sandeep Krishna, *Indian Institute of Science, Bangalore*

25. Mr. Sanjeev Kumar, *Harish-Chandra Research Institute, Allahabad*

26. Dr. Anjan Kundu, *Saha Institute of Nuclear Physics, Kolkata*

27. Mr. Siddhartha Lal, *Indian Institute of Science, Bangalore*

28. Ms. Tulika Maitra, *Indian Institute of Technology, Kharagpur*

29. Mr. Joydeep Majumdar, *Harish-Chandra Research Institute, Allahabad*

30. Ms. Poonam Mehta, *Delhi University, New Delhi,*

31. Mr. Pramod Kumar Mishra, *Banaras Hindu University, Varanasi*

32. Mr. Pradeep K. Mohanty, *Harish-Chandra Research Institute, Allahabad*

33. Mr. Partha Mukhopadhyay, *Harish-Chandra Research Institute, Allahabad*

34. Mr. Swagatam Mukhopadyay, *Indian Institute of Technology, Kanpur*

35. Mr. Nandan Pakhira, *Indian Institute of Science, Bangalore*

36. Ms. Debjani Pal, *Indian Statistical Institute, Kolkata*

37. Mr. Arun Paramekanti, *Tata Institute of Fundamental Research, Mumbai*

38. Mr. Rajarshi Ray, *Institute of Physics, Bhubaneswar*

39. Mr. Suryadeep Ray, *Harish-Chandra Research Institute, Allahabad*

40. Ms. Sumithra Sankararaman, *Institute of Mathematical Sciences, Chennai*

41. Mr. Prabudha Sanyal, *Indian Institute of Science, Bangalore*

42. Ms. Soma Sanyal, *Institute of Physics, Bhubaneswar*

43. Dr. Asok Kumar Sen, *Saha Institute of Nuclear Physics, Kolkata*

44. Dr. Parongama Sen, *Department of Physics, Calcutta University, Kolkata*

45. Mr. S. Shankaranarayan, *Inter University Centre for Astronomy and Astrophysics, Pune*

46. Mr. V. Venkatesh Shenoi, *Institute of Physics, Bhubaneswar*

47. Mr. Naveen Singh, *Banaras Hindu University, Varanasi*

48. Mr. Subrata Sur, *Harish-Chandra Research Institute, Allahabad*

49. Mr. Naveen Surendran, *Institute of Mathematical Sciences, Chennai*

50. Mr. Aseem Talukdar, *Delhi University, New Delhi*

51. Dr. Arghya Taraphdar, *Indian Institute of Technology, Kharagpur*

52. Mr. Ashoke Kumar Verma, *Bhabha Atomic Research Centre, Mumbai*

Introduction

Sumathi Rao

Harish-Chandra Research Institute
Chhatnag Road, Jhusi
Allahabad – 211 019

Condensed matter physics is a rich and diverse field; to understand most phenomena, several different techniques need to be applied and with different levels of sophistication. The subject can be approached by looking at different phenomena and learning the techniques that are relevant in that particular area, or by learning a few general techniques that are usually applicable to most phenomena.

This book, based on lectures that were given at a SERC school held a year ago at the Harish-Chandra Research Institute, emphasizes quantum field theory techniques that were developed in parallel in high energy physics and condensed matter physics. Well-known examples are the concept of broken symmetry, and the ideas of scale invariance, renormalisation group and fixed points. More recently, topological concepts have also been shown to have wide applications in many areas including condensed matter physics, particle physics and even astrophysics. Other more specialised techniques like bosonization and conformal field theories have also found applications both in physical problems in condensed matter physics like impurity problems and the the Kondo problem, and in more esoteric areas like string theory and D-branes. In these lecture notes, however, the emphasis has been mainly on developing the techniques and applying it to condensed matter systems.

The first chapter on **'quantum many particle systems'** by **Pinaki Majumdar** *(Harish-Chandra Research Institute, Allahabad)* explains the relevance of quantum field theory methods developed (mainly) in the context of high energy physics, to many particle systems, even though, in principle, quantum mechanics may suffice to understand these systems. It also gives a general overview of the kinds of phenomena one might expect in many-particle fermion or boson systems, and generally serves as an introduction for the various more specialised techniques studied in the other chapters. The main phenomena it covers are Bose-Einstein condensation (as a paradigm for interacting boson models) and superconductivity (as a paradigm for interacting fermion models).

List of errata in Chapter 3: Phase Transitions and Critical Phenomena: Renormalisation Group Method, *by Deepak Kumar*

1. Eq. (3.49): Replace by
$$U_f = \sum_{r_2} J(\vec{r}_1 - \vec{r}_2) G(\vec{r}_1 - \vec{r}_2) = \sum_{q_1} J_{-q_1} G(\vec{q}_1).$$

2. Eq. (3.50): Replace first equality by $U_f = J_0 \sum_q G(\vec{q})$.

3. Eq. (3.74): Replace by $F(K) = \sum_{n=2} \frac{1}{2^{n-1}} \ln\left(\frac{2C_n^2}{C_{n+1}}\right)$.

4. Eq. (3.121): Replace 3rd equation by $y_h = 1.37$.

5. Eq. (3.182): Replace last equality by $m(t, u) = \ldots\ldots = \left(\frac{|t|}{u}\right)^{\frac{1}{2}}$.

6. Eq. (3.202): Limits on all integrals are $\int_{\Lambda/b}^{\Lambda}$.

7. Eq. (3.210): Interchange δr and δu.

8. Eqs. (3.212) and (3.213): Replace β by B.

List of errata in Chapter 5: An Introduction to Bosonization and Some of its Applications, *by Sumathi Rao and Diptiman Sen*

1. Eq. (5.27): The expression $\alpha + i(x - x')$ should be replaced by $\alpha - i(x - x')$ in the second and third lines.

2. Eq. (5.28): In the first line, the expressions $\alpha - i(x - x')$ in the numerator and $\alpha + i(x - x')$ in the denominator should be interchanged. In the second line, the minus sign in front of $i/4$ should be removed.

3. Eq. (5.42): Replace ϕ by ϕ_R.

4. Eq. (5.53): On the right-hand side, the minus sign in front of $i\nu/4$ should be removed.

5. Eq. (5.55): On the right-hand side, a minus sign should be inserted in front of $i/2$.

6. In the first sentence after Eq. (5.55), the second equality should read as $\rho_\nu = -\partial_x \phi_\nu / \sqrt{\pi}$.

7. Eq. (5.86): In both lines, the sign $\sim$ should be replaced by $=$. In the first line, the expression $vt - x - i\alpha\text{sign}(t)$ on the right-hand side should be replaced by $\alpha - i(x - vt)\text{sign}(t)$. In the second line, the expression $vt + x - i\alpha\text{sign}(t)$ on the right-hand side should be replaced by $\alpha + i(x + vt)\text{sign}(t)$.

8. Eq. (5.88): The sign $\sim$ should be replaced by $=$. On the right-hand side, the expressions $vt - x - i\alpha(s_t)$ and $vt + x - i\alpha(s_t)$ in the denominator should be replaced by $\alpha - i(x - vt)s_t$ and $\alpha + i(x + vt)s_t$ respectively.

9. Eq. (5.93): The sign $\sim$ should be replaced by $=$. On the right-hand side, the expression $4\pi^2((vt - i\alpha(s_t))^2 - x^2)^K$ in the denominator should be replaced by $2\pi^2[x^2 - (vt - i\alpha s_t)^2]^K$.

10. Eq. (5.96): The sign $\sim$ should be replaced by $=$. On the right-hand side, the expressions $vt - x - i\alpha(s_t)$ and $vt + x - i\alpha(s_t)$ in the denominator should be replaced by $\alpha - i(x - vt)s_t$ and $\alpha + i(x + vt)s_t$ respectively.

11. In the sentence beginning after Eq. (5.149), the expression $(-/+)$ on the right-hand side of the first equality should be replaced by $(+/-)$.

12. Eq. (5.215): $\phi(0), \phi_{R/L}(0)$ should be replaced by $\phi(x = 0, \tau), \phi_{R/L}(x = 0, \tau)$ respectively.

13. Eq. (5.220): The second term on the right-hand side should be replaced by $(\lambda/2\pi\alpha) \int d\tau \cos 2\sqrt{\pi}\phi(0, \tau)$.

14. Eq. (5.226): In each of the four exponentials on the right-hand side, there should be a factor of $2\sqrt{\pi}$ immediately after the $-i$. In the next sentence, the equation $\phi(0) = \phi_R(0) + \phi_L(0)$ should be replaced by $2\sqrt{\pi}\phi(0) = 2\sqrt{\pi}(\phi_R(0) + \phi_L(0))$.

15. Eq. (5.227): On the right-hand side, $\cos(\theta_> - \theta_<)$ should be replaced by $\cos 2\sqrt{\pi}(\theta_> - \theta_<)$. In the next sentence, the equation $\phi_i(x = 0) = \pi/2$ sould be replaced by $2\sqrt{\pi}\phi_i(x = 0) = \pi/2$.

List of errata in Chapter 7: Low-dimensional Quantum Spin Systems, by Indrani Bose

1. Page 371: In the last line, Eq. (7.41) should be Eq. (7.47).

2. Page 391: Eighth line, omit 'and' before 'Eq. (7.107)'.

3. Page 402: Fig. 7.9 should be corrected. The spins in the dotted boxes should both be down. The spins adjacent to these spins should be up. Elsewhere, the spin arrangement should be up, down, up, down.

Introduction

The second chapter is on **'critical phenomena: a modern perspective'** by **Somendra M. Bhattacharjee** *(Institute of Physics, Bhubaneswar)*. Starting from thermodynamics, it leads to the idea of phase transitions through divergence of thermodynamic quantities and problems with extensivity. It then builds up to scaling and the emergence of divergent length scales in the problem. Exponents are introduced through the anomalous scaling of various correlation lengths and concepts such as universality are introduced. The ideas are illustrated through the examples of a polymer model (for a first order transition) and the Gaussian and ϕ^4 model (for continuous transition).

The third chapter is on **'phase transitions and critical phenomena: renormalisation group method'** by **Deepak Kumar** *(Jawaharlal Nehru University, New Delhi)*. Although, there is some overlap of the subject with the previous chapter, the examples chosen and the presentation of phase transitions is completely different. The Landau theory of phase transitions is studied using the example of the Ising model. Renormalisation group (RG) ideas are introduced and real space RG procedure is illustrated through the examples of the Ising chain and the 2 dimensional Ising model on a triangular lattice. Also the Wilson-Fisher momentum-space RG method is studied for the ϕ^4 theory and the generalisation to the n-component model is discussed.

It should be mentioned that both the second and third chapters emphasize the modern perspective of phase transitions and use a length scale based approach, which is also emphasized in the first chapter.

The fourth chapter on **'topological effects in condensed matter physics'** by **Ajit M. Srivastava** *(Institute of Physics, Bhubaneswar)* emphasizes the idea that topological concepts play an important role in a wide range of phenomena. Starting with the concept of order parameter and order parameter space, the subject of topological defects is introduced; and illustrated with examples from spin systems, superfluidity, superconductivity and nematic liquid crystals in condensed matter physics; and from grand unified theories (GUT), electroweak theories and chiral sigma models in particle physics. Mathematical ideas like homotopy group, based homotopy group and the classification of defects based on homotopy theory are also discussed in some detail. The chapter ends with a brief discussion of the formation of defects.

The fifth chapter is on **'introduction to bosonization and some of its applications'** by **Sumathi Rao** *(Harish-Chandra Research Institute, Allahabad)* and **Diptiman Sen** *(Indian Institute of Science, Bangalore)*. Starting with a general discussion of many particle systems in one-dimension, the concept of Luttinger liquids is introduced. It then shows how the idea of bosonization - rewriting a fermion model in terms of appropriate bosons - can be used as a non-perturbative technique to solve the interacting many-fermion model and calculate relevant correlation functions. The second half of the chapter deals with the application of the ideas of bosonization in simple one-dimensional models such as spin models and the Hubbard model, and also in the newly developed field of quantum wires, where the calculational tools of bosonization and RG are used to compute physically relevant quantities like conductances.

The next two chapters deal with low dimensional phenomena; however, quantum field theory is not the only method to study them. Both the lecturers have emphasized the phenomena, and described all methods used to understand them.

The sixth chapter on **'quantum Hall effect'** by **R. Rajaraman** *(Jawaharlal Nehru University, New Delhi)* starts from the basic Hall effect, and the Landau level problem in quantum mechanics, and builds up to the fractional quantum Hall effect, via Laughlin's variational wave-function approach and the plasma analogy. The use of quantum field theory in understanding this phenomenon is illustrated through (a variant of) the Chern-Simons theory. The chapter ends with its relation to Jain's theory of composite fermions.

The seventh chapter is on **'low dimensional quantum spin systems'** by **Indrani Bose** *(Jagdish Chandra Bose Institute, Kolkata)*. The Heisenberg ferromagnetic and antiferromagnetic spin systems are introduced and several theorems and rigorous results are proved. The techniques that are illustrated to solve several model problems include the variational technique and the method of constructing exact solutions. Finally, the Bethe ansatz solution is explained in detail using the example of the linear spin 1/2 Heisenberg antiferromagnetic spin chain.

There was also a set of introductory lectures on quantum many-particle physics (for students unfamiliar with quantum field theory methods or many-body techniques) by **R. Shankar** *(Institute of Mathematical Sciences, Chennai)*, a set of lectures on advanced topics in fractional quantum Hall effect and a special lecture on biology and physics by **G. Baskaran** *(Institute of Mathematical Sciences, Chennai)*, a set of lectures on conformal field theories with special emphasis on the application to the Kondo problem by **P. Durganandini** *(Pune University)* and a couple of lectures on superconductivity by **T. V. Ramakrishnan** *(Indian Institute of Science, Bangalore)*; unfortunately those lectures were not available and could not be included here.

Chapter 1

Quantum Many Particle Physics

Pinaki Majumdar

   Harish-Chandra Research Institute
   Chhatnag Road, Jhusi
   Allahabad – 211 019

This set of lectures focuses on some problems in itinerant Bose and Fermi systems. The first lecture will be an introduction to many-particle physics. In the second we will discuss phase transitions and symmetry breaking and the mean field approach to condensation in Bose systems. In the third we will look at strong interactions in 'normal' Fermi systems. The final lecture will be on strong coupling superconductivity which involves both strong interactions and symmetry breaking.

1.1 Preamble

'Statistical field theory' encompasses all the applications of quantum field theory to statistical physics. In the context of these notes, it is interpreted as the study of 'statistical mechanics of quantum many particle systems, employing the methods of field theory'. In this interpretation we are closer to the classic text of Abrikosov, Gorkov and Dzyaloshinski [1], than to the more modern approaches of Parisi [2] or Itzykson and Drouffe [3]. These lectures cover only quantum systems, fermions and bosons; other lectures in the school covered applications of field theory to problems in classical statistical mechanics and spin systems.

These lectures are meant to provide an introduction to itinerant quantum systems and discuss a few model problems. There is no claim at originality! The material is readily available in the literature [4, 5, 6, 7] and anyone interested should consult the references for a more comprehensive and rigorous coverage.

1.2 Introduction

We begin our survey of methods available for handling interacting many particle systems. There is no single method that is valid in all cases. We may tentatively classify the approaches as:

1. Exact solution!

1.2. Introduction

2. 'Mean field' theory for symmetry broken states.

3. Perturbation theory; weak coupling.

4. Strong coupling expansion (if/when possible).

5. Renormalisation group calculations.

6. Exact numerical calculations.

7. Large dimension expansion.

This is not an exhaustive list. Special techniques exist, for instance in one dimensional problems, as discussed elsewhere in the text. The emphasis here is on general applicability.

Of these methods 'exact solutions' are not generic [8]. For itinerant fermions only the simplest interacting model (the Hubbard model) can be solved in one dimension (1d). There are no exactly solvable models in two or higher dimensions. Renormalisation group calculations are discussed elsewhere, in the context of classical systems and for 1d fermions and quantum impurity models. Exact numerical calculations, for instance Quantum Monte Carlo (QMC) or exact diagonalisation methods, are powerful and quite generally applicable but lie beyond the scope of our discussion.

The large dimension, $d \to \infty$, approach to lattice models is a promising new method for strongly interacting systems but there was no time to discuss it. Introductory references are [9, 10].

This leaves us with the second, third and fourth topics on the list; which form the core of our discussion. Instead of dealing with them in the abstract, we draw on the few controlled calculations that one can do to illustrate both the technique and some physics. Examples can be drawn from

- The electron gas at small r_s (*i.e.*, high density),

- The 'hardcore' interacting fermion model at low density,

- Migdal-Eliashberg theory of the electron-phonon system,

- And the superfluid bose gas with weak repulsive interaction.

Unlike 'exact' numerical schemes, all analytical approaches to interacting systems involve some approximation, hopefully controlled, governed by the existence of a 'small' parameter. Most often this small parameter is not simply the interaction strength, unlike the way one learns to do perturbation calculations in a ϕ^4 theory or QED, say. For all the cases cited above one can do 'semi-analytic' calculations, quite controlled in a certain regime. The parameter defining this regime is, however, rather subtle. The small parameter in the high density electron gas is r_s, the mean interparticle separation measured in units of the Bohr radius, and not the Coulomb interaction energy V_{coul}! For the hardcore dilute fermi gas, it is $k_F a$ where a is the s-wave scattering length and k_F is the Fermi wavevector. The interaction potential is *infinite* at short distance and all finite order perturbation calculations diverge. The Migdal-Eliashberg theory of electron-phonon interactions is controlled by ω_D/E_F, the ratio of characteristic phonon/electron energy scales. The examples above essentially exhaust the 'solvable' limits of the classic quantum many body problems.

The motive behind this short discussion is to indicate that often calculations are controlled by a hidden but physically relevant small parameter, and not the 'smallness' of the bare coupling constant. The approximations are guided by intuition. Sometimes (most often!), however, there is *no* natural small parameter whatsoever, and one has to *introduce* an artificial small parameter to simplify the problem. In this class we have the 'large N' and 'large d' approximations, which employ $1/N$ and $1/d$ respectively as small parameters (N refers to an 'internal' quantum number, the number of orbitals, say, and d is spatial dimensionality).

1.3 Introduction to many particle physics

1.3.1 Phases of many particle systems

When approaching a vast subject it is helpful to have some crude organising principle, to avoid getting hopelessly lost. In terms of

1.3. Introduction to many particle physics

many particle systems you can classify them (trivially!) as

- 'free', or
- interacting

A more pertinent classification of the 'phases' of many particle systems is

- 'normal', *i.e.*, without any kind of long range order (LRO), or
- 'symmetry broken', *i.e.*, with some form of LRO.

Maybe we could also think of the the system as being in a temperature regime where quantum effects are negligible, *i.e.*,

- the system could be in a 'classical' regime, or
- the system is in a regime where the effects of quantum statistics is crucial.

Free systems, as the name implies, do not have any interparticle interaction. They constitute solvable 'model' problems where the dynamics and the statistical mechanics can be worked out exactly. Such systems, with the major exception of the free Bose gas, do not exhibit any 'ordering' phenomena at any temperature.

Interacting systems, by contrast, exhibit a rich variety of phases: liquid like, crystalline, magnetic, superfluid, etc. Most of these phases break some symmetry of the basic Hamiltonian. Our next lecture is devoted to symmetry breaking; here it suffices to note that the Hamiltonian of a system, modelling interparticle interactions etc, usually has the full translation and rotation symmetry of free space. This is forced by the fundamental conservation laws. The phases, however, usually exhibit lower symmetry, particularly in the low temperature regime. For instance, the periodic density variation in a crystal does not create a homogeneous and isotropic environment. Magnetisation in a ferromagnet singles out a particular direction in space. These 'broken symmetry' phases arise due to ordering tendencies inherent in the interaction term.

The temperature below which quantum effects are relevant is set, at least in free systems, by the particle density. It corresponds to the condition: de Broglie wavelength $\simeq$ interparticle separation. Since the energy of a free system is purely kinetic, and the only lengthscale in the problem is set by density, $i.e.$, $n^{-1/3}$, this energy scale, T_Q is $\sim \hbar^2 n^{2/3}/2m$, *for both bosons and fermions*. Dimensional arguments apart, such an energy scale actually arises in Fermi and Bose systems: it is the Fermi energy E_F for a Fermi system, and the Bose-Einstein condensation temperature T_{BEC} for a Bose system.

Interactions bring in additional energy scales. Most systems, gaseous at high temperature, tend to condense into a liquid and finally a solid at lower temperature. These transitions involve the competition of internal energy, which is gained by ordering, and entropy, which is lost. Quantum mechanics may or may not be relevant to the transition. The condensation of a gas, driven by interatomic forces, can occur at a temperature $T_{ord} \gg T_Q$. In that case the transition can be described purely classically, as in most gas-liquid and liquid-solid transitions. However, consider a situation where the particles are relatively light, $e.g.$, He^4 (or electrons!). T_Q will be large, and phase transitions can occur in a regime $T_{ord} \ll T_Q$. Quantum effects will be crucial in such transitions. The superfluid transition in He^4 and superconductivity in fermions are examples in this category. Sometimes quantum fluctuations can be strong enough to prevent any ordering transition at all.

1.3.2 Quantities of physical interest

What are the questions we might want to ask about a physical system? Let us look at the free Fermi gas for a hint. For instance, we say we can 'completely solve' the free Fermi gas because we can calculate all quantities of interest for this system.

These are, principally,

1.3. Introduction to many particle physics

- thermodynamic properties: internal energy, free energy, specific heat, compressibility and susceptibility.

- 'spectral properties', *i.e.*, the density of states (DOS) and spectral functions, and the momentum distribution function $\langle n_{\mathbf{k}} \rangle$.

- response to time dependent perturbations, and the transport coefficients, via linear response theory (assuming weak scattering effects).

We can calculate all these because we know all the eigenvalues and eigenfunctions of the free (or Bloch) electron gas. We will be interested in the same quantities for interacting systems as well but typically we will not have any clue about the many particle eigenvalues and eigenfunctions. This is where 'field theoretic' methods come in useful. They allow us to calculate the above quantities, approximately for interacting systems, without knowing the solutions of the many particle Schrodinger equation. These methods employ approximation schemes for the Green's functions of the system, and recover physically relevant quantities from them.

To be more specific, we will principally want to know about

- thermodynamic properties, as before, phase transitions and the nature of the broken symmetry phase.

- the spectral function: $A(\mathbf{k}, \omega)$, *i.e.*, the energy (ω) distribution of a particle added with momentum $\mathbf{k}$, and the "single particle" density of states, $N(\omega) = \sum_{\mathbf{k}} A(\mathbf{k}, \omega)$. A is calculated from the single particle Green's function as $A(\mathbf{k}, \omega) = -Im G(\mathbf{k}, \omega)/\pi$. Physically, $A(\mathbf{k}, \omega)$ contains information about the existence and spectrum of 'quasiparticles'.

- The transport properties of the system, *i.e.*, the conductivity $\sigma_{\alpha\beta}(\mathbf{q}, \omega)$, defined through $J_\alpha = \sigma_{\alpha\beta} E_\beta$. The conductivity involves a current-current correlation function. This is most easily seen by coupling in an external field through

$$\int \mathbf{j}(\mathbf{r},t)\cdot\mathbf{A}(\mathbf{r},t)d\mathbf{r},$$

and evaluating $\langle \mathbf{j} \rangle$ from the first order change in the wavefunctions due to $\mathbf{A}$. The resulting expression is related to $\langle \mathbf{j}(\mathbf{r},t)\mathbf{j}(\mathbf{r}',t') \rangle$, which is the *equilibrium* current-current correlation. This general connection is formalised in 'linear response theory' [12, 13], and is widely used in calculating response functions.

- Like the example above, correlation functions like density-density correlation $D(\mathbf{Q},\Omega)$, or spin-spin correlation $\chi(\mathbf{Q},\Omega)$, govern the response to x-ray or neutron beams and yield information about low energy fluctuations in the system.

1.3.3 Fermi and Bose liquids

The rest of the lectures will deal with specific models for interacting fermions and bosons; here let us try to build some intuition about the general features of Fermi and Bose systems.

(a) Free fermions:

The simplest reference point for fermions is of course the free Fermi gas (particles in free space). The Hamiltonian is

$$H = -\sum_i \frac{\hbar^2 \nabla_i^2}{2m} \tag{1.1}$$

where i runs over the particle coordinates. The Hamiltonian is a sum of single particle terms, so it is sufficient to solve the one particle problem. The single particle wavefunctions are solutions of $(-\hbar^2\nabla^2/2m)\psi_n = \epsilon_n\psi_n$. These are plane wave states, $\psi_\mathbf{k}(\mathbf{r}) \sim e^{i\mathbf{k}\cdot\mathbf{r}}/\sqrt{V}$. The many body state is a Slater determinant of such states (and not a simple product) to satisfy requirements of antisymmetry. In the second quantised representation we can write the same Hamiltonian as

$$H = \sum_\mathbf{k} \epsilon_\mathbf{k} c_\mathbf{k}^\dagger c_\mathbf{k} \tag{1.2}$$

1.3. Introduction to many particle physics

where we have dropped spin indices and the solution of the Schrodinger equation is embedded in $\epsilon_k = \hbar^2 k^2/2m$. The many particle eigenfunctions of this system are of the form

$$|\Psi\rangle = \{\prod_{n=1}^{N} c^{\dagger}_{\mathbf{k}_n}\}|0\rangle . \qquad (1.3)$$

The energy of such a state is $\sum_{n=1}^{N} \epsilon_{\mathbf{k}_n}$.

The ground state $|\Psi_0\rangle$ corresponds to filling up the N lowest single particle states and, since ϵ is a monotonic function of k, these are the states with the lowest momentum $|\mathbf{k}|$. This set of single particle states, for which $\langle n_{\mathbf{k}}\rangle = \langle c^{\dagger}_{\mathbf{k}} c_{\mathbf{k}}\rangle = 1$ constitute the *Fermi sea*. The 'surface' in $\mathbf{k}$ space enclosing these occupied states at $T = 0$ is called the Fermi surface (FS). The momentum distribution function $n_{\mathbf{k}}$ has unit discontinuity across the Fermi surface.

Excitations correspond to transferring a particle(s) from within the Fermi surface to higher energy states. Notice, as promised, we know *all* the eigenvalues and eigenfunctions of this trivial many particle system. The statistical mechanics of this system is easy to work out and is standard material (*e.g.* see [11]). Bypassing that we make a few remarks about the various properties of this system.

- As noted, there is a 'Fermi surface', denoted by the discontinuity in $\langle n_{\mathbf{k}}\rangle$, at $T = 0$. A sharp change survives as long as $T \ll E_F$ where E_F is the energy of the highest filled state at $T = 0$.

- The internal energy varies as $U(T) \sim U(0) + \mathcal{O}(T^2/E_F)$, and $C_V \sim T/E_F$. The Pauli spin susceptibility varies as $(\mu^2/E_F)(1 - \mathcal{O}(T^2/E_F^2))$.

- The 'spectral function' denoting the energy distribution of a particle propagating with momentum $\mathbf{k}$ is $\delta(\omega - \epsilon_{\mathbf{k}})$. The single particle DOS is given by $N(\omega) = \sum_{\mathbf{k}} \delta(\omega - \epsilon_{\mathbf{k}})$ and is independent of temperature or the number of particles.

- There are no 'collisions' in this system, so no relaxation; or equivalently, the relaxation time $\tau \to \infty$.

The free Fermi gas is always 'normal', it has no ordering tendencies; and its response to all external fields is bounded at all temperatures.

(b) Free bosons:

Just as for the free Fermi system, the statistical mechanics of the free Bose gas can be completely worked out. However, unlike the free Fermi gas, there is a singularity [11] in the thermodynamic functions of a (infinite, three dimensional) Bose system at a finite temperature which we denote as T_{BEC}^0. The chemical potential μ, which is $\sim -T\ln(T/T_{BEC})$ at high temperature increases towards $\mu = 0$ as T is lowered and beomes zero at a temperature defined by

$$\int d^d k \frac{1}{e^{\beta_c \epsilon_{\mathbf{k}}} - 1} = n. \quad (1.4)$$

Solving this equation, by scaling out T, we find $T_{BEC}^0 \sim \hbar^2 n^{2/3}/2m$, which is a 'quantum' scale (like E_F). There is no such energy scale in a free classical gas. For $T < T_{BEC}^0$, $\mu = 0$, and the ground state picks up a macroscopic occupancy. The ground state occupancy, N_0, is given by $N_0(T) = N(1 - (T/T_{BEC}^0)^{3/2})$. We use the notation T_{BEC} and T_{BEC}^0 interchangeably at the moment, T_{BEC}^0 refers to the condensation temperature in the 'free' system while T_{BEC} is supposed to incorporate interaction effects as well.

There is a phase transition, as evident from the singularity in the thermodynamic functions. Since the ground state occupation $\langle a_{k=0}^\dagger a_{k=0} \rangle \sim N \gg 1$, we may treat the operators $a_{k=0}^\dagger$ and $a_{k=0}$ as c numbers, neglecting their commutator. The symmetry breaking is related to $a_{k=0}^\dagger, a_{k=0}$ picking up an expectation value $\sim \sqrt{N}$ with *arbitrary phase*. In terms of the operator $\phi^\dagger(\mathbf{x})$ which 'creates' a particle at a point $\mathbf{x}$, we have,

$$\phi^\dagger(\mathbf{x}) = \phi_0 + \frac{1}{\sqrt{V}} \sum_{\mathbf{k} \neq 0} a_{\mathbf{k}}^\dagger e^{i\mathbf{k}\cdot\mathbf{x}} \quad (1.5)$$

1.3. Introduction to many particle physics

where $\phi_0 \sim \sqrt{N_0/V}$. There is a spatially uniform amplitude, $\phi_0 \sim \mathcal{O}(1)$ for $T \leq T_{BEC}$. ϕ_0 vanishes above the condensation temperature. In our notation, the condensate particle density $\phi_0^2 = n_0$.

We may summarise the low temperature properties of free bosons as follows:

- For $d = 3$ the free Bose gas condenses at $T = T_{BEC} \sim \hbar^2 n^{2/3}/2m$. The transition shows up as a derivative discontinuity in C_V.

- For $T < T_{BEC}$ a macroscopic fraction of particles are in the $k = 0$ state. The momentum distribution function is

$$n_{\mathbf{k}} = N_0 \delta(\mathbf{k}) + \frac{1}{e^{\beta \epsilon_{\mathbf{k}}} - 1} . \qquad (1.6)$$

- The excitation spectrum of the condensed system is still $\epsilon_{\mathbf{k}} \sim k^2/2m$, i.e., free particle like.

- The system is 'superfluid' but with critical velocity $v_c = 0$. This follows from an argument due to Landau which is worth repeating in detail: Consider the flow of a liquid. Its momentum can be dissipated if the friction between the liquid and the wall can create excitations in the liquid, with momentum $\mathbf{p}$ opposite the flow velocity $\mathbf{v}$. This excitation has energy $\epsilon_{\mathbf{p}}$ so, in a stationary frame, the energy of the liquid changes from $Mv^2/2$ to $\epsilon_{\mathbf{p}} + \mathbf{p} \cdot \mathbf{v} + Mv^2/2$. We would need $\epsilon_{\mathbf{p}} + \mathbf{p} \cdot \mathbf{v} < 0$ for this excitation to be created. Since $\epsilon_{\mathbf{p}} + \mathbf{p} \cdot \mathbf{v}$ takes its minimum value for $\mathbf{p}$ and $\mathbf{v}$ opposite, we need $\epsilon_p - pv < 0$, or $v > min(\epsilon_p/p)$. This relates v_c to a feature in the dispersion curve. For free bosons, $(\epsilon_p/p)_{min} = 0$ since $\epsilon_p \sim p^2$, so there is no superflow.

Exercise 1.1 *Plot out a dispersion and check the above condition.*

- Since there is no interparticle repulsion, or any fermion like exclusion, the free Bose gas is infinitely compressible.

(c) Interacting fermions:

We will discuss interacting fermions/bosons later using the full panoply of 'field theoretic' techniques. Can we say anything meaningful about the 'normal' phase of an interacting Fermi system or the low temperature properties of a Bose liquid without launching into a detailed calculation? For fermions such a description is provided by Landau Fermi liquid theory (FLT). A one paragraph summary of FLT might run as follows [14, 15, 16]:

- In a Fermi liquid the low energy excitations can be described as a *weakly interacting* gas of 'quasi-particles', irrespective of the detailed structure of the interacting ground state. FLT makes no comment on the nature of the ground state itself, *e.g.*, its energy or wavefunction.

- The quasiparticle states are a combination of a 'bare' particle (of momentum **k** say) and particle-hole excitations. They are in one to one correspondence with the single particle excitations of a free Fermi gas (they can be labelled by the same quantum numbers).

- The low temperature ($T \ll E_F^*$) thermodynamics and the low frequency ($\omega \ll E_F^*$), long wavelength ($q \ll k_F$) response of the system can be understood in terms of these quasi-particles. E_F^* is the effective Fermi energy of the system and can be much lower than the bare E_F.

- The response is essentially free Fermi like, with renormalised parameters. For example the specific heat, $C_V \sim \gamma T$, and χ tends to a constant as $T \to 0$.

- The paramaters defining the quasi-particle mass/interactions can be experimentally determined, and from microscopic theories too in simple cases.

Without reference to any specific Hamiltonian, we can identify the following 'microscopic' condition for Fermi liquid behaviour:

1.3. Introduction to many particle physics

- A system would exhibit Fermi liquid behaviour (in the sense above), when the momentum distribution $\langle n_{\mathbf{k}} \rangle$ of the interacting system still has a discontinuity $z_{\mathbf{k}}$ at k_F, i.e., the Fermi surface survives.

$$\langle n_{\mathbf{k}} \rangle|_{k \to k_F} = z_{\mathbf{k}} \theta(k_F - k) + \tilde{n}_{\mathbf{k}}$$

where $\tilde{n}_{\mathbf{k}}$ is smooth across the Fermi surface.

- This implies that for $k \to k_F$ the spectral function has a component $z_{\mathbf{k}} \delta(\omega - \tilde{\epsilon}_{\mathbf{k}})$. These coherent (long lived) 'quasiparticles' dominate the low energy physics. Free particles correspond to $z_{\mathbf{k}} = 1$.

Qualitatively, the presence of a Fermi surface, and the consequent phase space restrictions, lead to long lifetime for the low lying excitations. These quasiparticles, similar to excitations in the free Fermi gas, dominate the low temperature/low energy properties.

Since we are not able to directly calculate the actual renormalisations within Fermi liquid theory, nor, often, predict whether a given microscopic model will have a Fermi liquid ground state, one might be sceptical about the usefulness of this scheme. Fermi liquid theory is valuable because it creates a *common understanding* of a large number of physical properties in terms of a few experimentally determined parameters. At a more basic level, it emphasises the concept of weakly interacting quasiparticles as the relevant elementary excitations in strongly interacting systems.

'Proving' the existence of a Fermi liquid ground state (or otherwise) in a model, a topic of current research, is extremely difficult. However when such a state exists, as evidenced by experiments, it allows us to use a comprehensive phenomenology to understand all aspects of the low energy physics.

The most celebrated, and 'clean' instance of a Fermi liquid is liquid He^3. The behaviour of electrons in metals too, in most cases, can be understood within the Fermi liquid picture. Exceptions are one dimensional systems, discussed in the school, and the much studied high T_c materials. A more obvious breakdown of the Fermi liquid picture occurs when interactions lead to phase transitions, as in superconductivity.

(d) Interacting bosons:

What about the 'generic' interacting Bose liquid?
We might want to know if an interacting Bose system shares some features of the free Bose gas. Since a 'weakly' interacting system has a better chance of being described by the free system, let us proceed as follows:
• How is the weakly interacting Bose gas different from the free Bose gas? How is it similar?
• Does the strongly interacting Bose system bear any resemblance to the weakly interacting/free Bose gas?
Recap: The free Bose gas has ground state occupation $N_0 \sim N$, the excitations have single particle character, $\epsilon_\mathbf{k} \sim k^2/2m$, and there is no superfluidity (the critical velocity for 'superflow' is zero). The compressibility is infinite since there are no interactions nor any fermion like exclusion.
Interacting Bose systems have the following features at low temperature. We will look at some of these in more detail in the next lecture.

- In the presence of interactions (and at $T = 0$) the condensate fraction $\langle \hat{N}_0 \rangle / N \sim \mathcal{O}(1)$ although somewhat depleted. The momentum occupation (at $T = 0$) has the form:

$$\langle n_\mathbf{k} \rangle \sim N_0 \delta(\mathbf{k}) + \tilde{n}(\mathbf{k}) \qquad (1.7)$$

where $\tilde{n}(\mathbf{k})$ corresponds to the distribution in $\mathbf{k} \neq 0$ states.

- The excitation spectrum, at long wavelength, looks phonon like: $\epsilon_\mathbf{k} \sim v_{ph} k$, where v_{ph} is the phonon velocity. The excitations do *not* correspond to single particle excitations, but are actually density fluctuations. This arises from a combination of interaction effects and macroscopic N_0 and is discussed in detail in the next section.

- Superfluidity: The interacting system, in the condensed phase, sustains superflow if the flow velocity $v < v_c \approx v_{ph}$.

1.3. Introduction to many particle physics

Is this consistent with the Landau argument? In real systems (He^4, say) the critical velocity is rather less than v_{ph} and is set by the 'roton' minimum in the dispersion which corresponds to the creation of vortex like excitations in the liquid.

- The compressibility of the system is finite, stabilised by weak (repulsive) interactions.

At the level of the present discussion, let us just remember that a (weakly) interacting Bose system has a condensate at low temperature, like the free Bose gas, but interaction effects dramatically modify the long wavelength transport properties. The free Bose gas provides a starting point but interactions are also essential. There is no simple theory of a strongly interacting dense Bose system, which is what we might imagine He^4 to be. In the case of a strongly interacting but *dilute* Bose system, the 'bare' potential can be replaced by the exact two particle scattering amplitude. The predictions of such a theory are similar to what we have discussed above. Unfortunately the only real Bose system, liquid He^4, lies outside the domain of validity of such a theory.

Nevertheless BEC does seem to play a role in He^4: the superfluid transition occurs at ~ 2.19K, compared to $T^0_{BEC} \sim 3.14$K (for the system's parameters). However, the condensate fraction is $\sim 8\%$ at $T = 0$, way below saturation. Direct applicability of 'free Bose' ideas are doubtful, but the condensate does exist, as seen in neutron scattering, as do the other qualitative features described above. In particular the long wavelength spectrum is phonon like, exactly as described for weakly interacting bosons.

The principle of adiabatic continuity:

The discussion about generic low temperature properties of bosons and fermions is motivated by what has been called *adiabatic continuity* by P. W. Anderson [17]. It is often observed that strongly interacting systems retain some signature of the 'free' system, *e.g.*, the Fermi surface in Fermi systems or Bose condensation in Bose systems. These features, *i.e.*, the FS or a condensate, in turn govern the low temperature thermodynamic and transport properties of the system. The interacting system is seen to be a 'continuation' of the free system, in the sense that the properties of the interacting system can be accessed through some form of perturbation expansion about the free limit. Landau's Fermi liquid theory is explicitly in this spirit, and even in the Bose system, such an approach provides a useful starting point. 'Continuity' does not hold when there is a symmetry change induced by interactions, which is what we discuss in the next section.

1.4 Phase transitions/broken symmetry and Bose systems

1.4.1 Phase transitions and symmetry breaking

The identification of the symmetry broken in a phase with long range order (LRO) allows us to construct simple 'mean field' theories which provide a qualitative description of the system. Fluctuations of the order parameter, about the mean field state, usually define the low energy excitations in the system and dominate the low energy/low temperature properties. The mean field approach is, of course, not usable when there is no long range order, (like in the normal electron liquid, say) and there is no obvious reference state to expand about. Such systems require special, and specific, methods which we will discuss in the next section.

Phase transitions and broken symmetry [19, 20] have been covered in detail in the other chapters. To minimise repetition, here, we only highlight the key issues. We will work out one particular instance, Bose condensation in the interacting Bose gas, in some detail.

- A *phase transition* shows up as discontinuous behaviour in the thermodynamic properties of a system. Often, it leads to the emergence of long range order in the low temperature phase.

1.4. Phase transitions and broken symmetry

- The *order parameter* is a thermodynamic variable characterising the nature and extent of LRO. The most common, and oft-repeated, example is the magnetisation in a ferromagnet. The (vector) magnetisation, **M**, singles out a direction in space. The ordered state breaks the rotation invariance of the original Hamiltonian (see later). There are more subtle and non trivial examples of broken symmetry and order parameters, many of which are discussed by P. W. Anderson in [17] in the chapter on 'Broken Symmetry'.

    Exercise 1.2 *Can you identify the order parameter in a crystalline solid?*

- Why do systems 'order'? Consider an example of 'induced order' in a trivially solvable problem: a single spin (1/2) in a magnetic field. $H = -\mathbf{h} \cdot \mathbf{S} = -hS_z$. The partition function $Z = Tre^{-\beta H}$, and $m = \langle S_z \rangle \sim tanh(\beta h)$. The external field induces a magnetisation which grows with h/T. At $T = 0$ the spin aligns along the field. Increasing T leads to a decrease in m because the internal energy gained by aligning along the field is partially offset by the entropy lost.

- A system might want to order *spontaneously* if there are interactions which act as effective *internal fields*. In the model $H = -J \sum_{ij} \mathbf{S}_i . \mathbf{S}_j$, where $J > 0$ and the sum is carried out over nearest neighbours, the effective field on site 'i' is $J \sum_\delta \mathbf{S}_{i+\delta}$. In the simplest approximation one replaces $\mathbf{S}_{i+\delta}$ by $\langle \mathbf{S}_{i+\delta} \rangle \approx m$, where m is the bulk magnetisation. This field would tend to polarise the spin $\mathbf{S}_i$ parallel to it, and one can self consistently argue that the system will head towards a ferromagnetic state. This is also obvious looking at H itself which is minimised by polarising all spins parallel. Notice that H is invariant under global rotation of the spins but the order parameter picks out a specific direction in space.

- Here is a crude guiding principle. Look at the 'interaction' term in the Hamiltonian: for classical systems the ordered state arises from attempts to minimise this. We have a competetion between potential energy and entropy. Transitions in quantum systems are much more subtle and the kinetic energy plays a non trivial role.

Exercise 1.3 *Try to follow the above argument to explain why most liquids freeze at low temperature while liquid He^4 does not.*

- How good is the mean field picture? The mean field approximation suggests an uniform order parameter, *e.g.*, one in which $\mathbf{M}(\mathbf{r})$ is uniform in space. If the order parameter breaks a continuous symmetry of the Hamiltonian then long wavelength ($k \to 0$) variations of $\mathbf{M}$ would cost very little energy. In fact, the system should be gapless. The low energy thermodynamics and tranport properties will be defined by these (long lived, low energy) modes.

- Since we have approached phase transitions from the 'mean field' point of view it is necessary to add a cautionary note. Order parameter fluctuations can be large in a low dimensional system, particularly if a continuous symmmetry is involved. The mean field order might be completely suppressed by these fluctuations, pushing $T_c \to 0$. Quantum fluctuations have a similar effect.

Mean field theory often serves as the first approximation in handling broken symmetry. It involves replacing an interacting (and often unsolvable) problem with a non interacting problem in a *self consistent field*. The self consistent field is related to the order parameter and is itself some average over the systems variables. This leads to 'single spin' problems in magnetism, or quadratic fermion or boson problems in itinerant systems. Celebrated examples are Curie-Weiss theory in magnetism and the BCS theory of superconductivity.

You may have noticed that we have not invoked the 'smallness' of any parameter when discussing mean field theory. Did we need

1.4. Phase transitions and broken symmetry

to have a 'small J' to justify the mean field approximation in the spin model? The domain of validity of mean field theories has been discussed elsewhere in these lecture notes and is reviewed extensively in [20]. The role of spatial fluctuations, which we just hinted at above, is crucial in constructing a correct theory of phase transitions. Our attitude, in these lectures, is the following: mean field theories provide a starting point for understanding the qualitative behaviour of phases, even in the *absence of any small parameter*. The presence of long range order simplifies the behaviour of the system. In its absence, and in the presence of strong interactions, there are very few general methods. We will begin to appreciate the complications when we contrast the treatment of the Bose condensate, in the next subsection, with that of Fermi systems in the next section.

1.4.2 Symmetry breaking and interactions in BEC

The following short discussion will be our only encounter with Bose systems. It will highlight some of the issues discussed in the previous section.

The Bose condensed phase corresponds to macroscopic $\mathcal{O}(N)$ occupation of a single quantum state. For a translation invariant system this is the $\mathbf{k} = 0$ state. We have a satisfactory theory of interacting bosons in only two limits [18] -

- bosons with weak short range repulsive interaction, and

- a *dilute* Bose gas with arbitrarily strong repulsive interaction.

Neither of these corresponds to the case of liquid He^4, where the system is dense and the short range repulsion very strong. Nevertheless these model problems highlight how the macroscopic occupation of a single quantum state (the essential $T \to 0$ Bose character) controls the excitation spectrum and the physical properties of this many particle system.

It will be useful for you to remember the physics of the free Bose gas, *e.g.*, the temperature dependence of chemical potential,

$\mu(T)$, and properties of the condensed phase [11]. Here we consider how these properties are modified in the weakly interacting Bose gas. We will focus mainly on $T = 0$ properties. The Hamiltonian is given by

$$H = \int d\mathbf{x} \phi^\dagger(\mathbf{x})(-\nabla^2 - \mu)\phi(\mathbf{x})$$
$$+ \int d\mathbf{x} d\mathbf{y} \phi^\dagger(\mathbf{x})\phi^\dagger(\mathbf{y})V(\mathbf{x}-\mathbf{y})\phi(\mathbf{y})\phi(\mathbf{x}) \qquad (1.8)$$

which, Fourier transformed, looks like:

$$H = \sum_\mathbf{p} \epsilon_\mathbf{p}^0 a_\mathbf{p}^\dagger a_\mathbf{p} + \frac{1}{V} \sum_{\mathbf{pkq}} V_\mathbf{q} a_{\mathbf{p+q}}^\dagger a_{\mathbf{k-q}}^\dagger a_\mathbf{k} a_\mathbf{p} \qquad (1.9)$$

where the kinetic energy term $\epsilon_\mathbf{p}^0 \sim (p^2/2m) - \mu$, includes the chemical potential, and $V_\mathbf{q}$ is the Fourier transform of the interaction potential. We will examine low temperature properties, $T \ll T_{BEC}$, assuming $\langle n_{\mathbf{k}=0}\rangle = N_0 \sim \mathcal{O}(N) \gg 1$. The number of particles in the condensate at $T = 0$ is reduced from N due to interactions but still remains macroscopic.

The free problem, $V = 0$, has a Bose condensed ground state and even if V were weak we cannot do straightforward perturbation theory [5, 18]. Some degree of self consistency is required in handling interaction effects on the condensate. Here comes the crucial approximation. The boson operators for $\mathbf{k} = 0$ satisfy $[a_0, a_0^\dagger] = 1$. However, in the condensed phase $\langle a_0^\dagger a_0\rangle \gg 1$, so

$$\langle [a_0, a_0^\dagger]\rangle \ll \langle a_0^\dagger a_0\rangle, \langle a_0 a_0^\dagger\rangle . \qquad (1.10)$$

If we neglect the commutator, and treat $a_0, a_0^\dagger$ as 'c-numbers', then operators corresponding to the ground state disappear from the problem, replaced by $a_0 = \sqrt{N_0}$, etc. This was first suggested by Bogolyubov, and the following treatment, from [18], is essentially due to him.

In terms of the operator $\phi^\dagger(\mathbf{x})$ which 'creates' a particle at a point $\mathbf{x}$, the separation between the 'c' number and the operators leads to:

$$\phi^\dagger(\mathbf{x}) = \phi_0 + \frac{1}{\sqrt{V}} \sum_{\mathbf{k}\neq 0} \phi^\dagger(\mathbf{k}) e^{i\mathbf{k}\cdot\mathbf{x}} \qquad (1.11)$$

1.4. Phase transitions and broken symmetry

where $\phi_0 \sim \sqrt{N_0/V}$. We have replaced the operator, $a_0^\dagger$, by its expectation value, assumed to be spatially independent. See the connection with what we had discussed on mean field approximations.

Exercise 1.4 *If the order parameter is $\langle a_0^\dagger \rangle$, can you identify what symmetry is broken?*

If we rewrite the Hamiltonian using this separation of operators for $\mathbf{k} = 0$ and $\mathbf{k} \neq 0$, and use, $\phi^\dagger(\mathbf{x}) = \phi_0 + \frac{1}{\sqrt{V}} \sum_{\mathbf{k} \neq 0} \phi^\dagger(\mathbf{k}) e^{i\mathbf{k}\cdot\mathbf{x}}$, etc, the terms in the Hamiltonian can be regrouped in terms of powers of ϕ_0, where $\phi_0^2 = n_0$. The Hamiltonian looks like:

$$\begin{aligned} H = & \sum_{\mathbf{p} \neq 0} (\epsilon_\mathbf{p}^0 - \mu + n_0 V_0 + n_0 V_\mathbf{p}) a_\mathbf{p}^\dagger a_\mathbf{p} \\ & + \sum_{\mathbf{p} \neq 0} \frac{n_0 V_\mathbf{p}}{2} (a_\mathbf{p}^\dagger a_{-\mathbf{p}}^\dagger + a_\mathbf{p} a_{-\mathbf{p}}) + \frac{1}{2} n_0 N_0 V_0 \\ & + \sum_{\mathbf{pq} \neq 0} \sqrt{n_0} \frac{V_\mathbf{q}}{2} (a_{\mathbf{p}+\mathbf{q}}^\dagger a_\mathbf{q} a_\mathbf{p} + a_{\mathbf{p}+\mathbf{q}}^\dagger a_{-\mathbf{q}}^\dagger a_\mathbf{p}) \\ & + \sum_{\mathbf{pkq} \neq 0} \frac{V_\mathbf{q}}{2} a_{\mathbf{p}+\mathbf{q}}^\dagger a_{\mathbf{k}-\mathbf{q}}^\dagger a_\mathbf{k} a_\mathbf{p}. \end{aligned} \quad (1.12)$$

Exercise 1.5 *a) Why do we not have a term with only one $a_\mathbf{p}$ or $a_\mathbf{p}^\dagger$?*
b) Try to draw the Feynman graphs for the various vertices in H [5].

The salient features of H are:

- The Hamiltonian conserves total momentum.

- The number of bosons (in finite $\mathbf{k}$ states) is *not* conserved. Treating $a_0^\dagger$ as a c number has led to this situation. $N'_{op} = \sum_{\mathbf{p} \neq 0} a_\mathbf{p}^\dagger a_\mathbf{p}$ does not commute with H. Since $N = N_0 + \langle N'_{op} \rangle$, this implies that the total number of particles is not exactly conserved.

- The terms are classified accoring to powers of n_0. To obtain the spectrum of the system, to lowest order in V and retaining highest power of n_0, it is sufficient to work with the first two terms in H.

Since we are interested in the 'weak coupling' regime

$$H_{eff} = \sum_{\mathbf{p}\neq 0} \epsilon_{\mathbf{p}} a_{\mathbf{p}}^\dagger a_{\mathbf{p}} + \sum_{\mathbf{p}\neq 0} V'_{\mathbf{p}}(a_{\mathbf{p}}^\dagger a_{-\mathbf{p}}^\dagger + a_{\mathbf{p}} a_{-\mathbf{p}}) + \frac{1}{2} n_0 N_0 V_0 \quad (1.13)$$

where $\epsilon_{\mathbf{p}} = \epsilon_{\mathbf{p}}^0 - \mu + n_0 V_0 + n_0 V_{\mathbf{p}}$, and $V'_{\mathbf{p}} = n_0 V_{\mathbf{p}}/2$ and we have dropped terms involving interaction between particles outside the condensate. H_{eff} is a bilinear in boson operators and can be diagonalised by a canonical transformation. The number non-conserving terms $a^\dagger a^\dagger$, etc, conserve total momentum, so the eigenoperators have to be combinations of the form $g_{\mathbf{k}} a_{\mathbf{k}}^\dagger + h_{\mathbf{k}} a_{-\mathbf{k}}$. The transformation to the new (α) basis is given by

$$\begin{aligned} a_{\mathbf{p}} &= u_{\mathbf{p}} \alpha_{\mathbf{p}} - v_{\mathbf{p}} \alpha_{-\mathbf{p}}^\dagger \\ a_{\mathbf{p}}^\dagger &= u_{\mathbf{p}} \alpha_{\mathbf{p}}^\dagger - v_{\mathbf{p}} \alpha_{-\mathbf{p}} \end{aligned} \quad (1.14)$$

The transformation will be canonical if the new bosons obey the usual commutation rules, and preserve $[a_{\mathbf{p}}, a_{\mathbf{p}}^\dagger] = 1$. This requires

$$u_{\mathbf{p}}^2 [\alpha_{\mathbf{p}}, \alpha_{\mathbf{p}}^\dagger] - v_{\mathbf{p}}^2 [\alpha_{-\mathbf{p}}, \alpha_{-\mathbf{p}}^\dagger] = 1$$

i.e.,

$$u_{\mathbf{p}}^2 - v_{\mathbf{p}}^2 = 1 . \quad (1.15)$$

We can still choose the ratio $u_{\mathbf{p}}/v_{\mathbf{p}}$, which we use to eliminate 'off diagonal' terms in H_{eff}. This can be done if

$$\begin{aligned} u_{\mathbf{p}}^2 &= \frac{1}{2}(1 + \frac{\epsilon_{\mathbf{p}}}{\omega_{\mathbf{p}}}) \\ \text{and } v_{\mathbf{p}}^2 &= \frac{1}{2}(\frac{\epsilon_{\mathbf{p}}}{\omega_{\mathbf{p}}} - 1) , \end{aligned} \quad (1.16)$$

where $\omega_{\mathbf{p}}^2 = \epsilon_{\mathbf{p}}^2 - (2V'_{\mathbf{p}})^2$. We have to fix the two parameters, μ and n_0, to arrive at the spectrum.

1.4. Phase transitions and broken symmetry

The chemical potential is defined as $\partial E_0/\partial N$. The lowest order result, in V, is obtained by replacing all $a, a^\dagger$, by $N_0^{1/2} \sim N^{1/2}$. In this limit, of weak interaction and $T \to 0$, $\mu = n_0 V_0$. The order parameter, $n_0^{1/2}$, should be self-consistently calculated, but again in the low T, weak interaction, limit, $n_0 \approx n$. At finite T self consistency is essential. Whatever may be the self-consistent value of n_0, the spectrum has the following form. Substituting μ, $\epsilon_{\mathbf{p}} = \epsilon_{\mathbf{p}}^0 + n_0 V_{\mathbf{p}}$, we get

$$\omega_{\mathbf{p}} = (\frac{p^2}{m} n_0 V_p + \frac{p^4}{4m^2})^{1/2}. \qquad (1.17)$$

Since V_p is regular for $p \to 0$, at long wavelengths, $\omega_{\mathbf{p}} \sim v_s p$, where $v_s = \sqrt{n_0 V_0/m}$. *The elementary excitations of the system are phonons*, i.e., density fluctuations, and not particle like excitations as in the free Bose gas.

This essentially 'solves' the problem, and we have $H = \sum_{\mathbf{p}} \omega_{\mathbf{p}} \alpha_{\mathbf{p}}^\dagger \alpha_{\mathbf{p}}$. The spectrum involves the condensate fraction and interparticle interactions in an essential way. The condensate was present in the free system as well, so the free system is a sensible starting point, but interactions completely modify the nature of low lying states.

Exercise 1.6 *What is the ground state wavefunction, i.e., the wavefunction corresponding to the quasiparticle vacuum: $|\psi_0\rangle$ such that $\alpha_{\mathbf{p}}|\psi_0\rangle = 0$?*

We summarize below and mention some issues [21] which we have not been able to discuss:

- Ground state energy: the free Bose system has ground state energy zero ($k = 0$ state and no interactions). The ground state energy of the interacting system has an expansion of the form

$$E_0/N \sim \frac{n f_0}{m}(1 + \mathcal{O}(\sqrt{n f_0^3})) \qquad (1.18)$$

where n is the particle density and f_0 the s-wave scattering length $\sim V_0$. This reveals that the 'expansion parameter' in the Bose problem is $\sqrt{n f_0^3}$. This would be small for weak interactions (as in the Bogolyubov case), or for a dilute strongly repulsive system. This result goes beyond the mean field theory that we have discussed and actually quantifies the effect of the terms we neglected.

- Nature of the quasiparticle: Look at the long wavelength quasiparticle. We have $\alpha_{\mathbf{q}}^\dagger = u_\mathbf{q} a_\mathbf{q}^\dagger + v_\mathbf{q} a_{-\mathbf{q}}$ (by inverting the transformation). In the limit $q \to 0$, $u_\mathbf{q} = v_\mathbf{q} \sim 1/\sqrt{q}$. So, for $q \ll mv_s$, the quasiparticle is a superposition of a 'bare' particle of momemtum $\mathbf{q}$ and a 'bare' hole of momentum $-\mathbf{q}$. For a free system the quasiparticle would be a 'single particle'. At larger momenta the spectrum reverts to the single particle form.

- Superfluidity: the Bose system is superfluid, with critical velocity v_s.

- How well does all this describe He^4? The above calculation was explicitly for weak short range interactions. A microscopic model for He^4 would have to start with *strong* short range repulsion. The dilute Bose gas model [5] is a more appropriate starting point but even there the parameter regime appropriate to He^4 (high density) cannot be handled. Most of the *qualitative features* above, the phonon like spectrum, superfluidity, etc, are true of He^4, but the renormalisations, of the condensate say, are very strong. We still do not have a quantitative analytic theory of He^4. The best analytic approach to the excitation spectrum is the variational treatment of Feynman and Cohen [18].

- To look at the spatial fluctuations/dynamics of the order parameter and explore non equilibrium phenomena, it is better to define the broken symmetry state via $\langle \phi^\dagger(\mathbf{r}) \rangle \neq 0 \sim \sqrt{n_0} e^{i\theta(\mathbf{r})}$. This is a local intensive variable, akin to $\mathbf{m}(\mathbf{r})$ in a ferromagnet. One can write down an equation of motion for this 'classical field' within a Ginzburg-Landau scheme. This leads to the Gross-Pitaevskii equation [22].

1.5. Normal Fermi systems: model problems

- BEC in trapped atomic gases: Achieving BEC in trapped atomic gases (see [22] for a review) is one of the major experimental breakthroughs in recent years. The gas, with a large but *finite* number of atoms, condenses in a spatially inhomogeneous state, dictated by the trap potential. This is not condensation in a $k = 0$ state. The gases are dilute but due to the confinement, interaction effects are significant. A host of effects, not easily seen or poorly understood in He^4, have now been observed in these systems. These provide a truly new dimension to BEC, and deserve an entire school for their description!

1.5 Normal Fermi systems: model problems

When a Fermi system has long range order, *e.g.*, magnetic or superconducting, we can construct a mean field theory with an appropriate order parameter to obtain a qualitative picture of the phase. Technically, the mean-field problem is 'quadratic', *i.e.*, free particle like, and exactly solvable. When there is no LRO it is not clear what quadratic approximation to make, or if such approximations make any sense at all. In such cases, which we will call 'normal' Fermi systems, as opposed to 'ordered' systems, we can adopt either a phenomenological approach or a microscopic one.

If we are not interested in a detailed microscopic understanding, Landau Fermi liquid theory [14, 15, 16] often provides a consistent description of the low energy excitations in interacting Fermi systems. The shortcomings in this phenomenological approach are twofold:

- We never know *a priori* whether a system will be Fermi liquid: it may not, and even if it were we have no method of calculating the Fermi liquid corrections. However, when it does work, the Fermi liquid approach affords us an understanding of the 'low energy physics' within a simple conceptual framework.

- Even if the system were a Fermi liquid we can make no statements about its ground state energy (say), or excitations at high frequency or large momenta. Sometimes these features are interesting, and we have to adopt a more microscopic approach.

Microscopic calculations, in a realistic parameter regime, are difficult (if not impossible) to do. We will discuss two microscopic calculations in regimes where they are relatively straightforward. Before that, let us list the 'normal' fermion systems for which we might want a detailed microscopic understanding. These are:

- Electrons in metals! both the traditional Fermi liquids as well as the more recent high T_c cuprates. Also, low electron density systems in two dimensions, since there are interesting recent experiments in these systems.

- Liquid He^3 above its superfluid transition [16].

- Astrophysical objects: white dwarfs, neutron stars, etc.

- The fractional quantum Hall state (FQHE).

What would be the minimal models for describing these systems? In the original order:

- The electron gas: charged particles interacting through the Coulomb interaction. Sometimes the background periodic potential is relevant. Interactions with phonons and the effects of disorder could also be important.

- Dense fermion liquid of neutral particles, interacting via strong short range repulsion.

- High density (neutral?) system (forces?).

- The electron gas again, now in a quantising magnetic field, in the presence of disorder.

1.5. Normal Fermi systems: model problems

Let us postpone writing down the detailed models and discuss what is actually 'solvable' in interacting Fermi systems. We will then try to relate results on these 'minimal' models to real systems.

There are only two non trivial interacting fermion problems 'solvable' in dimension $d > 1$, in a restricted region of parameter space. These are

1. The strongly repulsive (short range, hardcore) dilute Fermi gas, and

2. The electron gas, interacting via long range Coulomb potential, in the limit of high density.

The model with weak short range repulsion can of course be handled through simple perturbation theory. The hardcore gas is supposed to describe a system with strong (singular) short range repulsion. The interatomic potential in He^3 can serve as an illustration of such an interaction. The model is parametrised by the 'scattering length' a, which for a 'hardcore' gas equals the particle diameter, and the Fermi wavenumber k_F. Unfortunately, in its regime of applicability, the solution does not really describe any experimentally realised system. The condition $k_F a \ll 1$ excludes the only possible relevant system liquid He^3; for He^3, $a \approx k_F^{-1}$. However, it serves as an important benchmark, particularly for lattice models with strong repulsion and low particle density. It also provides an explicit verification of the predictions of Fermi liquid theory in a specific microscopic context. Notice that this was also a solvable model for Bose systems.

The electron gas, describing particles in free space interacting via the Coulomb potential, can be thought of as a starting model for understanding the behaviour of conduction electrons in metals. It is usual to use the dimensionless mean interparticle separation r_s as a measure of the (inverse) density in these systems: $r_s = r_0/a_B$, where r_0 is the true interparticle separation and $a_B = \hbar^2/me^2$ is the Bohr radius. High density corresponds to $r_s \to 0$. It is in this regime that we have controlled calculations. It does not quite pertain to metallic densities ($r_s \sim 2-6$), neither does it incorporate 'solid state' periodic potential effects.

Despite these limitations these two very different models, hardcore-low density and long range-high density, serve as useful reference points for understanding the behaviour of real Fermi systems. They also illustrate how one needs to go beyond 'naive' perturbation theory to obtain physically meaningful answers.

1.5.1 Neutral fermions: dilute hardcore Fermi gas

The key to the solvability of the dilute problem is the rareness of multiparticle collisions. The particles have a finite range of interaction and the dominant effects are expected to arise from 'binary collisions'. Because of the low density, it is unlikely that a third particle will be simultaneously involved in this scattering process. The two particle collisions however cannot be treated within perturbation theory, because of the singular short range interaction, and have to be handled exactly. Treatment of multiple collisions between the same pair of particles is essential. For two low energy particles such multiple scattering is completely specified by the 's-wave' scattering length 'a'. The 'many body' nature of the system enters as an exclusion constraint for the intermediate states in scattering. It seems reasonable that we should have a controlled expansion for $a \ll 1/n^{1/3} \approx k_F^{-1}$. Let us look at the two particle problem in vacuum first, then put in the Fermi background.

Suppose the particles interact via a repulsive hard-core potential of strength V_0 and range a_0. The Schrodinger equation in the center of mass frame, for the relative wavefunction, satisfies

$$(\nabla^2 + k^2)\psi(\mathbf{x}) = v(\mathbf{x})\psi(\mathbf{x}) \tag{1.19}$$

where $k^2 = E$, the energy and $v(x)$ is just the hard-core potential mentioned above. The repulsive potential will tend to reduce the amplitude of ψ for $r < a_0$. When considering scattering, the Schrodinger equation can be written as an integral equation for the outgoing wave. Adding on the incident plane wave, the outgoing wave satisfies:

$$\psi_{\mathbf{k}}^{(+)}(\mathbf{x}) = e^{i\mathbf{k}\cdot\mathbf{x}} - \int d^3y\, G^{(+)}(\mathbf{x} - \mathbf{y}) v(\mathbf{y}) \psi_{\mathbf{k}}^{(+)}(\mathbf{y}) \ . \tag{1.20}$$

1.5. Normal Fermi systems: model problems

The Greens function, $G^{(+)}$, is defined via

$$(\nabla^2 + k^2)G^{(+)}(\mathbf{x} - \mathbf{y}) = -\delta(\mathbf{x} - \mathbf{y}) \tag{1.21}$$

and is solved by

$$G^{(+)}(\mathbf{x} - \mathbf{y}) = (1/4\pi)e^{ik|\mathbf{x}-\mathbf{y}|}/|\mathbf{x} - \mathbf{y}| . \tag{1.22}$$

Using this, the asymptotic form of the wavefunction can be written as

$$\psi_{\mathbf{k}}^{(+)}(\mathbf{x})_{x \to \infty} \sim e^{i\mathbf{k}\cdot\mathbf{x}} + f(\mathbf{k}, \mathbf{k}')\frac{e^{ikx}}{x} . \tag{1.23}$$

This identifies the scattering amplitude, which can also be written as

$$\begin{aligned} f(\mathbf{k}, \mathbf{k}') &= \int d^3y\, e^{-i\mathbf{k}'\cdot\mathbf{y}} v(\mathbf{y}) \psi_{\mathbf{k}}^{(+)}(\mathbf{y}) \\ &= \langle \psi_0(\mathbf{k}')|V|\psi(\mathbf{k})\rangle , \end{aligned} \tag{1.24}$$

where $|\psi_0(\mathbf{k}')\rangle$ is the incident plane wave state and $|\psi(\mathbf{k})\rangle$ is the *exact* outgoing state.

From our integral equation for $\psi_{\mathbf{k}}^{(+)}$ it is obvious that the wavefunction has to vanish wherever $v(\mathbf{x}) \to \infty$. Due to that, even for a singular potential (*i.e.*, one without a Fourier transform) the scattering amplitude f defined above remains finite. The potential *drastically alters* the wavefunction from its unperturbed form, making it vanish within the hardcore radius, an effect completely absent in the Born approximation. To obtain this effect, and evaluate the scattering amplitude, the integral equation for $\psi^{(+)}$ has to be solved *exactly*. The result is equivalent to summing all orders in perturbation theory. For hard sphere scattering, this calculation of the phase shifts and scattering amplitude is a standard problem in quantum mechanics [23].

The intent of the foregoing discussion was to demonstrate that multiple scattering is essential to handle hardcore repulsion between a pair of particles. In a *dilute* many particle system such pairwise scattering, summed to infinite order, should be the leading effect. The result differs somewhat from the simple two particle case due to the Fermi background.

We can phrase this in field theory language by classifying the Feynman graphs for the many particle system (see the paper by Galitskii in [4]). When a pair of particles is excited outside the Fermi sea, the dominant effect is the multiple scattering of these particles, compared to, say, the excitation of additional particle-hole pairs. The excitation of a particle-hole pair would involve the Fermi volume and is suppressed for small k_F. You can check that this is a remarkable simplification by trying to draw all the two particle scattering graphs to low order, V^3 say. The parameter which controls the relative magnitude of the neglected processes is $k_F a$, which is small in a low density system. Having selected out a class of processes (called 'ladder diagrams') from an infinite number of possibilities, the task is to sum this set. This is done via the 't-matrix' which sums the two particle scatterings and defines the effective vertex with which to do perturbation theory.

The actual calculation is reproduced in detail in [4, 5]. Here we outline the scheme, skipping the details. The effective vertex, Γ, for two particle scattering is defined by the sum of ladder diagrams. This is best seen graphically, see Fig.1.1.

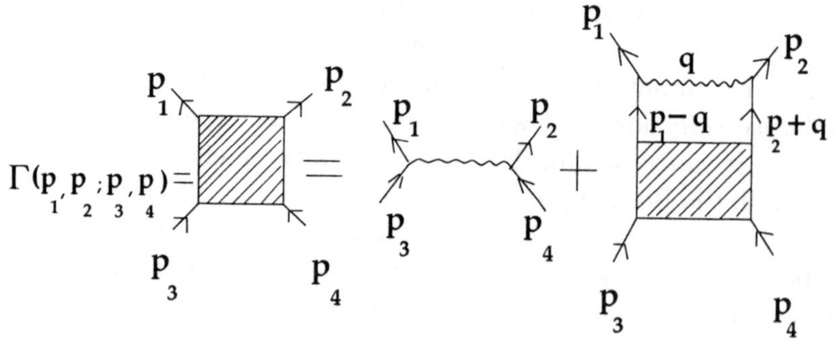

Figure 1.1: Ladder graphs for two particle scattering; integral equation for the effective interaction.

1.5. Normal Fermi systems: model problems

The integral equation for Γ is:

$$\Gamma(p_1, p_2 : p_3, p_4) = V(p_1 - p_3)$$
$$+ \int d^4q V(q) G_0(p_1 - q) G_0(p_2 + q) \Gamma(p_1 - q, p_2 + q : p_3, p_4) \,. \tag{1.25}$$

Here the p_i's are four momenta, and the G_0's are the bare Greens function. This equation encodes the same physics as our earlier integral equation for $\psi_{\mathbf{k}}^{(+)}(\mathbf{x})$ except that now we are able to handle the many particle background. The equation suggests an iterative structure: if you replace Γ on the right hand side with the expression defining the left hand side then,

$$\Gamma(p_1, p_2 : p_3, p_4) = V(p_1 - p_3)$$
$$+ \int d^4q V(q) G_0(p_1 - q) G_0(p_2 + q) V(p_1 - q - p_3) + \mathcal{O}(V^3). \tag{1.26}$$

Iteration generates the successive Born approximations. The special requirement for a hardcore V is that this integral equation be solved exactly. It should be obvious that the Fourier transform $V(q)$ does not even exist for such a potential, so term by term evaluation makes no sense.

Galitskii's method was to phrase the solution of this equation in terms of the *two particle scattering amplitude in free space*. In that case the singular potential drops out of the problem and we just calculate the renormalisation of the vertex due to fermion exclusion. These are corrections of $\mathcal{O}(k_F a)$ and higher to the bare scattering amplitude.

After solving for this 'two particle amplitude' most other quantities can be calculated as integrals over Γ. For instance, the self energy, Fig.1.2, can be calculated as

$$\Sigma(p) = 2 \int d^4k G_0(k) \Gamma(p, k : p, k) - \int d^4k G_0(k) \Gamma(k, p : p, k) \,. \tag{1.27}$$

Here the relative factor arises from spin. The Greens function is $G(\mathbf{p}, \omega) = (\omega - \epsilon_\mathbf{p} - \Sigma(\mathbf{p}, \omega))^{-1}$. From Σ, the energy and damping of quasiparticle excitations can be worked out.

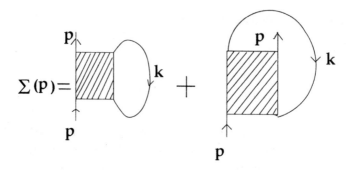

Figure 1.2: Self energy in the dilute 'hard-core' model.

Let us close this section with a look at the specific results obtained using this technique, and a few comments.

- The ground state energy has an expansion in powers of $k_F a$ as expected:

$$\frac{E_0}{N} = \frac{3k_F^2}{10m}(1 + \frac{10}{9\pi}k_F a + \mathcal{O}(k_F^2 a^2)) \ . \tag{1.28}$$

This is calculated from a knowledge of G and Σ.

- Single particle properties have the Fermi liquid form. Excitations near k_F are well defined, they are quasiparticles in the Fermi liquid sense. Specifically:

$$\text{Damping}: \quad \gamma_{\mathbf{p}} \sim E_F(k_F a)^2((p/k_F) - 1)^2$$

$$\text{Effective mass}: \quad m^*/m \sim 1 + k_F^2 a^2 \ . \tag{1.29}$$

- Apart from solvability, part of our motivation for studying this problem was the hope that this 'neutral fermion' model might describe some of the physics of He^3. In the form discussed here, we have only a low density theory. There were early attempts by Brueckner and coworkers to go beyond the low density regime by defining a set of self-consistent equations [4]. This has not proved to be a fruitful method.

1.5. Normal Fermi systems: model problems

Modern approaches to He^3 are of two kinds: (i) using variational wavefunctions which build in the short range correlations, reviewed in [6], or (ii) semi-microscopic theories, variously called 'paramagnon theory' or 'spin-fluctuation theory', where the coupling constant is a (small) phenomenological parameter. The reprint of P. W. Anderson and W. F. Brinkman in [17] discusses this approach. But a true 'field theoretic' first principles theory of liquid He^3, or of liquid He^4, has not proved possible.

1.5.2 Charged fermions: the electron gas

Let us directly write down the Hamiltonian for the 'electron gas'. We will always assume that there is a smeared positive background to ensure overall charge neutrality -

$$\begin{aligned} H &= \int d\mathbf{x} \psi^\dagger(\mathbf{x})(-\nabla^2)\psi(\mathbf{x}) \\ &+ \int d\mathbf{x} d\mathbf{y} \psi^\dagger(\mathbf{x})\psi^\dagger(\mathbf{y}) \frac{e^2}{|\mathbf{x}-\mathbf{y}|} \psi(\mathbf{y})\psi(\mathbf{x}) \,. \end{aligned} \quad (1.30)$$

The basic lengthscale in the problem is set by the particle density; this is $r_0 \sim 1/n^{1/3} \sim k_F^{-1}$. The kinetic energy, when we look at the free electron limit, varies as $k_F^2 \sim 1/r_0^2$. The Coulomb energy, on dimensional grounds, will vary as $\sim 1/r_0$. Here we are not concerned with the correct prefactors but just want to highlight the r_0 dependence. Obviously as $r_0 \to 0$, i.e., the 'high density' limit, the kinetic energy will dominate over the interaction energy. The behaviour should be free Fermi like. We should have a weakly interacting Fermi gas; probably a Fermi liquid.

As we lower the density, by increasing r_0, how long will this description work? At some point the $1/r_0^2$ dependence of the kinetic term will be overtaken by the $1/r_0$ in the potential energy, but that does not tell us where the system is headed. Here an argument due to Wigner comes in useful. Suppose you started from the opposite limit, of zero kinetic energy. The electrons would form a lattice to minimise their electrostatic energy. The energy of such a state will be $\sim 1/r_0$ (again!). What is the effect of kinetic energy, treated as a perturbation, on this solid. As you can try to check, the harmonic vibrations of this lattice lead to a kinetic energy $\sim 1/r_0^{3/2}$. This will be smaller than the electrostatic energy at sufficiently large r_0. The 'Wigner crystal' should be the correct ground state at large r_0.

It would seem that there is a transition in the electron gas, from a 'liquid state' to a localised (Wigner) phase, with decreasing density. What is the critical r_0? and what is the description of the electron liquid? We will try to answer the second question first.

Let us formalise our argument about kinetic versus potential energy and set up a perturbation theory. Rewriting H in the momentum basis:

$$H = \sum_{\mathbf{k},\sigma} \frac{\hbar^2 k^2}{2m} c^\dagger_{\mathbf{k}\sigma} c_{\mathbf{k}\sigma} + \frac{e^2}{2V} \sum_{\mathbf{kpq}} \sum_{\sigma\sigma'} \frac{4\pi}{q^2} c^\dagger_{\mathbf{k+q}\sigma} c^\dagger_{\mathbf{p-q}\sigma'} c_{\mathbf{p}\sigma'} c_{\mathbf{k}\sigma} \quad (1.31)$$

where we have Fourier transformed the $1/r$ potential assuming that we are in three dimensions. There are actually *two* length-scales in the problem. One is r_0, fixed by $4\pi r_0^3 N/3 = V$; the other is the Bohr radius (as in the hydrogen atom): $a_B = \hbar^2/me^2$. The relative magnitude of these two allows us to specify what we mean by 'large' or 'small' r_0, i.e., low or high density, and define the dimensionless 'lengthscale' $r_s \equiv r_0/a_B$. We can write H in terms of dimensionless variables as follows. With r_0 as the unit of length the dimensionless momenta are, $\bar{\mathbf{k}} = r_0 \mathbf{k}$, etc. Setting the reference energy scale $E_{ref} = e^2/a_0$,

$$\frac{H}{E_{ref}} = \frac{1}{r_s^2} \Big(\sum_{\bar{\mathbf{k}},\sigma} \frac{\bar{k}^2}{2} c^\dagger_{\bar{\mathbf{k}}\sigma} c_{\bar{\mathbf{k}}\sigma} + \frac{r_s}{2\bar{V}} \sum_{\bar{\mathbf{k}}\bar{\mathbf{p}}\bar{\mathbf{q}}} \sum_{\sigma\sigma'} \frac{4\pi}{\bar{q}^2} c^\dagger_{\bar{\mathbf{k}}+\bar{\mathbf{q}}\sigma} c^\dagger_{\bar{\mathbf{p}}-\bar{\mathbf{q}}\sigma'} c_{\bar{\mathbf{p}}\sigma'} c_{\bar{\mathbf{k}}\sigma} \Big) .$$
(1.32)

This form shows that in the limit $r_s \to 0$ the potential energy becomes a small perturbation. We seem to have located a control parameter and as long as r_s is small we should be able to do perturbation theory. Notice however that even though the prefactor

1.5. Normal Fermi systems: model problems

r_s may be small, as at high density, the potential is neither weak nor short range.

Let us set up the perturbation theory. First, the Hartree-Fock approximation: this is first order perturbation theory in V_q, the bare Coulomb interaction. We are interested in the electron self energy and the ground state energy. The Hartree contribution vanishes due to overall charge neutrality, $V_{q=0} = 0$. The exchange (Fock) correction to the ground state energy is

$$\langle F|H_{Coul}|F\rangle = -\sum_{\mathbf{q}} V_{\mathbf{q}} \sum_{k,\sigma} \langle F|n_{\mathbf{k}\sigma} n_{\mathbf{k}+\mathbf{q}\sigma}|F\rangle \approx -\frac{0.916}{r_s}. \quad (1.33)$$

Here $|F\rangle$ is the free fermion ground state. The free fermion kinetic energy is $2.21/r_s^2$ so the leading terms in the ground state energy per particle are

$$\left(\frac{E}{NE_{ref}}\right)_{r_s \to 0} \sim \frac{2.21}{r_s^2} - \frac{0.916}{r_s} + \ldots \quad (1.34)$$

and the corrections beyond Hartree-Fock constitute the correlation energy E_{corr}.

While the leading order Coulomb energy meets our expectation, in being small compared to the kinetic energy, there is trouble with the quasiparticle spectrum even at this level. Calculation of the Fock *self energy* (which is frequency independent) reveals that $\partial \Sigma_k / \partial k \to \infty$ for $k \to k_F$. Since the renormalised quasiparticle energy is $\epsilon_k = (k^2/2m) + \Sigma_k$, this implies that the 'effective mass' $(\partial \epsilon_k / \partial k)^{-1}$ vanishes at k_F. This, if true, would imply low temperature anomalies in C_V which have not been experimentally observed. The divergence is an artifact of our approximation; it arises from the bare Coulomb interaction, and hints that screening of the long range interaction is essential to obtain a well behaved theory.

Let the problem worsen somewhat: look at the results for the 'correlation energy' and the higher order corrections to Σ. Try to draw all the second order graphs for Σ. Some examples are shown in Fig.1.3. These involve terms of $\mathcal{O}(V^2)$ and beyond.

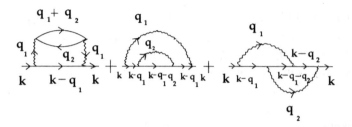

Figure 1.3: Some of the second order self energy graphs for the electron gas.

Concentrate on the term where the two Coulomb lines have the *same momentum q flowing through them*. The expression for this self energy, call it Σ'_2, is given by

$$\Sigma'_2(\mathbf{k}, ik_n) = \int d^3q V_q^2 (\frac{1}{\beta}) \sum_{q_n} \Pi_0(\mathbf{q}, iq_n) G_0(\mathbf{k}+\mathbf{q}, ik_n + iq_n)$$
$$\sim \int \frac{d^3q}{q^4}, \qquad (1.35)$$

where Π_0 is the free electron particle-hole bubble (polarisability). Actual evaluation of the integral confirms that it is indeed logarithmically divergent at small q. This is also true of the contribution to E_{corr} computed from this graph. The problem gets more severe with higher order self energies, and contributions to E_{corr}, where interaction lines of the form V_q^n are involved.

The problem suggests its own remedy as pointed out by Gell-Man and Brueckner (see [4]). The argument runs as follows: (i) the correlation energy of the electron gas *is* finite, since it is the difference between the exact ground state energy and the Hartree-Fock result, both of which are finite. (ii) As demonstrated even earlier by Pines [4] collective electron motion screens the Coulomb field beyond a distance $r_{max} \sim r_0^{1/2} a_B^{1/2} \sim r_0/\sqrt{r_s}$. This suggests that V_q gets modified from the $1/q^2$ form below $q_{min} \sim 1/r_{max} \sim \sqrt{r_s}/r_0$. The system spontaneously generates a lengthscale. This should emerge in a proper calculation of the 'dielectric function'.

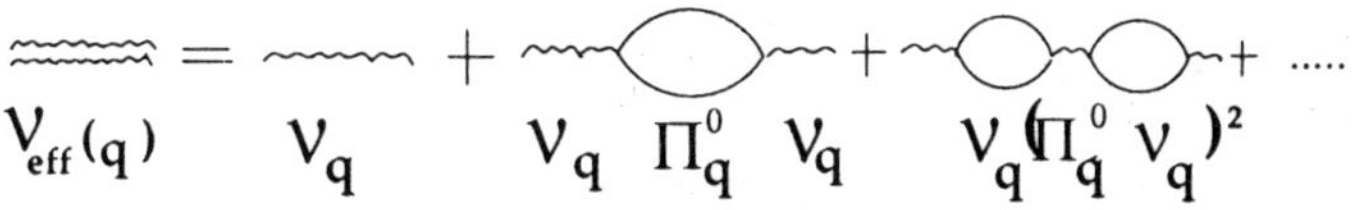

1.5. Normal Fermi systems: model problems

$$V_{eff}(q) = V_q + V_q \Pi^0_q V_q + V_q(\Pi^0_q V_q)^2 + \ldots$$

Figure 1.4: Screening, at RPA level, in the electron gas.

It emerges that the most divergent terms (in the low r_s limit) arise from interaction lines of the form, V_q or $V_q \Pi_0 V_q$, ..., i.e., generally $V_q^{n+1} \Pi_0^n$. The divergence is removed and a physically sensible theory is obtained when one replaces V_q by the *effective interaction*

$$V_{eff}(q) = V_q + V_q \Pi_0 V_q + V_q \Pi_0 V_q \Pi_0 V_q + \ldots = \frac{V_q}{1 - \Pi_0 V_q} . \quad (1.36)$$

This is the famous 'random phase approximation' (RPA), Fig.1.4. The screened interaction can be calculated from a knowledge of $\Pi_0(\mathbf{q}, \omega)$. This can be evaluated analytically [5, 6] though the algebra is tedious. To check the spatial behaviour of the screened interaction let us examine the static $V_{eff}(q)$. This has the form:

$$V_{eff}(q) = \frac{e^2}{q^2 + \alpha r_s k_F^2 g(q/k_F)} , \quad (1.37)$$

where α is a numerical factor, and $g(x)$ is related to the Lindhard function [5]. Since we are interested in the $q \to 0$ behaviour we just use $g(0)$, which is 1, and write $V_{eff}(q) \sim e^2/(q^2 + \alpha r_s k_F^2)$. The effective potential is *finite* at $q = 0$. This form of V_{eff} is valid only at small q; nevertheless let us Fourier transform this to get a feel for the spatial variation of V_{eff}. This leads to

$$V_{eff}(r) \approx (e^2/r) e^{-r/\lambda} \quad (1.38)$$

where $\lambda \sim 1/(k_F \sqrt{r_s})$. This λ is the 'Thomas-Fermi' screening length (and is related to r_{max} discussed earlier). Within this (somewhat crude) approximation the $1/r$ potential seems to get converted to the Yukawa form. Using the full q dependence of

$g(q/k_F)$ it emerges that, at zero temperature, the decay is actually an oscillatory power law: $V_{eff}(r) \sim (1/r^3)\cos(2k_F r)$.

Here is a summary of the results within RPA:

- For the ground state energy we may have naively expected an expansion of the form

$$\frac{E}{N}_{r_s \to 0} \sim \frac{e^2}{a_0 r_s^2}(a + br_s + cr_s^2 + \ldots) \qquad (1.39)$$

The energy actually has an expansion of the form

$$\frac{E}{N}_{r_s \to 0} \sim \frac{e^2}{a_0 r_s^2}(a + br_s + cr_s^2 \ln r_s + \ldots) \qquad (1.40)$$

We have seen the first two terms. The $\ln r_s$ contribution comes from the RPA. The expansion is not analytic even as $r_s \to 0$.

- We have not looked at the self energy calculation. This is done using V_{eff}, and the answers for the single particle excitations are:
excitation energy:

$$\epsilon_{\mathbf{k}} \sim (k_F^2/2m)((k^2/k_F^2) - Ar_s[(k/k_F) + \text{const}]) , \qquad (1.41)$$

and damping:

$$\gamma_{\mathbf{k}} = (k_F^2/2m) B \sqrt{r_s}((k/k_F) - 1)^2 ; \qquad (1.42)$$

here A and B are numerical factors [4]. The damping reveals that the system is a Fermi liquid.

A few comments:

- The small q divergence of $V(q)$ precludes any straightforward expansion. The task is to screen the long range interaction first! This defines the effective interaction V_{eff} with which to do perturbation theory.

1.5. Normal Fermi systems: model problems

- We have built up a 'picture' of the high density electron liquid: the effective interaction is screened due to particle motion, and the quasiparticles are a combination of the bare electron and its 'screening cloud'. The RPA level description is adequate at small r_s. Real metals, however, are at intermediate density, $r_s \sim 4 - 6$. The RPA, pushed to this regime exhibits various pathologies: *e.g.*, beyond $r_s \sim \mathcal{O}(1)$, the calculated pair distribution function $g(r)$ turns negative. The physics of screening is still relevant but the 'polarisability' Π_0 has to be modified to include charge correlations. These are called 'local field corrections' [6]. While some of these seem to work, *there is really no controlled calculation for the electron liquid at large r_s*.

- Where is the Wigner transition? This seems to occur at very large r_s. In three dimension the critical radius $r_s^c \sim 110$, while in two dimension $r_s^c \sim 40$. The transition from the electron liquid to the Wigner phase is a fascinating problem but there are few analytical tools for studying it. A Quantum Monte Carlo study is given in Ref.[24].

- Interplay with electron-phonon interactions: this takes us to the physics of real metals. We will discuss some of these in the next lecture.

- Disorder + Coulomb interactions: the interplay of these two effects is crucial for an understanding of insulator-metal transitions. This is a vast subject; one reference is [25].

In conclusion:

For the two fermion problems we studied, with short range hardcore repulsion, and long range Coulomb interaction, straightforward perturbation theory is useless. The models, however, should be well behaved, or so one expects on physical grounds. Formally, resummation methods remove the singularity. In the hardcore case it ensures that the wavefunction vanishes within the hardcore radius, imposed via the repeated scattering. For the charged (but overall neutral) system, the first task is to screen the interaction through the RPA.

These initial steps are usually not enough to guarantee satisfactory quantitative results over a wide parameter regime. Their value lies in removing the divergence and making the theory well behaved. They define the new 'renormalised' model on which further approximations can be made.

We studied two very different models, which turned out to be solvable in rather distinct parameter regimes. Is there a common feature? In both of them, the accessible regime corresponded to kinetic energy $\gg$ potential energy ($k_F a$ or r_s small). Sometimes in the opposite regime, K.E $\ll$ P.E, the problem is also solvable. It is the intermediate region which is hardest.

1.6 Electrons and phonons: Migdal-Eliashberg theory

The electron-phonon interaction dominates the physics of metals, except in the oxides where direct electron-electron interaction seems to be the crucial ingredient. Most of the finite temperature resistivity in a metal arises from the scattering of electrons by phonons. The same interaction, as a virtual process instead of real scattering, leads to 'pairing' of electrons and superconductivity in many of the traditional, low T_c, systems.

The full problem in a real metal is extremely complicated as the Hamiltonian will shortly show. In the lowest approximation, electrons propagate in the background of a periodic potential, while the phonons are elementary excitations, $i.e.$, normal mode vibrations, of this periodic structure. Not surprisingly, the electrons 'sense' the lattice vibration; that is the origin of electron-phonon coupling. Additionally there is the direct Coulomb interaction, V_{ee}, among the conduction electrons. It is important to incorporate this because in real materials the effective electron-electron interaction will arise from a combination of V_{ee} and the phonon induced interaction. It is a complex coupled field theory.

We will begin at the beginning, with the simple free electron

1.6. Electrons and phonons: Migdal-Eliashberg theory

gas interacting via the BCS interaction, and build up finally to the full Migdal-Eliashberg theory. The sequence is:

- The free electron gas and the BCS instability.
- The 'real' metal: Migdal approximation for the normal state.
- BCS theory: Greens function formulation.
- Strong interactions and superconductivity: Nambu-Eliashberg scheme.

The Hamiltonian for a metal, in full glory, is:

$$
\begin{aligned}
H &= \sum_{\mathbf{p},\sigma} \frac{p^2}{2m} c^\dagger_{\mathbf{p}\sigma} c_{\mathbf{p}\sigma} + \sum_{\mathbf{q}\lambda} \Omega_{\mathbf{q}\lambda}(b^\dagger_{\mathbf{q}\lambda} b_{\mathbf{q}\lambda} + \frac{1}{2}) \\
&+ \sum_{\mathbf{q}\mathbf{p}\sigma} V(\mathbf{q}) S(\mathbf{q}) c^\dagger_{\mathbf{p}+\mathbf{q}\sigma} c_{\mathbf{p}\sigma} + \sum_{\mathbf{p}\mathbf{k}\mathbf{q}\sigma\sigma'} \frac{4\pi e^2}{q^2} c^\dagger_{\mathbf{p}+\mathbf{q}\sigma} c^\dagger_{\mathbf{k}-\mathbf{q}\sigma'} c_{\mathbf{k}\sigma'} c_{\mathbf{p}\sigma} \\
&+ \sum_{\mathbf{p}\mathbf{p}'\lambda\sigma} g_{\mathbf{p}\mathbf{p}'\lambda} \phi_{\mathbf{p}'-\mathbf{p},\lambda} c^\dagger_{\mathbf{p}'\sigma} c_{\mathbf{p}\sigma} .
\end{aligned}
\qquad (1.43)
$$

The first two terms are 'bare' electron and phonon energies, the third term corresponds to the periodic crystal potential, the structure factor $S(\mathbf{q}) = N\delta_{\mathbf{q},\mathbf{G}}$, where $\mathbf{G}'s$ are reciprocal lattice vectors. $V(\mathbf{q})$ is the (pseudo)potential between the ion core and the electron. Terms 1 and 3 generate the electronic 'band structure'. The fourth term corresponds to electron-electron interaction and the last one is the electron-phonon interaction, with the displacement field

$$\phi_{\mathbf{q}\lambda} = b_{\mathbf{q}\lambda} + b^\dagger_{-\mathbf{q}\lambda}$$

and the coupling constant

$$g_{\mathbf{p}\mathbf{p}'\lambda} = -i\left(\frac{N}{2M\Omega_{\mathbf{p}'-\mathbf{p},\lambda}}\right)^{1/2} (\mathbf{p}' - \mathbf{p}) \cdot \vec{\epsilon}_{\mathbf{p}'-\mathbf{p},\lambda} V(\mathbf{p}' - \mathbf{p}) . \qquad (1.44)$$

Here M is the ionic mass and $\vec{\epsilon}$ the polarisation vector for the phonon.

1.6.1 Weak coupling theory: BCS

It is not easy to see the familiar BCS reduced Hamiltonian in the complicated set of terms above. In this section we will not recapitulate the BCS derivation in detail but point out the innovation that was involved in their solution. This feature is something that we would want to retain when we construct a theory of superconductivity based on the full Hamiltonian.

The BCS approach to superconductivity was partly motivated by Cooper's discovery [26] that, in the presence of a Fermi surface, a pair of electrons added in the state $\mathbf{k}\uparrow, -\mathbf{k}\downarrow$, (with $k > k_F$) form a bound state even for an *arbitrarily weak attractive interaction*. The Pauli principle restricts the intermediate states $(\mathbf{k}'\uparrow, -\mathbf{k}'\downarrow)$ in the scattering to $k' > k_F$. In the absence of interactions, the lowest energy for adding these two particles would have been $2\epsilon_F$, i.e., the energy of the system would have increased by $\Delta E = 2\epsilon_F$. Bound state formation leads to $\Delta E = 2\epsilon_F - 2\hbar\omega e^{-2/N(0)V}$. Here V is a model interaction potential (discussed later), not to be confused with the electron-ion potential. ω is related to the range of V in $\mathbf{k}$ space, and $N(0)$ is the DOS at ϵ_F. Since the interaction actually exists between *all pairs* of electrons in the system, and not just the added pair, this lowering of energy hinted at a possible instability of the Fermi sea towards some unusual paired state. The nature of the ordered state and an estimate of the transition temperature had to await the full BCS calculation.

The only direct interaction between electrons in our starting Hamiltonian is via the Coulomb term. This is repulsive, so the attraction, if any exists, has to come from the phonons. If we ignore the Coulomb term for the moment, and assume that the electron-phonon coupling is weak, then we can 'integrate out' the phonons to $\mathcal{O}(g^2)$. This will generate an effective electron-electron interaction of the form $\sim g^2 D(\mathbf{q}, \omega)$ where $D(\mathbf{q}, \omega)$ is the propagator for the phonons.

In the BCS case we consider this interaction for $\omega = \epsilon_\mathbf{k} - \epsilon_{\mathbf{k}'}$ (the phonon scatters an electron from $\mathbf{k}$ to $\mathbf{k}'$) and only for $\epsilon_\mathbf{k}$, $\epsilon_{\mathbf{k}'}$ within ω_D of ϵ_F. Furthermore we selectively consider only

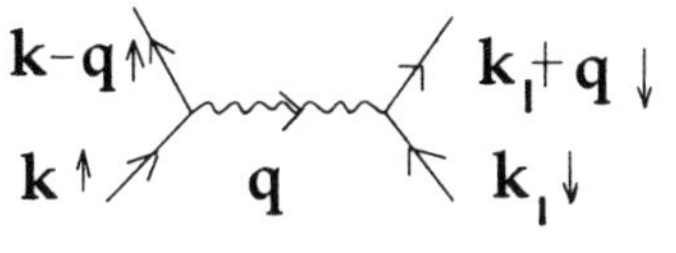

1.6. Electrons and phonons: Migdal-Eliashberg theory

Figure 1.5: The phonon mediated effective electron-electron interaction involved in the BCS theory. The wavy line is $g^2 D(\mathbf{q}, \Omega = \epsilon_k - \epsilon_{k-q})$, where $D(\mathbf{q}, \Omega)$ is the phonon propagator.

pairwise scattering from $|\mathbf{k} \uparrow, -\mathbf{k} \downarrow\rangle$ to states $|\mathbf{k}' \uparrow, -\mathbf{k}' \downarrow\rangle$ and not the more general process $|\mathbf{k} \uparrow, \mathbf{k}_1 \downarrow\rangle \rightarrow |\mathbf{k} - \mathbf{q} \uparrow, \mathbf{k}_1 + \mathbf{q} \downarrow\rangle$, as shown in Fig.1.5. This leads to the BCS 'pairing' Hamiltonian (below). Having seen the 'actual' Hamiltonian, many of these approximations might appear *ad hoc*. Here, we will learn our lessons from the BCS theory, and then we will gradually put back the complexity of the full Hamiltonian.

The BCS 'pairing' Hamiltonian is:

$$H_p = \sum_{\mathbf{k},\sigma} \epsilon_\mathbf{k} c^\dagger_{\mathbf{k}\sigma} c_{\mathbf{k}\sigma} + \sum_{\mathbf{k},\mathbf{k}'} V_{\mathbf{k},\mathbf{k}'} c^\dagger_{\mathbf{k}\uparrow} c^\dagger_{-\mathbf{k}\downarrow} c_{-\mathbf{k}'\downarrow} c_{\mathbf{k}'\uparrow} \quad (1.45)$$

where the interaction $V_{\mathbf{k},\mathbf{k}'} = -V$ for $|\epsilon_\mathbf{k}|, |\epsilon_{\mathbf{k}'}| < \hbar \omega_{ph}$ and vanishes otherwise. ω_{ph} is a typical phonon energy $\approx \omega_D$, the Debye frequency. Given this Hamiltonian, which only involves scattering of pairs $\mathbf{k} \uparrow, -\mathbf{k} \downarrow$, the BCS calculation is a mean field theory in terms of the order parameter $\Delta_\mathbf{k} = -\sum_{\mathbf{k}'} V_{\mathbf{k},\mathbf{k}'} \langle c_{-\mathbf{k}'\downarrow} c_{\mathbf{k}'\uparrow} \rangle$.

The 'mean field' Hamiltonian is:

$$H_m = \sum_{\mathbf{k},\sigma} \epsilon_\mathbf{k} c^\dagger_{\mathbf{k}\sigma} c_{\mathbf{k}\sigma} - \sum_\mathbf{k} (\Delta_\mathbf{k} c^\dagger_{\mathbf{k}\uparrow} c^\dagger_{-\mathbf{k}\downarrow} + \Delta^*_\mathbf{k} c_{-\mathbf{k}\downarrow} c_{\mathbf{k}\uparrow}) \quad (1.46)$$

upto a constant term. The mean field amplitude is to be self-consistently calculated from $\langle c_{-\mathbf{k}'\downarrow} c_{\mathbf{k}'\uparrow} \rangle$ in the eigenstates of H_m. This bilinear Hamiltonian can be easily diagonalised and the quasiparticle energy is given by:

$$E_\mathbf{k} = \sqrt{\epsilon_\mathbf{k}^2 + \Delta_\mathbf{k}^2} \quad (1.47)$$

where $\Delta_{\mathbf{k}}$ is the self-consistent amplitude. In the quasiparticle basis H_m looks like

$$H_m = \sum_{\mathbf{k}} E_{\mathbf{k}} (\gamma^\dagger_{\mathbf{k}0} \gamma_{\mathbf{k}0} + \gamma^\dagger_{\mathbf{k}1} \gamma_{\mathbf{k}1}) \qquad (1.48)$$

where the operators γ, etc, are a linear combination of $c, c^\dagger$. Notice the similarity with the Bogolyubov scheme for bosons. All the excitation properties and responses of the superconductor can be worked out from this quasiparticle Hamiltonian. For reference, the BCS prediction for the transition temperature is

$$T_c = 1.14 \hbar \omega_D e^{-\frac{1}{N(0)V}} . \qquad (1.49)$$

Notice that T_c is non-analytic in V. The superconducting transition cannot be accessed via perturbation theory.

Beyond the specific calculations with a model Hamiltonian and explaining a vast range of phenomena (!), the most significant achievement of BCS was the identification of the 'anomalous amplitude' $\Delta_{\mathbf{k}}$ *which specifies the nature of superconducting order.* Even where the detailed predictions of BCS theory do not hold, the essential nature of the superconducting state is still defined by this 'pairing amplitude', akin to magnetisation in a ferromagnet, but much more subtle.

Let us note the assumptions underlying BCS theory:

- The starting Hamiltonian H_p for BCS starts essentially with free particles which interact via $V_{\mathbf{k},\mathbf{k}'}$. *The normal state excitations are supposed to be Fermi liquid quasiparticles.*

- There is an instantaneous effective interaction, arising from a combination of the screened V_{ee} and phonon induced attraction. (We dropped V_{ee} but it is always present in any real system.)

- The pairing correlations are severely restricted by the requirement of correlated occupancy of 'time reversed' ($\mathbf{k} \uparrow$, $-\mathbf{k} \downarrow$) quasi-particle states. In this way, the correlations specific to the superconducting phase are isolated and taken into account non perturbatively.

1.6. Electrons and phonons: Migdal-Eliashberg theory

The underlying physical model, Landau Fermi liquid theory, is valid for $T, \epsilon_\mathbf{k} \ll \omega_{ph} \sim \omega_D$. For an electron with enough energy to emit 'real' phonons the lifetime is extremely short, so the quasi-particle picture fails. One expects deviations from the BCS model in cases where T_c is high, and ω_D is low.

The deviations from BCS theory can arise in at least three ways:

- the quasi-particle description can become inadequate if the damping rate should become equal to the excitation energy.

- the assumption of an effective two body instantaneous interaction between the quasi-particles may not adequately represent the physics of the retarded, phonon induced, interaction.

- the pairing hypothesis may break down!

Since the attractive interaction arises from phonon exchange, it is resonant for energy transfers $\sim \omega_D$. This means that *an important part of the pairing energy arises from virtually excited quasi-particles of energy $\sim \omega_D$ relative to E_F*. For strong coupling superconductors, the lifetime of a quasi-particle of energy $\sim \omega_D$ is so short that the 'long lived quasi-particle' picture fails. In addition the detailed space-time dependence of the effective electron-electron interaction is significant in these materials.

Empirically, in real metals, the first two BCS assumptions are often violated. The T_c is 'large', the electron lifetimes (at T_c) are short and the effective interaction has significant $\mathbf{q}, \omega$ dependence. These are 'strong coupling' superconductors, and to be able to handle them we need to look at electron-phonon physics in the normal state in more detail. To get a feel for the numbers, T_c/ω_D is approximately as follows in some elements: 0.25% in Al, 1.7% in In, 3.5% in Pb and 5.7% in Hg. Al and In are reasonably well described by BCS theory, while Pb and Hg are in the strong coupling regime.

We need a better starting point to understand superconductivity in metals with strong electron-phonon coupling. The next section discusses such a theory.

1.6.2 The normal state: Migdal theory

What is the physics of electron-phonon coupling in the normal state? Let us consider, in a qualitative fashion, some of the consequences of this coupling in a metal. There are two 'real' processes arising from the electron-phonon interaction:

- At finite T the most obvious effect is electrical resistivity due to scattering of electrons by phonons.

- A related phenomena is phonon attenuation, the absorption of sound wave by an electron gas.

There are also three 'virtual' processes:

- in presence of a phonon field the quasi-particle properties of the electron gas will be changed (encoded in the electron self-energy Σ_{el}).

- the phonon field polarises the electron gas; this in turn modifies the effective interaction between the ions and shifts the phonon frequencies (arises from the phonon self-energy Π_{ph}).

- the phonon field gives a new mechanism of interaction between the electrons (an effective 'vertex').

How do we proceed with a calculation and quantify these effects?

We do not have a 'small parameter' in hand yet, so let us proceed formally. Suppose we ignore the e-e interaction for the time being, as well as band structure effects. The electron self energy $\Sigma(p)$ and the phonon self energy $\Pi(p)$ are given exactly by

$$\begin{aligned}\Sigma(p) &= \int d^4q g_q G(p+q) D(q) \Gamma(p,q,p+q) \\ \Pi(p) &= \int d^4q g_q G(p+q) G(q) \Gamma(q,p,p+q)\ . \end{aligned} \quad (1.50)$$

Here G is the exact electron propagator and D is the exact phonon propagator, neither of which we know *a priori*. Γ is the vertex

1.6. Electrons and phonons: Migdal-Eliashberg theory

function, which encodes how the effective electron-phonon interaction is modified from the 'bare' coupling g. The variables q, etc, refer to four momenta $(\mathbf{q}, \omega)$. The four momenta in the vertex are labelled as follows: (i) the phonon momentum is chosen to flow *out* of the vertex, (ii) the momenta sequence is, outgoing electron, outgoing phonon, incoming electron. The convention for the vertex and the skeleton graphs for the electron and phonon self energies are shown in Fig.1.6. The expressions for Σ and Π are *not perturbative*. They include all possible renormalisations in G, D and Γ.

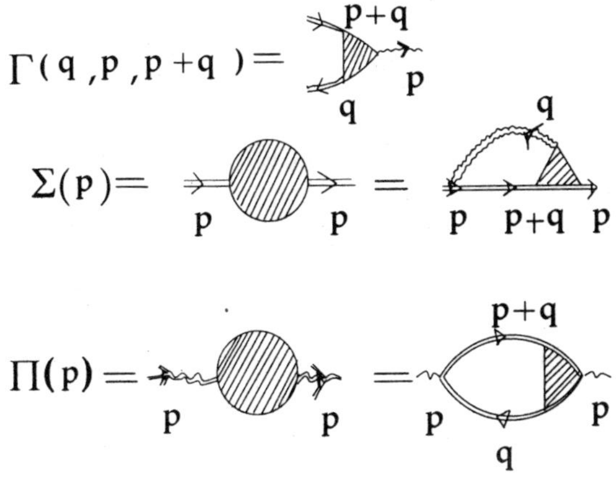

Figure 1.6: Convention for the vertex function Γ, and the normal state electron and phonon self energy.

We also have:

$$\begin{aligned} G^{-1}(p) &= G_0^{-1}(p) - \Sigma(p) \\ D^{-1}(p) &= D_0^{-1}(p) - \Pi(p) \ . \end{aligned} \quad (1.51)$$

These equations are the Dyson equations for the electron and the phonon respectively.

Let us count the number of unknowns in the four equations above. They are G, D, Σ, Π and Γ. The bare propagators G_0 and D_0 are known. We have 5 unknowns and 4 coupled equations for them. *No approximations* have been made till now, nor have any assumptions been made about parameters or coupling constants. Our set of equations is exact, and also quite useless at the moment!

This is where an argument due to Migdal allows a breakthrough. Migdal's work formalises what has been known as the Born-Oppenheimer theorem. This theorem is based on the separation of timescales between the electrons (with $\tau^{-1} \sim$ bandwidth) and the lattice vibrations (with $\tau^{-1} \sim \omega_D$) and states: "the conduction electron wavefunction can be accurately described as following the motion of the ion cores *adiabatically*, upto terms $\sim \mathcal{O}(\sqrt{m/M})$". We can write an electronic wavefunction *as a functional of the ionic coordinates*, $\psi_{el}\{\mathbf{R}_i\}$, though the ions themselves are quantum degrees of freedom. Migdal's theorem, which we will discuss soon, casts this in the language of Green's functions and allows an approximation for the vertex Γ. This enables us to close the set of equations above.

Let us look at the physical content of the approximation first. In this coupled problem one should start by discussing the phonons using electron states which do not contain phonon effects. Why? Although the phonons affect the electrons, they alter electron states only within a Debye energy of the Fermi surface. These are a small fraction of the total number of electrons. However, in considering the effect of electrons on the phonons we shall average over *all* the occupied states of the electron system. This average is affected negligibly by those few electrons which are themselves affected by the phonons. So: solve for the phonons first, using electronic states which have no phonon corrections (this is the adiabatic approximation), then recompute the electronic spectrum using these phonons.

How does all this simplify the vertex function? Expanding order by order in g, the vertex Γ involves the series shown in Fig.1.7. Actual evaluation of low order graphs indicates that

$$\Gamma \sim \Gamma_0 (1 + \mathcal{O}(\sqrt{\frac{m}{M}})) , \qquad (1.52)$$

where Γ_0 is the bare vertex g. So, even if the dimensionless electron-phonon coupling constant is $\mathcal{O}(1)$, the small parameter

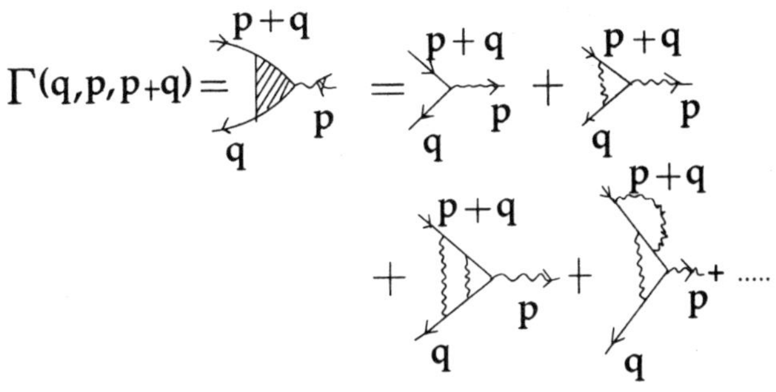

1.6. Electrons and phonons: Migdal-Eliashberg theory

$$\Gamma(q,p,p+q) = \text{(diagram)} = \text{(diagram)} + \text{(diagram)} + \text{(diagram)} + \text{(diagram)} + \cdots$$

Figure 1.7: Low order expansion of the electron-phonon vertex.

m/M allows for a controlled approximation to the vertex. This is Migdal's theorem. Since the vertex is specified, now we have a closed set of equations for the electron and phonon spectra.

Using $\Gamma = \Gamma_0 = g$ the electron spectral properties can be computed from the equations

$$\begin{align} \Sigma(p) &= \int d^4q \, g_q^2 G(p+q) D(q) \\ \Pi(p) &= \int d^4q \, g_q^2 G(p+q) G(q) \, . \end{align} \quad (1.53)$$

The Dyson equations for the electron and phonon (Eq.1.51), and these one loop self energies define *coupled non linear integral equations* for G and D. In actually solving these equations one could do the following: (i) solve the set of four equations self-consistently, (ii) calculate Π using G_0 and substitute the corresponding D into the equation for Σ. (iii) drop Π, replace D by D_0 and solve the integral equation for G. Notice that this already involves solving a self-consistent equation although no order parameter is involved at this stage. These diagrammatic self-consistent approximations in terms of 'dressed' propagators are called *skeleton expansions*, in contrast to a simple perturbative expansion.

What if we do retain the Coulomb term? We have some experience in handling Coulomb effects in the electron gas (Sec. 1.5) albeit in the high density limit. The extension to metallic densities is somewhat *ad hoc*, and the simultaneous inclusion of electron-phonon interaction brings in a host of effects whose relative importance has to be carefully judged [27]. To quote the final result, one replaces the bare electron-electron interaction V_{ee} by the screened interaction $V_{ee}(\mathbf{q})/\epsilon(\mathbf{q})$ (ϵ is the dielectric function) and, when computing the electron self energy, uses a 'total' interaction:

$$V_{tot} = \frac{V_{ee}(\mathbf{q})}{\epsilon(\mathbf{q})} + g^2 D(\mathbf{q},\omega) . \tag{1.54}$$

The solutions of the coupled equations which arise from the Migdal approximation are discussed in [26, 27].

Let us summarise this part. All this was supposed to impress on you that as far as electron-phonon interaction in a metal is concerned we have a believable theory which allows calculations accurate to $\mathcal{O}(\sqrt{m/M})$, remarkable in a strong coupling many body theory. There has been no 'quasiparticle' assumption. However, there is one question which should bother you. If we have such an accurate theory of the metal, can we also predict the superconducting transition from it?! It turns out that the processes which lead to the superconducting instability (the 'Cooper channel') are not incorporated in the Migdal theory! The Migdal equations have to be expanded to include the possibility of pairing. Before we look at that theory, which was formulated by Eliashberg, let us take a look at BCS theory in the Greens function formulation. This will ease the transition to Eliashberg.

1.6.3 BCS theory: Greens function approach

Before starting to draw Feynman graphs it is worthwhile to look again at the mean field form of the BCS Hamiltonian given by

$$H_m = \sum_{\mathbf{k},\sigma} \epsilon_\mathbf{k} c^\dagger_{\mathbf{k}\sigma} c_{\mathbf{k}\sigma} - \sum_\mathbf{k} (\Delta^+_\mathbf{k} c^\dagger_{\mathbf{k}\uparrow} c^\dagger_{-\mathbf{k}\downarrow} + \Delta^-_\mathbf{k} c_{-\mathbf{k}\downarrow} c_{\mathbf{k}\uparrow}) . \tag{1.55}$$

We have used a slightly different notation in H_m here; to compare with our earlier expression $\Delta^+_\mathbf{k} = \Delta_\mathbf{k}$ and $\Delta^-_\mathbf{k} = \Delta^*_\mathbf{k}$. What are the processes implied by H_m? (*i*) There is propagation of an

1.6. Electrons and phonons: Migdal-Eliashberg theory

electron with energy $\epsilon_{\bf k}$, (ii) There can be 'creation' of a pair, $c^\dagger_{{\bf k}\uparrow} c^\dagger_{-{\bf k}\downarrow}$ with amplitude $-\Delta^+_{\bf k}$, similarly 'annihilation' with amplitude $-\Delta^-_{\bf k}$. (iii) There are no 'many body' (four fermion) terms, nor is there any scattering of the form $c^\dagger_{{\bf k}+{\bf q}} c_{\bf k}$, i.e., no 'normal self energy'.

Why recount this in detail? By using the 'amplitudes' described above we can write down the generalised Dyson equation for this system and directly 'diagonalise' H_m. Here is how:

Since there are amplitudes for propagation ($c^\dagger c$) as well as pair creation and annihilation, we have to bring in both the 'normal' Greens function, $G_{\bf k} = \langle T c^\dagger_{{\bf k},\uparrow}(\tau) c_{{\bf k},\uparrow}(0) \rangle$, and the anomalous Greens function $F^+_{\bf k} = \langle T c^\dagger_{{\bf k},\uparrow}(\tau) c^\dagger_{-{\bf k},\downarrow}(0) \rangle$ and its conjugate $F^-_{\bf k}$. These are the 'propagators' while the vertices are defined by $\Delta^\pm_{\bf k}$, shown in Fig.1.8. This is a 'one body potential' just as you would have for impurity scattering, except, here you *conserve momentum but do not conserve particle number*. The Dyson equation will now be a matrix equation involving G, F^+, F^-, etc. Writing out the equations for G and F^+ by using our 'vertices' we have

$$G_{\bf k} = G^0_{\bf k} - G^0_{\bf k} \Delta^-_{\bf k} F^+_{\bf k}$$

$$\text{and } F^+_{\bf k} = -G^0_{-{\bf k}} \Delta^+_{\bf k} G_{\bf k} \,. \tag{1.56}$$

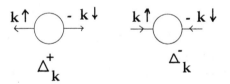

Figure 1.8: 'Vertices' in the Green's function version of BCS theory.

Combining the two equations we have

$$G_{\bf k} = G^0_{\bf k} \{ 1 - G^0_{\bf k} G^0_{-{\bf k}} |\Delta_{\bf k}|^2 \}^{-1} \,, \tag{1.57}$$

and on using $G_{\mathbf{k}}^0 = (\omega - \epsilon_{\mathbf{k}})^{-1}$ we have

$$G_{\mathbf{k}} = \frac{\omega + \epsilon_{\mathbf{k}}}{(\omega - E_{\mathbf{k}})(\omega + E_{\mathbf{k}})}$$

$$\text{and } F_{\mathbf{k}}^+ = -\frac{\Delta_{\mathbf{k}}^+}{(\omega - E_{\mathbf{k}})(\omega + E_{\mathbf{k}})} \quad (1.58)$$

where $E_{\mathbf{k}} = \sqrt{\epsilon_{\mathbf{k}}^2 + |\Delta_{\mathbf{k}}|^2}$. We need $\Delta_{\mathbf{k}}$ to complete the calculation. The self consistency condition fixes this amplitude. Thus:

$$\begin{aligned}
\Delta_{\mathbf{k}} &= -\sum_{\mathbf{k}'} V_{\mathbf{k},\mathbf{k}'} \Delta_{\mathbf{k}'} \\
&= -\sum_{\mathbf{k}'} V_{\mathbf{k},\mathbf{k}'} F_{\mathbf{k}'}^+(\tau = 0) \\
&= -\sum_{\mathbf{k}'} V_{\mathbf{k},\mathbf{k}'} (\frac{1}{\beta}) \sum_n F_{\mathbf{k}'}^+(i\omega_n) \, . \quad (1.59)
\end{aligned}$$

F^+ depends on Δ through the Dyson equation. The Dyson equation combined with the equation above is the Greens function version of BCS theory.

Let us now see how a proper strong coupling theory of the superconducting phase looks.

1.6.4 Strong coupling superconductivity: Eliashberg theory

We have seen a strong coupling theory for the metal, as well as the BCS theory, formulated in the language of Greens functions The Migdal theory does not describe the superconductor, while BCS oversimplifies the normal state. The Eliashberg theory combines the strengths of these two approaches. It is 'mean field like' in that it calculates some amplitudes self consistently. However, it goes beyond mean field in that it does not assume a single particle picture. It is formulated not in terms of a static self-consistent 'potential' but the Greens functions and self energy of the system, with damping and retardation effects built in. For most metals it comes close to being an 'exact solution' for superconductivity within the defined Hamiltonian.

1.6. Electrons and phonons: Migdal-Eliashberg theory

The theory can be cast in a formal and elegant form proposed by Nambu [26, 27] using spinor notation. This approach defines the operator $\Psi_\mathbf{k} = (c_{\mathbf{k}\uparrow}, c^\dagger_{-\mathbf{k}\downarrow})^T$ and its adjoint $\Psi^\dagger_\mathbf{k}$ and sets up the theory in terms of the matrix Greens function connecting these. We will avoid this matrix notation and work explicitly with the G's and $F^\pm$. As in the last section we are still interested in .

$$\begin{aligned}
G_\mathbf{k} &= \langle T c_{\mathbf{k}\uparrow}(\tau) c^\dagger_{\mathbf{k}\uparrow}(0) \rangle \\
F^-_\mathbf{k} &= \langle T c_{\mathbf{k}\uparrow}(\tau) c_{-\mathbf{k}\downarrow}(0) \rangle \\
F^+_\mathbf{k} &= \langle T c^\dagger_{-\mathbf{k}\downarrow}(\tau) c^\dagger_{\mathbf{k}\uparrow}(0) \rangle \\
\tilde{G}_{-\mathbf{k}} &= \langle T c^\dagger_{-\mathbf{k}\downarrow}(\tau) c_{-\mathbf{k}\downarrow}(0) \rangle .
\end{aligned} \quad (1.60)$$

G and $\tilde{G}$ are 'normal' Greens function for up spin electrons and down spin holes respectively (they are 'normal' in the sense that they survive in the normal phase). The $F^\pm$ are the 'anomalous' propagators associated with the order parameter. At 'equal time', $\tau = 0$, they can be related to the BCS order parameter as we have seen in the previous subsection.

Let us first write the 'Dyson equation' for the G's and F's. Fig.1.9 shows the diagrammatic form for the Dyson equation, now involving both normal and 'anomalous' propagators and self energies -

$$G_\mathbf{k} = G^0_\mathbf{k} + G^0_\mathbf{k} \Sigma^n_\mathbf{k} G_\mathbf{k} + G^0_\mathbf{k} \Sigma^-_\mathbf{k} F^+_\mathbf{k}$$

$$\text{and} \quad F^+_\mathbf{k} = G^0_{-\mathbf{k}} \Sigma^n_{-\mathbf{k}} F^+_\mathbf{k} + G^0_{-\mathbf{k}} \Sigma^+_\mathbf{k} G_\mathbf{k} . \quad (1.61)$$

Here Σ^n is the 'normal' self energy and the $\Sigma^\pm$ are the 'anomalous' self energies, or generalised versions of $\Delta^\pm$ in BCS theory. We can write the equations for $\tilde{G}$ and F^- as well but they are related by symmetry to the equations above. All the functions above depend on frequency ω; we have suppressed it for notational convenience.

What is the principle involved in writing these equations?

(i) G conserves particle number (overall), so it has to be built from 'blocks' of the form $\Sigma^n G$ or $\Sigma^- F^+$ or $\Sigma^+ F^-$. Since an iterative structure is implied by using the full propagators G, F, etc, on the

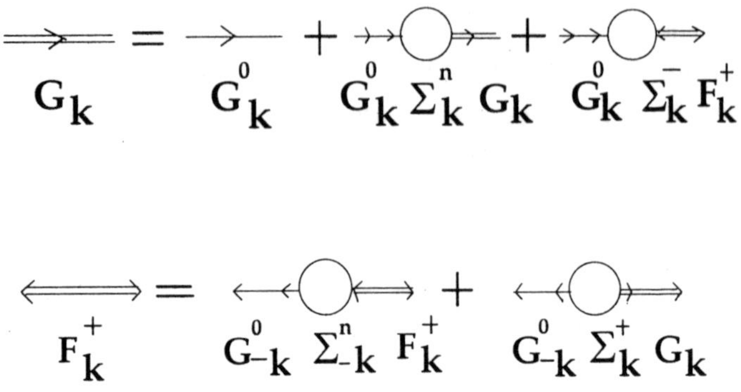

Figure 1.9: Diagrammatic representation for the Dyson equation within Migdal-Eliashberg theory.

right hand side, it is sufficient to use G^0 and the two other terms given above in Eq.1.61. The others are generated by iteration or coupling via the second equation. G can propagate as in the normal state, or via anomalous amplitudes in the intermediate states.

(ii) There is no bare propagator for F^+ since F^+ does not survive in the unordered phase. Unlike G, both the terms contributing to F^+ involve anomalous amplitudes which, self consistently, appear only for $T \leq T_c$.

The Dyson equation can be inverted to write the G's and F's in terms of Σ^n and $\Sigma^\pm$:

$$\begin{aligned}G_{\mathbf{k}} &= \frac{G_{\mathbf{k}}^0(1 - G_{-\mathbf{k}}^0 \Sigma_{-\mathbf{k}}^n)}{(1 - G_{\mathbf{k}}^0 \Sigma_{\mathbf{k}}^n)(1 - G_{-\mathbf{k}}^0 \Sigma_{-\mathbf{k}}^n) - G_{\mathbf{k}}^0 G_{-\mathbf{k}}^0 \Sigma_{\mathbf{k}}^+ \Sigma_{\mathbf{k}}^-} \\ &= \frac{G_{-\mathbf{k}}^{0\,-1} - \Sigma_{-\mathbf{k}}^n}{(G_{\mathbf{k}}^{0\,-1} - \Sigma_{\mathbf{k}}^n)(G_{-\mathbf{k}}^{0\,-1} - \Sigma_{-\mathbf{k}}^n) - \Sigma_{\mathbf{k}}^+ \Sigma_{\mathbf{k}}^-} \; ; \end{aligned} \quad (1.62)$$

this is a generalised version of the BCS equation for G, which corresponds to $\Sigma^n = 0$ and $\Sigma^\pm = \Delta$. (The equations for F can be similarly found). One can also check that the normal state Dyson equation is recovered for $\Sigma^\pm = 0$.

To complete the theory we need a prescription for Σ^n and

1.6. Electrons and phonons: Migdal-Eliashberg theory

$\Sigma^{\pm}$. We are not making a mean field approximation so we will not factorise the interaction terms in our Hamiltonian. We follow the Migdal prescription for our Σ's. According to this scheme a self-consistent one loop approximation suffices in the normal phase, with $\Gamma = g$. We use the same prescription here, shown in Fig.1.10, except that (i) the G which enters Σ^n itself satisfies a more complicated Dyson equation, and (ii) now there are also anomalous, one loop, self energies, $\Sigma^{\pm}$, constructed out of the $F's$. The full set of equations is shown in the next page.

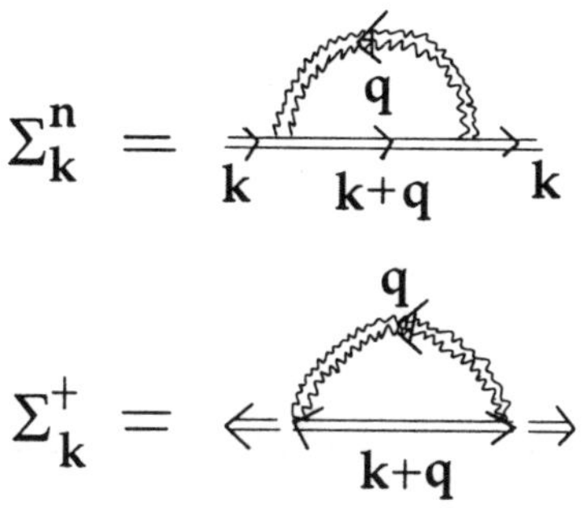

$$\Sigma_k^n = \quad \underset{k \quad\quad k+q \quad\quad k}{\overset{q}{\longrightarrow}}$$

$$\Sigma_k^+ = \quad \underset{k+q}{\overset{q}{\longleftarrow\longrightarrow}}$$

Figure 1.10: Normal and anomalous self energies within the Migdal-Eliashberg theory.

A few comments

1. Unlike BCS, the *vertices* here are not number non conserving. They are defined by the original Hamiltonian. However, the $\Sigma^{\pm}$ can be anomalous because they involve $F^{\pm}$. Here the $\Sigma^{\pm}$ assume the role of $\Delta^{\pm}$ in BCS.

2. The normal state self energy Σ^n is non trivial, this is what introduces damping. We had $\Sigma^n = 0$ within BCS.

3. The $\Sigma^{\pm}$ are *frequency dependent*, the order parameter is not a static amplitude anymore.

4. Solving the integral equations for the self energies is equivalent to the BCS self-consistency. There is no additional self

consistency involved (unlike the Greens function formulation of BCS).

Given below is the *full set* of Eliashberg equations (if you wanted to set eyes on them!). Limiting cases lead you to BCS or Migdal theory. We use four momentum notation.

$$\begin{align}
G_k &= G_k^0 + G_k^0 \Sigma_k^n G_k + G_k^0 \Sigma_k^- F_k^+ \\
F_k^+ &= G_{-k}^0 \Sigma_{-k}^n F_k^+ + G_{-k}^0 \Sigma_k^+ G_k \\
\Sigma_k^n &= \int d^4q \, G_{k+q} V_{tot}^q \\
\Sigma_k^\pm &= \int d^4q \, F_{k+q}^\pm V_{tot}^q \\
V_{tot}^q &= \frac{V_{ee}}{\epsilon_q} + g^2 D_q \\
D_k &= D_k^0 + D_k^0 \Pi_k D_k \\
\Pi_k &= \int d^4q \, g_q^2 G_{k+q} G_q \, .
\end{align}$$
(1.63)

Exercise 1.7 *Check that the limiting cases of the above set of equations lead you to the BCS or Migdal theory.*

Before we close:

- We have an intimidating set of coupled equations. Solving these is an art form and quite at the frontier even with the computational resources available today.

- The relation between Δ, ω_D and T_c is quite involved in strong coupling superconductors; there is no exact analytic form (as can be guessed from the equations above). Reference [27] discusses some limiting cases.

- The order parameter is complex and frequency dependent, in contrast to the real BCS 'gap'. This is reflected in the single particle spectrum and tunneling properties.

- Vertex corrections beyond Migdal: divergence of the ladder sum. There is a family of graphs not included in the Migdal scheme, being formally of order $\sqrt{m/M}$, whose (infinite) sum diverges at a certain temperature. The long range order implied by the divergence is taken into account by the $F^{\pm}$ in the Eliashberg generalisation. However, *superconducting fluctuation effects in the normal state are not covered by the Eliashberg approach.*

1.7 Conclusion: 'field theory' and many particle physics

We are not starting all over again! At the end of this long set of lectures maybe it is just worth asking 'what have 'field theoretic' methods contributed to our understanding of many particle physics?'

Historically, field theoretic methods were those developed by Feynman, Dyson, Schwinger, and others for understanding QED. They were 'field theoretic' in that one had to go beyond quantum mechanics, and the Schrodinger equation, to set up a theory. There is no way to write a 'many particle' Schrodinger equation for a quantum field, since often the particle number is not conserved. Technically, one has to work with Green's functions, employing the methods of diagrammatic perturbation theory.

Unlike quantum field theory (qft), there *is* a Schrodinger equation governing many particle phenomena in condensed matter. Particle number (of electrons, atoms, etc) is conserved at the energies of interest, and, if you wish, you could work with wavefunctions involving all their coordinates! In fact variational calculations, which predate field theoretic methods, employ precisely such an approach. The transition in the 1950's was from these early variational calculations to diagrammatic methods, pioneered by Gell-Man, Brueckner, etc. The advantage of this method was that one could access all the physical properties, and not just the ground state wavefunction. Most of the problems we discussed were 'solved' in 50's, employing the then newly available techniques of qft. These classic problems include the dilute Bose gas (Bogolyubov, Beliaev), the fermion problems (Gell-Man & Brueckner, Pines, Galitskii), electron-phonon physics (Migdal), superconductivity (Gorkov, Eliashberg). *These were all semi-analytic solutions*, depending on some small parameter for a controlled calculation using diagrammatic methods.

Our intuition today about these *model problems* dates back to these solutions. However, where *real systems* are concerned, *e.g.*, the He liquids, or the low density electron gas, the methods of field theory are not directly applicable. We are no longer in a perturbative regime. Here one falls back on variational calculations, or numerical simulations. So, the physical picture we have of interacting systems is based in large measure on perturbative field theoretic methods, but detailed quantitative answers often depend on numerical simulations.

Several modern problems in condensed matter also employ field theory; celebrated examples are Chern-Simons theory in the quantum Hall effect, non linear sigma models in the theory of electron localisation, and gauge theories for strongly interacting fermions. Some of these are discussed in [28, 29]. If you think you have mastered the classics, move on to them!

Bibliography

[1] A. A. Abrikosov, L. P. Gorkov and I. E. Dzyaloshinskii, *Methods of Quantum Field Theory in Statistical Physics*, Prentice-Hall, New York (1964).

[2] G. Parisi, *Statistical Field Theory*, Addison Wesley (1988).

[3] C. Itzykson and J. M. Drouffe, *Statistical Field Theory*, Cambridge (1989).

[4] D. Pines, *The Many Body Problem*, Benjamin, New York (1962). One of the clearest introductions to many body physics, and a collection of classic reprints.

[5] A. L. Fetter and J. D. Walecka, *Quantum Theory of Many Particle Systems*, McGraw Hill, New York (1971).

[6] G. D. Mahan, *Many Particle Physics*, Plenum (1990).

[7] J. W. Negele and H. Orland *Quantum Many Particle Systems*, Addison Wesley (1988).

[8] For a discussion on exactly solvable models, and a collection of reprints, see D. C. Mattis, *The Many Body Problem*, World Scientific (1993).

[9] A. Georges et al., Rev. Mod. Phys. **68**, 13 (1996).

[10] See the article by D. Vollhardt in *Correlated Electron Systems: Jerusalem Winter School on Theoretical Physics*, Ed. V. J. Emery, World Scientific (1993)

[11] K. Huang, *Statistical Mechanics*, Wiley (1963).

[12] D. Forster, *Hydrodynamic Fluctuations, Broken Symmetry and Correlation Functions*, Addison-Wesley (1990).

[13] R. Kubo, M. Toda, N. Hashitsume, *Statistical Physics II*, Springer-Verlag (1995)

[14] P. Nozieres, *Interacting Fermi Systems*, Benjamin, New York (1963).

[15] H. J. Schulz, Les Houches lectures, in *Mesoscopic Quantum Physics*, Ed: E. Akkermans *et al.*, North-Holland (1995).

[16] A. J. Leggett, Rev. Mod. Phys. **47**, 331 (1975).

[17] P. W. Anderson, *Basic Notions of Condensed Matter Physics*, Addison-Wesley (1984).

[18] P. Nozieres and D. Pines, *Quantum Liquids II*, Addison Wesley, (1990), (which we have faithfully followed).

[19] H. E. Stanley, *Introduction to Phase Transitions and Critical Phenomena*, Oxford (1971).

[20] P. M. Chaikin and T. C. Lubensky, *Principles of Condensed Matter Physics*, Cambridge (1995).

[21] *Bose-Einstein Condensation*, Ed. A. Griffin, D. W. Snoke and S. Stringari, Cambridge (1995), provides a recent (but pre atom trap BEC) update on the physics of BEC.

[22] There are two detailed reviews on BEC in trapped gases: F. Dalfovo *et al.*, Rev. Mod. Phys. **71** 463 (1999), and H. T. C. Stoof, Les Houches Lecture Notes 'Field Theory for Trapped Atomic Gases' (cond-mat archive: 9910441).

[23] J. J. Sakurai, *Modern Quantum Mechanics*, Addison-Wesley (1995).

[24] D. M. Ceperley and B. J. Alder, Phys. Rev. Lett. **45**, 567 (1980).

[25] *Electron-Electron Interactions in Disordered Systems*, Ed. A. L. Efros and M. Pollak, North-Holland (1985).

[26] See J. R. Schrieffer, *Theory of Superconductivity*, Addison-Wesley (1983), for a complete description of BCS theory and post BCS developments.

[27] See the article by D. J. Scalapino in *Superconductivity*, Vol I, Ed. R. D. Parks, Marcel Dekker (1969).

[28] Les Houches Lecture Notes, *Strongly Interacting Fermions and High T_c Superconductivity*, Ed. B. Doucot and J. Zinn Justin, North Holland (1995).

[29] *Correlated Electron Systems: Jerusalem Winter School on Theoretical Physics*, Ed. V. J. Emery, World Scientific (1993)

Chapter 2

Critical Phenomena: An Introduction from a Modern Perspective

Somendra M. Bhattacharjee

Institute of Physics
Sachivalaya Marg
Bhubaneswar – 751 005

Our aim in this set of lectures is to give an introduction to critical phenomena that emphasizes the emergence of and the role played by diverging length-scales. It is now accepted that renormalization group gives the basic understanding of these phenomena and so, instead of following the traditional historical trail, we try to develop the subject in a way that lays stress on the length-scale based approach.

2.1 Preamble

A phase transition is defined as a singularity in the free-energy or any thermodynamic property of a system. "A system" is to be understood in a general sense; it is defined by a Hamiltonian under various external macroscopic constraints. For example, an isolated collection of particles would be an example of a system, just as the same collection exchanging heat or work with surrounding. It will also be the same system even when it exchanges particles with surroundings. The three examples given are the standard microcanonical, canonical and grand canonical ensembles in statistical mechanics. There are many other types of ensembles. All these ensembles are expected to give the same description of the system in the thermodynamic limit; at least that is what says the principle of equivalence of ensembles.

Phase transitions are generally classified as first order, second order, ..., etc, or first order and continuous type. The behaviour of any system near or at a continuous transition (also called critical point) is distinctly different from the behaviour far away from it. This peculiarity near a critical point is known as critical phenomenon.

Power laws are distinctive features of critical phenomena and can be ascribed to diverging length-scales. This connection is now so well-characterized that any phenomena, equilibrium or nonequilibrium, thermal or nonthermal, showing power laws tend to be

2.1. Preamble

interpreted in the same fashion of equilibrium critical phenomena through the identification of relevant terms, scaling, and diverging lengths.

In these notes, we shall concentrate on the general aspects of critical phenomena. A few features of first order transitions will be touched upon at some point.

One of the most important contributions of the studies of critical phenomena is the shift in point of view to a length-scale based analysis and classifying terms of a Hamiltonian (or in any other description of a problem) as *relevant* or *irrelevant* rather than classifying them as numerically weak or strong. The approach we take focuses on this and related issues.

Our approach is to show that the class of singular behaviour seen near criticality can be characterized and understood via an emergent diverging length scale as the transition point is approached. This way of characterizing singularities find straightforward generalization to many phenomena beyond equilibrium phase transitions.

Defining via singularity is the most general way of characterizing a phase transition. For a large class of systems, singularities could occur due to "ordering" after a phase transition (or symmetry breaking) but that is not necessarily a requirement for all transitions. In other words, the existence of an "order-parameter" like quantity is not a prerequisite for understanding criticality though it might be useful in many contexts. We therefore deviate from the conventional historical approach to the subject via mean field theory, Landau theory and experimental observations leading to scaling.

Outline of the notes: The problem from a theoretical point of view is presented, with a little bit of recapitulation of familiar things, in Secs. 2.2 to 2.4. A possible resolution through scaling is introduced that leads to finite-size scaling. The consequences and physical interpretation leading to length-scale dependent parameters are discussed in Secs. 2.5 to 2.8. The Gaussian model and the ϕ^4 theory are used as examples in Sec. 2.9. The problems (exercises), even if not attempted, should be read for continuity.

Only elementary calculus is used throughout, but the reader is assumed to have the background of statistical mechanics at the level of Reif.

2.1.1 Large system: Thermodynamic limit

In the canonical ensemble, the properties of the system with Hamiltonian H are obtained from the free energy,

$$F = -k_B T \ln Z, \qquad (2.1)$$

where k_B is the Boltzmann constant, T the temperature and $Z = \sum e^{-\beta H}$ is the partition function with $\beta = 1/k_B T$. If a phase transition is a singularity of this free energy, then, for a simple well-defined H, it can occur only if Z has zeros for real values of the parameters. This cannot be the case for a finite-sized system because Z is a sum of a finite number of Boltzmann factors (which are always positive). The zeros of Z or singularities of F can only be for complex values of the parameters. These complex zeros might have real limits when $N \to \infty$ (infinite number of terms to be added in Z) . Then and only then can a phase transition occur.

> **Conclusion:** A phase transition cannot take place in a finite system. The thermodynamic limit, $N \to \infty$, has to be taken.

2.2 Where is the problem?

Statistical mechanics starts with the idea of a micro-canonical ensemble: the Boltzmann formula $S = k_B \ln \Omega$ where Ω is the degeneracy or the number of states accessible to the system under the given conditions of energy (E), volume (V), number of particles (N), etc.

Thermodynamics starts with the basic postulate of the existence of an 'entropy' function S, so that

$$E = E(V, S, N), \quad \text{or} \quad S = S(E, V, N). \qquad (2.2)$$

2.2. Where is the problem?

One may consider either a system of fixed number of particles in equilibrium with surroundings or a large isolated system and focus on a small part which is in equilibrium with the rest. By releasing constraints (like E =const, or V = const, or N = const, etc) one changes the ensemble. This is tantamount to changing the thermodynamic potential required to describe the system.

Mathematically, the change of one thermodynamic potential to another is achieved via Legendre transformation. For example, from entropy to temperature (micro-canonical to canonical) is done via

$$T = \left.\frac{\partial E(V, S, N)}{\partial S}\right|_{V,N}, \qquad (2.3)$$

$$F = E - TS, \qquad (2.4)$$

where F is now a function of (V, T, N). This is possible if and only if Eq.(2.3) can be inverted to eliminate S on the right hand side of Eq.(2.4) in favour of T. This inversion is *possible* if $\partial T/\partial S \neq 0$. Similarly, we require $\partial p/\partial V \neq 0$, $\partial h/\partial M \neq 0$, or $\partial \mu/\partial N \neq 0$, etc, where μ is the chemical potential, h the magnetic field and M the magnetization (for a magnetic system), for change of appropriate ensembles.

Since $\partial S/\partial T$ is related to the heat capacity of the system, we see trouble if the specific heat *diverges* for some values of the parameters of the system. This looks like an algebraic problem of the transformation but it is vividly reflected in the argument for the equivalence of ensembles. So let us recollect that.

Take the case of canonical and micro-canonical ensembles. The two ensembles are equivalent if the energy fluctuation in the canonical case is very small. First note that, if $F \propto N$, $\langle E \rangle$ and $\langle E^2 \rangle - \langle E \rangle^2$ are both proportional to N (more on this later). The angular brackets indicate statistical mechanical averaging. A straight forward manipulation, using derivatives of the partition function, yields (C_V: constant volume total heat capacity)

$$C_V = \frac{1}{k_B T^2}\left(\langle E^2\rangle - \langle E\rangle^2\right) \implies \frac{\Delta E}{E} = k_B T^2 \frac{\sqrt{c_V}}{\sqrt{N}}, \qquad (2.5)$$

where ΔE is the standard deviation of E and c_V is the specific heat (heat capacity per particle). For a thermodynamic system, $N \to \infty$, and this fluctuation goes to zero, provided c_V is finite. Hence the equivalence.

Well, the argument fails when $c_V \to \infty$.

By our definition of phase transitions, a diverging second derivative of the free energy implies a phase transition. A hypothetical possibility? No, it occurs in many systems in many different contexts. Such points will be classified as critical points. Wait for a better definition of a critical point.

Critical points seem to be at odds with the conventional wisdom of thermodynamics and statistical mechanics.

2.3 Recapitulation - A few formal stuff

A consistent thermodynamic description requires a few postulates, some important ones of which are:
(i) Existence of "entropy" (already mentioned).
(ii) Extensivity of E and S.
(iii) Convexity of the free energy.
Postulate (ii) is a simple additivity property, while (iii) is required for stability. The formulation of statistical mechanics guarantees (iii) but not (ii). We need to ensure (ii) to make contact with thermodynamics and this restricts the form of the Hamiltonians we need to consider. (See exercises 2.5 and 2.7 below).

2.3.1 Extensivity

From our experience of macroscopic system, we expect that under a rescaling $V \to bV$, $N \to bN$, and $S \to bS$, total energy should change accordingly, i.e.,

$$E(bV, bS, bN) = bE(V, S, N) \quad \text{(homogeneous function).} \quad (2.6)$$

By choosing $b = 1/N$, (b is arbitrary), we have

$$E(V, S, N) = N \, \mathcal{E}(V/N, S/N), \quad \text{where} \quad \mathcal{E}(x, y) = E(x, y, 1). \quad (2.7)$$

2.3. Recapitulation - A few formal stuff

Note that $\mathcal{E}(v, s)$ is the energy per particle and being a function of v, s (volume and entropy per particle), $\mathcal{E}(v, s)$ is independent of the overall size (e.g., N) of the system. This proportionality to the number of particles (or volume) is *extensivity*.

Exercise 2.1 *Additivity means* $E(V_1, S_1, N_1) + E(V_2, S_2, N_2) = E(V_1 + V_2, S_1 + S_2, N_1 + N_2)$. *Show that this implies Eq.(2.6).*

Exercise 2.2 *Sanctity of extensivity: Remember Gibbs paradox and its resolution?*

The classical thermodynamics is based on Eq.(2.6) which, via the Euler relation for homogeneous functions, gives

$$S\frac{\partial E}{\partial S} + V\frac{\partial E}{\partial V} + N\frac{\partial E}{\partial N} = E = TS - pV + \mu N, \qquad (2.8)$$

defining $T = \partial E/\partial S$, $-p = \partial E/\partial V$ and $\mu = \partial E/\partial N$, where the partial derivatives are taken keeping the other variables constant. Each of the three new parameters defined, T, p and μ, is a derivative of an extensive quantity with respect to another extensive variable. Therefore T, p, μ, and variables like these are independent of the scale factor b of Eq.(3). Such quantities are called *intensive quantities*. [1]

Exercise 2.3 *Take* $b = 1 + \delta l$ *and derive Eq.(2.8) for* $\delta l \to 0$.

Exercise 2.4 *From Eq.(2.6) show that*

$$F(bV, T, bN) = bF(V, T, N), \quad \text{and} \quad F(V, T, N) = N f(v, T). \qquad (2.9)$$

---

[1] Quantities like v, or s, an extensive quantity per unit volume or per particle, are also independent of the scale factor b. To distinguish these from the other type, we may call them "densities" or "fields" (highly confusing for field theorists). In most cases, these could be distinguished from the context.

2.3.2 Convexity: Stability

We recognize that the derivatives needed for the Legendre transforms (see Eqs. (2.3)) are derivatives of conjugate pairs: specific heat, compressibility, susceptibility, etc. These derivatives (second derivatives of free energy) are called *response functions*, because they measure the change in the extensive variable as the externally imposed conjugate intensive quantity is changed.

A generalization of the derivation of Eq.(2.5) for a pair (y, Φ) (Φ being the extensive variable and y the conjugate intensive variable) leads to

$$\left.\frac{\partial \langle \Phi \rangle}{\partial y}\right|_{y=0} = \frac{1}{k_B T}(\langle \Phi^2 \rangle - \langle \Phi \rangle^2)\Big|_{y=0} \qquad (2.10)$$

which connects the response function for Φ to the latter's variance. The latter being a positive definite quantity ensures that the response function is of a particular sign only. This in turn shows that the free energy as a function of y can have only one type of curvature (positive or negative) - a property known as convexity of the free energy. This positivity of the response function guarantees thermodynamic stability. This is the third postulate of thermodynamics mentioned above (quite often stated as the maximization/minimization principle). This connection between response and fluctuation plays a crucial role in the subsequent development, especially in developing a correlation function based approach.

The main points we need here are
(1) thermodynamic potentials are additive and therefore obey a simple scaling.
(2) There are conjugate pairs of variables (T, S), (p, V), (μ, N), etc, where the first one in each set is intensive (*i.e.*, independent of the size) while the second one is extensive (*i.e.*, proportional to the size). One may change variables (*e.g.*, by releasing a constraint) and this is the change of ensemble in statistical mechanics. Two bodies in contact in equilibrium need to have equality of the relevant intensive quantities (remember zeroth law?).

2.3. Recapitulation - A few formal stuff

Comments

- Equivalence of ensembles relies on the sharpness of the probability distribution for say E. The width of the distribution is related to the corresponding response function. For sharp distributions, the probability of E for the combined system can be taken as the product of the individual probabilities (i.e., $P_{V_1+V_2}(E_1 + E_2) = P_{V_1}(E_1)P_{V_2}(E_2)$), whence follows extensivity. Broad distributions would create problem here.

- Broad distributions imply large fluctuations. It is therefore expected that fluctuations would play an important role in critical phenomena. Fluctuations are responsible for the problem with naive extensivity.

- Let us write $\Phi = \int d^d x \, \phi(\mathbf{x})$, where $\phi(\mathbf{x})$ is a local quantity (density). We have written it as a d-dimensional integral, though it would be a sum for discrete systems. The width of the probability distribution for Φ around the mean $\langle \Phi \rangle$ is given by the corresponding "susceptibility" as given by Eq.(2.10). Assuming translational invariance, $\langle \phi(x) \rangle = \langle \phi \rangle$, and denoting the response function per particle by χ, the relative width of $P(\Phi)$ is given by $N^{-1/2}\sqrt{\chi}/\langle \phi \rangle$.

Exercise 2.5 *A very important model that is used extensively is the Ising model.*

$$H = -J \sum_{<ij>} s_i s_j - h \sum_i s_i, \quad J, h > 0, \qquad (2.11)$$

where the spins $s_i = \pm 1$ are situated on a d-dimensional lattice, the interaction could be restricted to nearest-neighbours only (denoted by $<ij>$), and h is an external magnetic field.
(1) Show that the free energy is extensive for any value of T and h, except possibly a particular point. (Difficult).
Show this extensivity at $T = 0$ and $T = \infty$. (easy)

(2) How do we define the dimensionality of the lattice? (Hint: How does the number of paths connecting two points grow with their separation?) For uniform hyper-cubic lattices (linear, square, cubic, ...) d can be uniquely defined by the number of nearest-neighbours.

Exercise 2.6 *Consider now more general Hamiltonians where each spin interacts with every-other:*

$$H_{\text{bogus}} = -J\sum_{i<j} s_i s_j, \quad \text{and} \quad H_{\text{mf}} = -\frac{J}{2N}\sum_{i,j} s_i s_j.$$

(N is total number of spins.) (2.12)

Show that there is a problem with extensivity for H_{bogus} but not for H_{mf} (for $N \to \infty$). Do this for $T = 0$. This model (H_{mf}) gives the mean-field theory as an infinite dimensional model. H_{bogus} is to be dumped.

Exercise 2.7 *Diamagnetism shows negative susceptibility. Any problem with convexity?*

2.4 Consequences of divergence - Problem with extensivity?

Choose a quantity say the specific heat or susceptibility, that diverges at the critical point $T = T_c$ in the thermodynamic limit. For simplicity we keep only T as the control parameter[2] (all others being kept at their respective critical values). Let us denote this quantity by $C(T, N)$ - a total quantity for the whole system - obtained from the partition function of an N-particle system.

Extensivity requires that $N^{-1}C(T, N)$ has an N-independent limit for $N \to \infty$, say $c(T)$, so that for large but finite N

$$C(T, N) = N\ c(T) + C_{\text{cor}}(T, N). \qquad (2.13)$$

[2] We have deliberately chosen an intensive variable. The case of the conjugate extensive variable is left as a problem. (See Ex. 2.14.)

2.4. Consequences of divergence

Here $C_{\text{cor}}(T, N)$ is the correction term in an asymptotic analysis. An obvious expectation is $\frac{1}{N}C_{\text{cor}}(T, N) \to 0$ as $N \to \infty$. Well, Eq.(2.13) cannot be valid at $T = T_c$, because the left hand side is finite while $c(T_c) = \infty$ (by hypothesis), *i.e.*, the correction term has to be as large as the main term. This is not the only problem. Right at $T = T_c$, N is the only parameter in hand, with all others at their respective critical values. If we want $\frac{1}{N}C(T_c, N)$ to diverge as $N \to \infty$, we need

$$\frac{C(T_c, N)}{N} = C_0 N^p, \quad (p > 0), \Longrightarrow \boxed{C(T_c, N) \sim N^{1+p}}, \quad (2.14)$$

which looks like a violation of Eq.(2.6). We have assumed a power law form in the above equation, but many other possibilities exist. Our choice is motivated by the fact that a large number of systems (real or models) do show such power laws.

Extensivity, ad nauseum, is a consequence of additivity: small pieces can be glued together to form a big piece with no change in property. In a sense boundaries can be ignored. This has to fail at the critical point. But how?

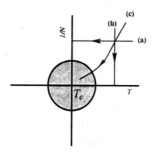

Figure 2.1: Schematic diagram showing the various paths in T vs N^{-1} plane, around the critical point $T = T_c$.

If we study the free energy, which cannot diverge, we might in principle write down an expansion of the type proposed in Eq.(2.13). This expansion is in N for any T. From such a free-energy, we might compute the specific heat by taking derivatives and face again the problem of infinities on the right hand side (rhs) at $T = T_c$. Another way of saying the same thing is that the free energy at the critical point behaves as N^{1+q} with $q \leq 0$ but the specific heat obtained by taking derivatives of the free energy with respect to temperature gets a different power of N. This means that N and temperature occur as a combination variable and not independently so that the limits $N \to \infty$ and $T \to T_c$ (taking derivatives) may not commute. The double limit needs to be considered carefully.

The dilemma is shown schematically in Fig.2.1. The expansion in N, Eq.(2.13), is valid along path (a) ($T \to T_c$ for a fixed $N < \infty$) and path (b) (thermodynamic limit for a fixed T), but the expansion is not valid in the neighbourhood of the critical point. The non-extensive behaviour gets reflected if the critical point $(T = T_c, N^{-1} = 0)$ is reached via a suitable path like (c) whose form is obtained in the next section. An isolated critical point *does* affect its neighbourhood as in any problem of non-uniform convergence. The neighbouring region (shaded in Fig.2.1) is called critical region.

Question remains: Can the specific heat be divergent?

Comments

- If the specific heat has the anomalous behaviour at $T = T_c$, then the energy-density is also expected to have so. In general if the response function for a variable $< \Phi >$ has a singular behaviour, the variable itself will also be singular. In the next section, we see how these are related.

- With Eq.(2.14) in hand, we need to redo the argument made in Sec. II regarding fluctuation. This is done in Sec. 2.8.

- Question of divergence of specific heat: The answer is obviously "no" for a single particle system with a simple Hamiltonian (perfect gas, oscillator, two-level system ..). For a noninteracting collection of such simple systems, the free energy is always proportional to N and no criticality may

2.5. Generalized scaling

occur even in the limit $N \to \infty$. Interaction is needed and essential for criticality.

2.5 Generalized scaling

The simple scaling of Eq.(2.9) needs to be modified to allow for the apparent non-extensive behaviour at a critical point. To show this we work in the ensemble with T as the variable (and all other parameters are kept fixed at their critical values). More variables will be taken up after that. We start with a generalization of scaling and show that it works. After working out a few consequences, we discuss the physical significance of the original scaling hypothesis.

2.5.1 One variable: Temperature

Let us define $t = (T - T_c)/T_c$. Remember that this is an intensive variable and would not have required any scale factor on rescaling as per Eq.(2.6) or Ex.(2.9). In order not to clutter the equations with extra symbols, we consider a lattice problem for which N and V can be interchanged (e.g., the Ising model of Eq. (2.34)). Defining $N = L^d$ with L as a linear dimension of a d-dimensional cube, we take a way out as

$$F(t, L) = b^{-d+y_f} F(b^{-y_t}t, bL), \quad \text{(generalized homogeneity)}, \tag{2.15}$$

from which Eq.(2.6) or Eq.(2.9) can be recovered for $y_f = y_t = 0$. We first show that this generalized homogeneous function works and then, look at its consequences. Physical significance is discussed after that.

Choosing $b = L^{-1}$,

$$F(t, L) = L^d L^{-y_f} \mathcal{F}\left(\frac{t}{L^{-y_t}}\right), \implies f(t, L) = b^{-y_f} f(b^{-y_t}t, bL), \tag{2.16}$$

where $f(t, L) \equiv F(t, L)/N$ is the free-energy per unit volume. We used $\mathcal{F}(x) = F(x, 1)$. Finiteness of free-energy requires $y_f > 0$.

For $t \neq 0$, demand extensivity. To get a linear dependence of F on N, we require[3] $\mathcal{F}(x) \sim |x|^q$ such that $-y_f + qy_t = 0$. This gives

$$f(t) \equiv \lim_{L \to \infty} f(t, L) \sim |t|^{y_f/y_t}, \quad \text{or} \quad f(t) = b^{-y_f} f(b^{y_t} t). \quad (2.17)$$

Since the free energy per particle cannot be infinite even at T_c, we need $y_f > 0$ and $y_t > 0$. The last expression could also be written as $f(t) = b^{-y_f/y_t} f(bt)$ by redefining b.

If we are right at the critical point, the free energy density behaves as

$$f(t = 0, L) \sim L^{-y_f}, \quad (2.18)$$

the "nonextensive" feature we are looking for. The generalized homogeneous function works.

> **Exercise 2.8** Why cannot we choose $F(T - T_c, L) = b^{-d+y_f} F(T - T_c, bL)$ as a simple generalization? Hint: non-extensive everywhere.

> **Exercise 2.9** Is there any other generalization (other than Eq.(2.15)) that could have been done?

Comments

A few things are to be noted in these manipulations.

- A singularity of the free energy in the thermodynamic limit is manifest in Eq.(2.17) and the singular behaviour is a power law in the deviation from the critical point. These powers are called exponents.

- We at the end recovered the much coveted extensivity, everywhere save the critical point, but at the cost of a scaling of the intensive variable.

---
[3]The sign $\sim$ is used to denote the functional dependence on the variables on the rhs, coefficients etc, may not be explicitly shown. Sign $\approx$ is to be used to denote the leading term or terms with all coefficients.

2.5. Generalized scaling

- What we are focusing on here is the singular part of the free-energy. There could (and will) be an analytic piece that will not show any such anomalous scaling.

- With a positive y_t, we realize that the effective temperature is further away from the critical temperature as the size is increased, if we want to keep the free energy per particle (or per unit volume) the same. This desire to keep the same free energy per unit volume is consistent with the content of Eq.(2.7).

- A more dramatic result is that an argument $L/|t|^{-1/y_t}$ can be thought of as a comparison of the length or size of the system with a characteristic scale ξ of the system. This scale ξ diverges when $t \to 0$ as $\boxed{\xi \sim |t|^{-\nu}}$ where $\boxed{\nu = 1/y_t}$.

- A diverging length scale is the hallmark of a critical point and in fact *a critical point or criticality is defined as a point with a diverging length scale.*

- No, we have neither devised a way of bypassing the Legendre transformation problem nor ignored it. We chose the right ensemble (*i.e.*, right thermodynamic potential) to do our analysis. It is conceptually helpful to use the ensemble that involves the intensive variables. For example, in first-order transitions (like, say, solid-liquid transition) there would be discontinuities in some extensive variables like the density, the energy density, the entropy (latent heat), etc, but the intensive variables remain continuous. It is this fact that prompts one to use vertex functions in field theoretic analyses - but that is beyond the scope of these notes.

- It still begs the question: What is this length ξ?

Specific heat: power laws and exponents

The specific heat is $c \sim \partial^2 f / \partial t^2$. From Eq. (2.17), choosing $b = |t|^{-1}$, it follows that $c \approx f''(\pm 1) |t|^{-2+y_f/y_t}$, where prime denotes derivatives and $\pm$ stands for $t \gtrless 0$. Note that the specific heat shows a power law singularity with an exponent

$$\alpha = 2 - \frac{y_f}{y_t}, \quad \text{defining } \alpha \text{ via } c \sim |t|^{-\alpha}, \text{ or } \alpha = - \lim_{t \to 0} \frac{\log c}{\log |t|}. \tag{2.19}$$

The exponent for divergence of c is same on both sides of the critical point at $t = 0$, though the amplitudes $f''(\pm 1)$ need not be the same.

For $L < \infty$, the specific heat behaviour is

$$\boxed{c(t, L) = L^{\alpha/\nu} \, \mathcal{C}(t L^{1/\nu})}, \tag{2.20}$$

making the L-dependence explicit as required by Eq.(2.14). We leave it to the reader to establish the relationship between $\mathcal{C}$ of Eq.(2.20) and $\mathcal{F}$ of Eq.(2.16).

Exercise 2.10 *The two-dimensional ferromagnetic Ising model, Eq.(2.11), in zero field ($h = 0$) has a logarithmic divergence of specific heat, $c \sim -\ln |t|$. How will the above formulation handle this? Note also that $\ln x = \lim_{n \to 0} (x^n - 1)/n$.*

On exponents: hyper-scaling

Since free energy and energy-density cannot diverge, we need to have $\alpha < 1$. This puts a limit

$$\frac{y_f}{y_t} \geq 1 \text{ or } y_f \geq y_t \text{ or } \nu \geq \frac{1}{y_f}. \tag{2.21}$$

At the critical point $k_B T_c$ is the energy scale, and therefore $F/(L^d k_B T_c)$ dimensionally behaves like inverse volume. In absence of any other scale, one might expect

$$\frac{F(t=0, L)}{k_B T_c N} \sim L^{-d}, \quad \Longrightarrow \quad \boxed{y_f = d}. \tag{2.22}$$

2.5. Generalized scaling

A similar argument for $L \to \infty$, and $t \neq 0$, would then give, with ξ as the important length-scale,

$$\frac{f}{k_B T_c} \sim \xi^{-d} \sim t^{+d\nu}, \implies \boxed{2 - \alpha = d\nu}. \qquad (2.23)$$

This relation between α and ν involving d is called *hyper-scaling*. Putting them all together

$$\frac{f}{k_B T_c} = L^{-d} \mathcal{F}\left(\frac{t}{L^{-1/\nu}}\right). \qquad (2.24)$$

In general

$$2 - \alpha = y_f \nu \qquad (2.25)$$

where y_f could be different from d if some other length-scale plays an important role at the critical point.

We did get the value of y_f easily under certain conditions, but it is not possible to obtain ν. Needless to say, with other variables, there will be more exponents. The values of these exponents are to be determined either from experiments or from statistical mechanical calculations with appropriate Hamiltonians.

Exercise 2.11 *If there is no singularity for a finite system, then how can there be a diverging length scale in say Eq.(2.24)? How to interpret this properly? Hint: asymptotic expansion (see Ex. 2.12 and Eq.(2.37)).*

Exercise 2.12 *A problem from school algebra: Consider the sum $S(x, N) = \sum_{n=0}^{n=N-1} x^n$ for $x \leq 1$. Show that for large N, and x close to 1, $S(x, N) = N\tilde{S}(Nt)$ where $t = 1-x$, and $\tilde{S}(z) = (1-e^{-z})/z$. Recover, from this asymptotic form, the exact result (i) for $N \to \infty$ for any x (Path b of Fig.2.1), and (ii) for $x \to 1$ for any finite N (Path a of Fig.2.1). Compute numerically $S(x, N)$ for various values of N and $x \leq 1$, plot $S(x, N)/N$ vs Nt and compare with $\tilde{S}(z)$ (Path c of*

Fig.2.1). Note the strong deviation from the asymptotic scaling for small N or large deviation of x from 1. This is actually a comparison of the partial sums of the infinite series around the singular point.

Role of fluctuations: upper critical dimension: I

From the form of the free-energy, the finite-size scaling of the energy at $T = T_c$ can be obtained, namely $E/N \sim L^{-(1-\alpha)/\nu}$. Let us now go back to Eq.(2.5). The relative width of the distribution for E is given by

$$\frac{\Delta E}{E} \sim L^{\alpha/2\nu} L^{(1-\alpha)/\nu} L^{-d/2} = L^{(y_f - d)/2}, \qquad (2.26)$$

remembering that $N = L^d$, and using Eq.(2.25). If now hyperscaling is valid, $y_f = d$, and the relative width does not vanish even in the limit of $L \to \infty$. Therefore for such a case, fluctuation plays a crucial role. If, however, $y_f < d$, then the relative width vanishes in the thermodynamic limit. We then get a situation where we have a critical point with a diverging length-scale and possibly diverging response functions, but the effect of fluctuations is not strong in the sense that the probability distribution remains sharp around the average value. Similar results are found for other extensive variables also as shown in Sec. 2.5.3.

One thing becomes clear: the dimensionality of the system is very important and the lower the dimensionality the stronger is the effect of fluctuations. We might expect that for large enough d, this condition $y_f < d$ will be satisfied. One may then try to understand such critical systems by considering the average system ignoring fluctuations altogether. This class of theories is called mean-field theory.

If there is a finite value of $d = d_u$ above which $y_f < d$, then fluctuations can be ignored for all $d > d_u$. This d_u is called the *upper critical dimension* of the system. Hope would be that the effect of fluctuations can be studied in a controlled manner around this d_u with $\epsilon = d_u - d$ as a small parameter. This is the basis of the ϵ-expansion for many systems.

2.5. Generalized scaling

We, however, caution the reader that the finite-size scaling in a mean-field theory (or $d > d_u$) is more complicated. We are concentrating only on the fluctuation dominated case.

2.5.2 Solidarity with thermodynamics

The problem with extensivity can now be understood in terms of a diverging length scale. In the strict sense, there is actually no violation of extensivity. The idea of adding small pieces to build a bigger one makes sense provided the smaller pieces themselves are representative of the bulk. For systems with short range interactions and small intrinsic or characteristic length scales, this is reasonable because the "small size" effects or boundary effects are perceptible only if the size is comparable to these lengths. These effects are small corrections that can be safely ignored. In case of a diverging length-scale or characteristic length-scales larger than the system size, smaller pieces cannot be added up. Thus at a critical point with diverging length-scales, the notion of adding up smaller pieces fails. The length-scale ξ is the appropriate scale for comparison and so at a critical point the whole sample is to be treated as a single one.

At any temperature $t \neq 0$, $\xi \sim |t|^{-\nu}$ is large but finite. If we take a block of size ξ^d as a unit, then extensivity requires that the free energy be proportional to the number of such blocks or blobs i.e., $F = (N/\xi^d) f_0 \sim N|t|^{d\nu} f_0$, with f_0 as the free-energy per blob. If f_0 is independent of ξ, then $f \sim f_0 |t|^{d\nu}$, recovering Eq. (2.23). In case f_0, the free energy of a blob of volume ξ^d, depends on ξ, then extra ξ contribution is expected and one gets Eq.(2.25) with $y_f \neq d$.

Exercise 2.13 *Anisotropic system: A particular two-dimensional system of size $L_x \times L_y$ shows the following scaling behaviour*

$$f(t, L_x, L_y) \approx X\mathsf{f}\left(L_x t^{\nu_x}, L_y t^{\nu_y}\right), \qquad (2.27)$$

X being the size dependent prefactor. There are now two different length scales in the two directions.

What are the length-scale exponents? Find the relation (hyper-scaling) among α, ν_x and ν_y. What would be the form of X if one takes $L_x \times \infty$ strips. (Ans: $X = L_x^{\alpha/\nu_x}$.) What would be X for $\infty \times L_y$ strips? We assume that the strips do not show any phase transition. What would be the right combination variable if both L_x and L_y are to be used in the prefactor? (Ans: $X = (L_x^{-1/\nu_x} + L_y^{-1/\nu_y})^{-\alpha}$. Why?)

Exercise 2.14 *Cross-over: Think of a cylindrical geometry. The system is finite in one direction (length L) but infinite in the remaining $d-1$ directions. There will be a critical behaviour in this geometry if d is not too small. Consider the $d-1$ to d dimensional crossover as the finite length L is made larger. Consider various situations: (a) $d < d_u$, (b) $d > d_u$ but $d-1 < d_u$, and (c) $d-1 > d_u$.*

2.5.3 More variables: Temperature and field

Generalization of the scaling of Eq.(2.15) to more variables is straight forward. Take a variable h which represents another intensive quantity like pressure, magnetic field, etc, measured from its critical value. The critical point is now at $t = 0, h = 0$. The free energy can be written as

$$\begin{aligned} F(t,h,L) &= b^{-d+y_f} \, F(b^{-y_t}t, b^{-y_h}h, bL), \\ &\stackrel{N\to\infty}{\Longrightarrow} f(t,h) = b^{y_f} f(b^{-y_t}t, b^{-y_h}h), \end{aligned} \quad (2.28)$$

so that the same series of manipulations done for $h = 0$ would lead to

$$f(T,h,L) \equiv \frac{1}{N} F(t,h,L) = \xi^{-y_f} \mathcal{F}_\pm \left(\frac{h}{\xi^{-y_h}}, \frac{L}{\xi} \right), \quad (2.29)$$

$$= L^{-y_f} \tilde{\mathcal{F}} \left(\frac{t}{L^{-1/\nu}}, \frac{h}{L^{-y_h}} \right), (2.30)$$

where $\mathcal{F}_\pm(x,y) = F(\pm 1, x, y)$.

2.5. Generalized scaling

If $tL^{y_t} = const$, then, as per Eqs. (2.30) and (2.24), extensivity is nowhere to be found - that is curve (c) in Fig.2.1. The scaling form of Eq.(2.24) or (2.30) is known as finite-size scaling.

Comments

- For a fixed N, if $t \to 0$, then $f \sim L^{-y_f}$.

- For a fixed t, if $N \to \infty$, then

$$f \sim |t|^{2-\alpha} \hat{\mathcal{F}}_\pm \left(\frac{h}{|t|^{y_h/y_t}} \right). \tag{2.31}$$

- We see the general feature:
  critical region $\iff$ "nonextensivity" $\iff$ finite size scaling $\iff$ Scaling.

- Once we take the diverging length-scale as the sole scale for the problem, the finite-size scaling form can be obtained from the bulk behaviour as well. A finite length L matters only when it is comparable to ξ. For $L \gg \xi$, the above-mentioned blob picture is valid and for $L < \xi$ the whole system needs to be treated as a critical one. If the bulk behaviour is like $c(t) \sim |t|^{-\alpha}$, then the size-dependent crossover is given by

$$c(t, L) \sim |t|^{-\alpha} \, \tilde{\mathcal{C}}(L/\xi, h/\xi^{y_h}) \sim L^{\alpha/\nu} \, \mathcal{C}(t/L^{-1/\nu}, h/L^{y_h}).$$

(See Eq.(2.20).) Historically, bulk scaling was proposed first and finite-size scaling came later on.

- We cannot overemphasize the fact that h and t are completely independent variables. Yet in Eq.(2.31) the free-energy in the thermodynamic limit depends on the single combination variable $h/t^{y_f/y_t}$, and not on h and t separately. Results of experiments (real or numerical) done at various values of h and t can be collapsed on to a single curve, unthinkable away from the critical region.

- It seems there is one exponent too many for the bulk free energy density f in Eq.(2.28), y_f/y_t and y_h/y_t would have sufficed. Yes, it would if we consider thermodynamics alone. But the path via finite-sized system showed us the necessity of another exponent, ν, and so we keep all the three in Eq.(2.28).

Exercise 2.15 *Take the Legendre transform of f in Eq.(2.28) with respect to t. Discuss its scaling property. Caution: The scaling is not for the total entropy or entropy density. The special scaling is for the deviation from the critical value of the entropy at the critical point. This shows the difference of the scaling around the critical point and the simple scaling of thermodynamics.*

More variables mean more exponents

The power laws we saw earlier tend to suggest similar behaviour for other physical quantities also. We define several such practically important exponents, but shall ultimately see that all of these can be expressed in terms of the three y_f, y_t, y_h introduced earlier.

For concreteness, the magnetic language of the Ising model of Eq. (2.11) is used. Let us assume that there is a critical point for the Ising model (and there is one) so that the free energy is given by Eq.(2.31). We have already defined α as the specific heat exponent. We define β for magnetization, γ for susceptibility, δ for critical isotherm. Apart from these thermodynamic exponents, two other exponents are needed ν (already introduced) for lengthscale and η for critical correlations.

Remembering that the magnetization is the derivative of the free-energy with respect to h, we get $m \approx |t|^{(y_f - y_h)/y_t} \mathcal{F}_\pm'(h/t^{y_h/y_f})$. From this, see that for $h = 0$, the magnetization vanishes at the critical point as

$$m \sim |t|^\beta, \quad \text{with} \quad \beta = (y_f - y_h)/y_t. \qquad (2.32)$$

2.5. Generalized scaling

For a ferromagnetic transition, there is no magnetization in zero magnetic field in the high temperature phase (paramagnet), and therefore $\mathcal{F}_+'(0) = 0$ but $\mathcal{F}_-'(0) \neq 0$. A quantity like m that describes the "ordering" of the system is called an order parameter and β is the order-parameter exponent. We repeat that phase transitions need not necessarily have an order parameter. (see Sec. 2.7.2).

Susceptibility χ is the response of magnetization, and we see in zero field ($h = 0$) ($\chi = \partial m/\partial h$)

$$\chi \sim |t|^{-\gamma}, \quad \text{with} \quad \gamma = (2y_h - y_f)/y_t. \tag{2.33}$$

At the critical point ($t = 0$) in presence of a field (critical isochore), we get

$$m \approx h^{1/\delta} m_0, \quad \delta = y_h/(y_f - y_h). \tag{2.34}$$

Here the amplitude $m_0 = \lim_{x \to \infty} x^{-(y_f - y_h)/y_h} \mathcal{F}_\pm'(x)$. In general, $\delta \neq 1$. One expects a linear relationship ("linear response") between "cause" (h) and "effect" (m), but that turns out not to be the case at criticality. A linear relation can never give an infinite χ!

A major consequence of the homogeneity of the free energy is the power law behaviour of various physical quantities (various derivatives of free energy) and all of these are obtained from the three basic exponents. In case of hyper-scaling (*i.e.*, $y_f = d$) we have a further reduction and only two exponents are needed for a complete description of the critical behaviour of a system.

For a thermodynamic description this looks enough, but we already saw the usefulness of a length-scale based analysis. We need to define the length-scale properly and in the process we will find a more useful critical exponent η.

Comments

- The blob picture can be used for susceptibility near the critical point. There are N/ξ^d blobs. Each blob of size ξ^d can be thought of as a critical object. The total susceptibility is given by $N\chi = (N/\xi^d)\chi_{\text{blob}}$, where χ_{blob} is the total susceptibility of a blob. Finite-size scaling predicts $\chi_{\text{blob}} = \xi^d \xi^{\gamma/\nu}$ so that we obtain $\chi \sim |t|^{-\gamma}$.

- Note that we require $y_h < y_f < 2y_h$.

Role of fluctuations: upper critical dimension : II

The role of fluctuations was analyzed earlier in the context of specific heat. Let us now reanalyze it for the probability distribution for magnetization M. The finite-size scaling behaviours of (total) magnetization and (total) susceptibility at $T = T_c$ are $\langle M \rangle \sim L^d L^{-\beta/\nu}$ and $N\chi \sim L^d L^{\gamma/\nu}$. The relative width $\Delta M/M$ of the probability distribution for M is $L^{\gamma/2\nu} L^{\beta/\nu} L^{-d/2} = L^{(y_f - d)/2}$ where Eqs. (2.35) and (2.25) have been used.

It is reassuring that the condition for sharpness (or broadness) of the probability distribution for M is the same as obtained for specific heat (see Sec. 2.5.1). The upper critical dimension is the same no matter which conjugate pair we use.

2.5.4 On exponent relations

Once the exponent identifications are made, with only two independent ones, it is possible to write down many relations involving the (in principle) experimentally measurable exponents. For example, by adding the exponents,

$$\alpha + 2\beta + \gamma = 2. \tag{2.35}$$

If we take the free-energy density $\sim hm$ and $m = \chi^{-1}h$, then it follows that $2\beta + \gamma = 2 - \alpha$, as in the above equation, without using any explicit formula. Another way of re-writing the above relation $(-\alpha = 2(\beta - 1) - \gamma)$ suggests that the behaviour of specific heat $(c_{h=0})$ is similar to $(\partial m/\partial t)^2 \chi^{-1}$. In fact, thermodynamics gives us the formula

$$c_h - c_m = T \left(\frac{\partial m}{\partial T}\right)_h^2 \chi_T^{-1}, \tag{2.36}$$

2.6. Relevance, irrelevance and universality

where c_x is the specific heat with x constant. Since specific heat is positive definite, it follows from Eq.(2.36) for c_h with the intensive variable h held constant at the critical value $h = 0$ that $\alpha + 2\beta + \gamma \geq 2$.

Exercise 2.16 *Thermodynamic argument seems to indicate inequality rather than equality in Eq.(2.35). Find out the conditions for which a strict inequality is expected.*

2.6 Relevance, irrelevance and universality

A few observations should not miss our attention. Since $y_t > 0$, the combination variable Lt^{1/y_t} in, say, Eq.(2.20) or (2.30) have different limits for $t = 0$ and $t \neq 0$ in the limit $L \to \infty$. This difference actually gave us the different L-dependent behaviour of the free-energy or the specific heat, or, as a matter of fact, any physical quantity we may calculate or observe. In the same way, if $y_h > 0$, then in the bulk case, if $t \to 0$, *i.e.*, as the critical point is approached, the combination or scaling variable $h/t^{y_h/y_t}$ goes on increasing if $h \neq 0$, no matter how small it is. The behaviour is different if h is strictly zero. Such a tendency of a parameter to grow also tells us that zero field and nonzero field behaviours are different. We call these variables *relevant variables*.

In case $y_h < 0$, then for $t \to 0$, the scaled variable is zero irrespective of its value. In such a case whether the system has $h \neq 0$ to start with is immaterial. No wonder these are to be called *irrelevant variables*.

There could also be *redundant* terms that do not matter at all, like a constant added to a Hamiltonian. We ignore them altogether.

To be at a critical point, the relevant variables must be tuned properly to be at their critical values (*e.g.*, $t = h = 0$ in the magnetic example) because they take us away from the critical point. Irrelevant variables do not matter as such but they do play a significant role if we want to go beyond the leading behaviour.

Special situations arise, if the scaled function shows a singularity as an irrelevant variable scales to zero. Such variables are then important for the critical behaviour, though they do not take the system away from criticality. Such a variable is called a *dangerous irrelevant variable*.

Critical points are classified by the number of relevant variables required to describe them. An ordinary critical point requires two (for the magnetic problem, temperature and magnetic field; for the liquid-gas transition, temperature and pressure). If three relevant parameters are needed it is called a tricritical point and so on.

Our analysis so far has been restricted to thermodynamic parameters like t, h, etc, but this can be extended to any parameter occurring in the Hamiltonian in a statistical mechanical approach. The starting Hamiltonian may have a large number of parameters based on the microscopic details of the system. But as we look at longer length-scales close to the critical point, all these parameters can be classified under the banner of relevance and irrelevance. By throwing away the irrelevant terms, for the leading behaviour, an enormous simplification ensues (e.g, all ordinary critical points will have only two relevant parameters, only difference may be in the numerical values of the exponents ν and η, etc.). It might then be expected that the numerical values of the exponents could be identified from certain basic symmetries, etc, of the Hamiltonian. This is the concept of Universality. A universality class would be described by the exponents and also the amplitude ratios of the various singular quantities on both sides of the critical point.

The idea of universality transcends the domain of critical phenomena. Whenever we are interested in properties on a scale much bigger than the underlying or microscopic scales, there seems to be a set of properties which are quantitatively same no matter what the microscopic details are. The exponents we have seen are just one such examples. Historically it was found that the shape of the coexistence curve near the liquid-vapour critical point is independent of the chemical composition and is also very similar to the critical behaviour of several magnets. Once the universality

2.7. Digression

class of a system is identified, the set of universal quantities can be obtained by studying a simpler model system than the original one with all details. The simpler model is expected to focus on the relevant variables only or at most a few irrelevant ones.

Universality is not just a set of exponents. In a scaling description, the amplitudes and even the scaling functions are independent of gross details except that the arguments of the functions may involve nonuniversal metric factors. In a finite size scaling there could be a dependence on the boundary conditions as well.

Whether a variable is relevant or not is determined by its scaling exponent (e.g., y_h in the previous case). In certain cases, one may get these by simple arguments (Gaussian model in Sec. 2.9) but in most cases, these are to be determined.

Exercise 2.17 *Can there be critical cases with only one relevant variable or no relevant variables at all?*

2.7 Digression: First-order transition and transition with no ordering

2.7.1 A first-order transition: $\alpha=1$

What happens at the borderline of $\alpha = 1$? The form of the free-energy tells us that the energy-density will have a discontinuity on the two sides of the singular point. Such a phase transition with a discontinuity in any first derivative of the free-energy is defined as a first-order transition. This value of α leads to $\nu = 1/d$.

We see the possibility of a first order transition in the same framework developed for the critical point. But first-order transitions can be of other types also, and they may not necessarily have any diverging length-scales associated with it. One needs to be careful about it.

As an example, let us consider a very simple configuration space for a magnet (a crude approximation to an Ising-type model). We replace the N-spin configurations by two types only. A zero-energy ground state – all spins up or all down – two-fold degenerate, and an excited state of energy $N\epsilon$ with degeneracy 2^N. (This is a one-dimensional ferro-electric six vertex model, in disguise.) The partition function for this model is $Z(x,N) = 2 + (2x)^N$ where $x = \exp(-\beta\epsilon)$. It is easy to see, by taking the $N \to \infty$ limit, that there is first-order transition at $x = 1/2$ with total energy going from zero for $x < 1/2$ to $N\epsilon$ for $x > 1/2$. There is no problem with extensivity for $x \neq 1/2$. In the limit $N \to \infty$, the specific heat is just a delta function at the transition point. For finite N, the specific heat can be written in the form

$$C(x,N) = N^2 \ln 2 \, \frac{2e^{Nt}}{(2+e^{Nt})^2}, \qquad (2.37)$$

where $t = 1-2x$. For a d-dimensional system, take $N = L^d$. What we now see from the scaling variable is that there is a diverging length scale with exponent $\nu = 1/d$.

Comments

- Fig.2.2 shows $C(x,N)$ for various values of x and N computed from the partition function and compared with the scaling function of Eq.(2.37). This is an example of *data collapse*.

- The thermodynamic limit for any given value of x (Path b of Fig.2.1) comes from the tail of the scaling function of Fig.2.2. This plot amplifies the critical region of Fig.2.1.

- The N^2 prefactor in Eq.(2.37) is consistent with an exponent α/ν with $\alpha = 1$. Remember that $N = L^d$.

- The peak of the scaling function in Fig.2.2 is not at the bulk transition point.

Exercise 2.18 *What is this length scale in the context of this model?*

This simple model calculation can be generalized. One way of studying a first-order transition is to compute the free energies of

2.7. Digression

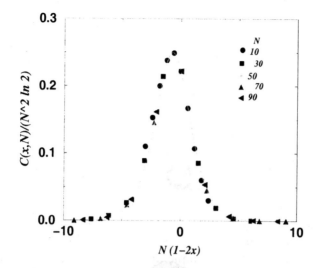

Figure 2.2: Data collapse of $C(x, N)/N^2$ vs Nt for various N. The solid line is the finite-size scaling function of Eq.(2.37).

the two phases independently (like *e.g.*, solid and liquid) and then finding the lower free energy curve. Taking these free energies per particle, f_1 and f_2, as coming from a restricted sum over states of the full partition function, the total partition function can be approximated by $Z = \exp(-N\beta f_1) + \exp(-N\beta f_2)$, with the free energy per particle $f = \min(f_1, f_2)$ for $N \to \infty$. In general, f_i's are analytic functions and the transition temperature is determined by $f_1(T_c) = f_2(T_c)$, so that $\Delta f = |f_1(T) - f_2(T)| \sim |T - T_c|$ from a Taylor series expansion around $T = T_c$. The partition function can therefore be written as $Z = \exp(-N\beta f) \left[1 + \exp(-N/\xi^d)\right]$, where $\xi^{-d} = \beta_c |\Delta f|$. This defines a length scale $\xi \sim |T - T_c|^{-1/d}$. For first-order transitions, $\boxed{\nu = 1/d}$.

2.7.2 Example: Polymers : no "ordering"

We give an example here of a phase transition for which there is no "ordering" unlike the magnetic case, and the interest in the critical point is because of finite-size behaviour.

Polymers are long linear objects, abundantly occurring in nature. Let us take a single flexible polymer in a solvent (a single polymer of length N in d-dimensional space of infinite volume). Depending on the nature of the solvent and the monomers constituting the polymer, there could be effective repulsion or attraction between the monomers. By changing temperature, one may go from a repulsion-dominated (monomer favouring solvent molecules) (self avoiding walk) to an attraction-dominated (monomer preferring monomer) phase. This is called a collapse transition. Such a transition for a single molecule can occur only in the thermodynamic limit of length $N \to \infty$. But that is of no interest because the polymers are necessarily of finite lengths. The two phases in this example are described by the overall size of the polymer, e.g., by the mean square end-to-end distance $\langle R^2 \rangle \sim N^{2\nu}$ In a collapse transition the value of this size exponent ν changes from $\nu \approx 3/(d+2)$ to $\nu = 1/d$ in d-dimensions. This is an example of a phase transition or critical point which may not be viewed as ordering (unlike the magnetic system) - it is no less important though. The main features of the transition are reflected in the "finite-size" behaviour (with respect to N) both at the transition and in the two phases, and the general analysis we have done so far is easily applicable here.

Exercise 2.19 *What sort of a critical point is the collapse transition? Hint: How many relevant parameters at a collapse transition? We need to make length going to infinity, we need to adjust temperature and if we take a solution of polymers, we need to make concentration equal to zero (i.e., single chain limit).*
Ans: tricritical point.

Exercise 2.20 *For a polymeric phase transition of the type mentioned above, show that the specific heat per unit length (or per monomer) can be written in a scaling form $c = N^{2\phi-1}\mathcal{C}((T - T_c)N^\phi)$.*

2.8 Exponents and correlations

2.8.1 Correlation function

Consider the Ising model of Eq.(2.11) or (2.12) and take $M = \sum_i \langle s_i \rangle$. The fluctuation-response theorem of Eq.(2.10) and translational invariance can be used to express the susceptibility (χ per particle) as

$$N\chi = (k_B T)^{-1} \sum_i \sum_j \langle (s_i - \langle s \rangle)(s_j - \langle s \rangle) \rangle = N(k_B T)^{-1} \sum_{\vec{x}} g(\vec{x}) \qquad (2.38)$$

where $g(\vec{x}) = \langle (s_0 - \langle s \rangle)(s_{\vec{x}} - \langle s \rangle) \rangle$ is the pair correlation function. The behaviour of the susceptibility is therefore dependent on the behaviour of the correlation function. This relation is quite general, as pointed out in Sec. 2.3.2. By expressing the total quantity $\Phi = \sum_\mathbf{x} \phi(\mathbf{x})$, in terms of a local density, we have $\langle \Phi^2 \rangle - \langle \Phi \rangle^2 = N \sum_\mathbf{x} g_{\phi\phi}(\mathbf{x})$, where $g_{\phi\phi}(\mathbf{x}) = \langle (\phi(0) - \langle \phi \rangle)(\phi(\mathbf{x}) - \langle \phi \rangle) \rangle$.

For simplicity, let us replace the sum in Eq.(2.38) by an integral so that $\chi = \int d^d x\, g(\vec{x})$. For the Ising case, s is bounded and so is g. We also know that χ diverges at least at T_c. The only way this can happen is from the divergence of the integral or the sum - it is the large distance property of $g(x)$ that controls the behaviour. We can conclude that, at least at T_c, $g(x)$ cannot be a short ranged function, but it has to decay to zero for infinitely large distances. In d dimensions, the decay at $T = T_c$ for large x is given by

$$g(\vec{x}) \sim x^{-(d-2+\eta)} \quad \text{with} \quad \eta \geq 0 \qquad (T = T_c). \qquad (2.39)$$

For $T \neq T_c$, convergence requires that $g(r)$ should decay sufficiently faster than this; hence, it better be of short-range character with a characteristic length scale ξ, e.g., $\exp(-x/\xi)$. The decay for any temperature can be written as

$$g(\vec{x}) = \frac{1}{x^{d-2+\eta}}\, \mathbf{g}\left(\frac{x}{\xi}\right) \qquad (T \neq T_c), \qquad \text{(see comments below)} \qquad (2.40)$$

which goes over to the $T = T_c$ case if this length-scale ξ diverges for $T \to T_c$. The length-scale we saw popping out naturally for comparisons of the length of the system can now be identified as the correlation length, the scale for correlations. A precise definition[4] of ξ would be from the second moment of $g(\vec{x})$ as

$$\xi^2 = \frac{\int d^d x \, x^2 g(\vec{x})}{\int d^d x \, g(\vec{x})} \sim |t|^{-2\nu}. \qquad (2.41)$$

(For an isotropic system, the first moment vanishes by symmetry.) It is possible to define such scales via higher moments also. Under the assumption of one scale, all of these will have similar divergences at a critical point. Nevertheless, it is worth keeping in mind that there are cases where one may need to study higher order moments and different moments defining different length-scales.

Exercise 2.21 *Can $\eta < 0$?*

Exercise 2.22 *Generalize Eq.(2.41) for anisotropic cases like in Ex. 2.14.*

Comments

- The on-site term, $i = j$, with prefactor $(k_B T)^{-1}$ in Eq. (2.38) is the susceptibility of individual spins (or isolated degrees of freedom), the Curie susceptibility. The correlation contribution is special to an interacting system. Note that the correlation vanishes for a noninteracting system.

- The fluctuation becomes long-ranged at the critical point. On the high temperature side $\langle s_i \rangle = 0$ (no magnetization), and so the fluctuation correlation is the same as the spin-spin correlation.

- For low temperatures ($T < T_c$), the spin-spin correlation approaches a constant ($\langle s \rangle^2 \neq 0$) for large separations. This approach to the constant value is generally short-ranged,

---
[4]There are in fact many ways of defining a length scale. We use a $g(x)$-based definition.

2.8. Exponents and correlations

i.e., very rapid, like exponential, but becomes long-ranged at T_c.

- We get a long-range correlation even though the interactions are just short ranged (*e.g.*, nearest neighbour interaction in the chosen Ising model).

- A finite size scaling for the correlation length itself would be $\xi \sim L$. Quite often it is possible to write it as an equality with the amplitude determined by the known exponents. The amplitude surely depends on which correlation function is used to determine ξ, boundary conditions, etc.

- For historical reasons Eq.(2.40) is generally written in a slightly different form as $g(\vec{x}) = \frac{1}{x^{d-2+\eta}} \, D\left(\frac{x}{\xi}\right) \exp(-x/\xi)$, with $D(z) \to 1$ for $z \to 0$, while $D(z) \sim z^{(d-3+2\eta)/2}$ for large z. Such a form is required to make correspondence with the Ornstein-Zernicke theory of correlations, which we shall not discuss in these notes.

2.8.2 Relations among the exponents

Let us think of the pair correlation function. It decays rapidly once we are on a scale greater than the correlation length ξ. Close to T_c, we can think of the system as blobs of highly correlated regions - the blobs are of size ξ^d in d dimensions. These are the blobs we introduced to salvage extensivity in Sec. 2.5.2. Inside a blob ($r \ll \xi$), the fluctuations are critical-like and at a simple level, a blob can be thought of as at T_c. On a bigger length scale $x \gg \xi$, the blobs are independent. Basically, we are arguing that it is the correlation length that matters - all other length scales are unimportant.

We use this simple picture for the susceptibility. We can cutoff the integral at $x \sim \xi$, and inside this region $g(x) \sim x^{-(d-2+\eta)}$. The integral $\int^\xi dx \, x^{1-\eta} \sim \xi^{2-\eta}$. Using the temperature dependence of ξ, we get the temperature dependence of χ as $|t|^{-\nu(2-\eta)}$. The

net result is
$$\gamma = \nu(2 - \eta). \quad (2.42)$$
This relation with the help of Eq.(2.33) gives
$$y_h = \frac{y_f + 2 - \eta}{2}. \quad (2.43)$$

This is a very important relation - it shows that how the external magnetic field or the nonthermal relevant variable scales away from the critical point is determined by the decay of correlations *at criticality*. It is straightforward to see that $\alpha = \nu(2 - \eta_E)$.

We now have the full identification of the three exponents:
$$\boxed{y_t = 1/\nu, \; y_f = d, \; y_h = \tfrac{d+2-\eta}{2}} \quad (2.44)$$

All the nuances of critical phenomena are qualitatively and quantitatively expressed in terms of the correlation function and the exponents needed for it. It is a shift from a thermodynamic description to a purely statistical mechanical one. *It is the correlations of the degrees of freedom that control completely the whole phenomenon.*

In this particular example of ordinary critical point, one of the exponents, η is really an exponent defined at criticality, while the other, ν, is an off-critical one. An off-critical exponent helps in the description of the approach to the criticality. This can however be traded for purely critical exponents.

Exercise 2.23 *What if $y_f \neq d$?*

Comments

- Now that we have emphasized the role of correlation functions and expressed y_h in terms of η, why are we not doing the same thing for specific heat? Eq.(2.5) tells us that specific heat is related to energy-energy correlation function, and so shouldn't we define an exponent η_E? We could if we want to. But convince yourself, by using hyper-scaling, that $y_t = 1/\nu = (d + 2 - \eta_E)/2$. We may use either ν or η_E.

2.8. Exponents and correlations

- One may generalize the analysis of Sec.2.8.1 for any local variable $L(\{\phi(x)\})$ and the corresponding response function will be determined by the critical correlation of $\langle L(\{\phi(x_1)\})L(\{\phi(x_2)\})\rangle$.

In a correlation based approach, we need to compute the critical correlations of all possible combinations of the degrees of freedom (e.g., in the Ising case, spin-spin, energy-energy and so on). From these, the relevant variables can be identified. In the Ising criticality case, there are only two. In addition, for a thermodynamic problem, we need to know the finite size behaviour of the free energy at the critical point (or assume hyperscaling). These three critical parameters then completely specify the leading behaviour even around the transition point.

To repeat, a thermodynamic description would focus on y_f, y_t, y_h for a critical point, while a statistical mechanical description would focus on the set of purely critical exponents y_f/y_t, η_E, and η. The scaling relations we derived show their equivalence.

Exercise 2.24 *Can this happen: the spin-spin correlation function is given by Eq.(2.40) and the energy-energy correlation is given by $g_{EE} \sim x^{-(d-2+\eta_E)} g_E(x/\xi_E)$ where $\xi_E \sim |t|^{-\nu_E}$, but $\nu_E \neq \nu$?*

2.8.3 What's b anyway? : Length-scale dependent parameters

If we look at the free energy of Eq.(2.17) or Eq.(2.28) in the thermodynamic limit of $L \to \infty$, a question may be asked, "what is b now?". Answer is in the correlation function. Under a rescaling $x \to x/b$, we see $g(x,t) = b^{-(d-2+\eta)} g(x/b, b^{y_t} t)$, where the t-dependence has been made explicit. This resembles the scaling of the free-energy. An interpretation of this equation is that if we change the scale of length measurements (actually the scale of microscopic details - though not apparent right now), the parameters of the problem and the concerned physical quantity get rescaled. Quantitatively, for $x \to x' = x/b$, we have $t \to t' = b^{y_t}t$, $h \to h' = b^{y_h}h$, and $f \to f' = b^{y_f}f$.

For a given problem (*i.e.*, $t, h, \ldots$, fixed) such a scale transformation leads to a transformation of the parameters and the physical quantities. Repeated applications of this transformation would land us on a set of parameters which are functions of b. These are the scale dependent parameters - criticality is best understood in terms of these scale-dependent parameters.

The scale-dependence of the parameters is best represented by infinitesimal scale transformation $b = 1 + \delta l$ (compare with Ex. 2.3). We may write down a differential equation for Eq.(2.17) as

$$\left(\zeta(t)\frac{\partial}{\partial t} - y_f\right)f = 0, \qquad (2.45)$$

where $\zeta(t) = \partial t/\partial l = y_t t$. This equation has the expected solution $f \sim t^{y_f/y_t}$.

This is a new way of looking at the problem. Instead of studying a system for various values of the parameters ("coupling constants"), we are studying it at various scales to see how it behaves in the long-scale limit, since after all thermodynamic or macroscopic behaviour is for a large system.

The equation derived just above, (Eq.(2.45), describes the flow of the free-energy as the scale is changed, and such equations are called flow equations. Any physical quantity will have a flow equation associated with it. However with only a few relevant parameters and a few independent exponents, not all flow equations are necessary. A fuller picture emerges from a renormalization group analysis. Actual flow equations turn out to be slightly more complicated and the simpler equation of Eq.(2.45) is obtained under special conditions of "fixed points". The critical behaviour and the universality classes are ultimately linked to the fixed points.

Exercise 2.25 *Possibility of a dangerous fixed point?:*
A fixed point by definition is a point that does not change under rescaling. If a parameter say u has a stable fixed point at $u = u^*$ then $u - u^*$ can be taken as

2.9 Models as examples: Gaussian and ϕ^4

an irrelevant variable. Can this variable be a "dangerous irrelevant variable", or, in other words, can there be a dangerous fixed point?

2.9 Models as examples: Gaussian and ϕ^4

At this point it is helpful to consider a few examples. The Ising model has been introduced in Ex. 2.5. Here we consider a continuum version of it. A naive continuum limit would give a Gaussian model for a field variable $-\infty < \phi < \infty$,

$$\begin{aligned}\frac{H}{k_B T} &= \int d^d x \, [\frac{1}{2}(\nabla\phi)^2 + \frac{1}{2}r\phi^2 - h\phi] \\ &= \int \frac{d^d q}{(2\pi)^d}[\frac{1}{2}(q^2 + r)\phi_k \phi_{-k}] - h\phi_{k=0}, \quad (2.46)\end{aligned}$$

with a cut-off in real space, i.e., $|x| > a$ or in momentum space $|q| < \Lambda \sim a^{-1}$, where a could be the lattice spacing. We shall assume a spherical Brillouin zone. A better continuum limit is the ϕ^4 model,

$$\frac{H}{k_B T} = \int d^d x \, [\frac{1}{2}(\nabla\phi)^2 + \frac{1}{2}r\phi^2 + u\phi^4]. \quad (2.47)$$

Exercise 2.26 *Starting from the Ising model on a hyper-cubic lattice of coordination number $q = 2d$, get the continuum form given above. Show that $r \sim T - T_m$ with $T_m = qJ$.*

'The reason for considering a continuum limit is to take advantage of dimensional analysis. Both sides of Eqs. (2.46) and (2.47) being dimensionless and the right hand side expressed as a volume integral helps us in introducing a length based analysis in a natural way.

That $r = 0$ is a singularity is obvious from Eq.(2.46) because of the instability at $r < 0$. One can derive all the thermodynamic properties for the Gaussian model and convince oneself that there is a singularity at $r = 0$. We do not go into that. Here let us accept that $r = 0$ corresponds to a critical point.

- In general, one would have a term of the type $\frac{1}{2}c(\nabla\phi)^2$ in Eqs. (2.46) and (2.47). However, if c does not change sign, one may absorb it in the definition of ϕ and redefine r and u. This has been done in those two equations. There are problems where c may change sign when external parameters are changed (Lifshitz point), and in such cases c needs to be kept explicitly. It would be necessary to keep c explicitly in case the field variable cannot be scaled arbitrarily, as e.g., if ϕ is like an angular variable.

2.9.1 Specific heat for the Gaussian model

Using Gaussian integrations, one can compute the zero-field specific heat for the Gaussian model. The leading term is given by the integral

$$c_{h=0} \approx \xi^{4-d}\frac{Ta_2}{2}\frac{K_d}{(2\pi)^d}\int_0^{\Lambda\xi}\frac{q^{d-1}}{(1+q^2)^2}\,dq + \ldots, \qquad (2.48)$$

where $\xi = r^{-1/2}$, $r = a_2(T-T_m)$ and K_d is the surface integral of a d–dimensional unit sphere. The length-scale exponent is $\nu = 1/2$.

Exercise 2.27 *Derive Eq.(2.48). Show that $\eta_E = d - 2$.*

A dimensional analysis gives

$$\begin{aligned}[\phi(x)] &= L^{(2-d)/2},\ [r] = L^{-2},\ [\Lambda] = L^{-1},\\ \text{and }[u] &= L^{d-4}, [h] = L^{-(2+d)/2},\end{aligned} \qquad (2.49)$$

where L is a length scale. This simple dimensional analysis already identifies a diverging length-scale $\xi = r^{-1/2}$, used in Eq. (2.48). Note also that the specific heat is a volume integral of the ϕ^2-ϕ^2 correlation function in real space. A dimensionally correct form is therefore

$$c = \xi^{4-d}\,\mathcal{C}(u\xi^{4-d}, h\xi^{(2+d)/2}, \Lambda\xi). \qquad (2.50)$$

We take the Gaussian model first ($u = 0$) and no external field, $h = 0$. Then $c = \xi^{4-d}\mathcal{C}(\Lambda\xi)$ and Eq.(2.48) is in this form. Take

2.9. Models as examples: Gaussian and ϕ^4

the limit $\xi \to \infty$. The behaviour of the specific heat now depends on $\mathcal{C}(z)$ as $z \to \infty$. If this limit is finite, then the cut-off can be completely forgotten and ξ plays the important role. In such a case the microscopic parameters are not important for the critical behaviour. This happens, as we see from the integral in Eq. (2.48) for $d < 4$, and $\alpha = (4-d)/2$, a value that satisfies hyper-scaling with $\nu = 1/2$. However for $d > 4$, the integral diverges at the upper limit and $\mathcal{C}(z) \sim z^{d-4}$, leaving us with a non-divergent specific heat, i.e., $\alpha = 0$. Hyper-scaling also gets violated. For $h \neq 0$, Gaussian integrations yield $y_h = (d+2)/2$, consistent with dimensional analysis. The exponents are

$$\boxed{y_f = d \text{ for } d < 4, \text{ but } y_f = 4 \text{ for } d > 4}$$
and $\boxed{\nu = 1/2 \text{ and } \eta = 0, \text{ for all } d}$. (2.51)

In the above example we chose the specific heat because of its interesting behaviour for $d < 4$ and $d > 4$. Take susceptibility. Dimensional analysis gives $\chi \sim r^{-1}$. Convince yourself, by using Gaussian integrals, that this is so for all d. However the finite-size scaling behaviour will be different for $d < 4$ and $d > 4$.

From the value of α or y_f, (see Eq.(2.51)) and the sharpness criterion of Sec. 2.5.1, we see that fluctuations can be ignored if $d > 4$. In a sense $d = 4$ turns out to be the upper-critical dimension for this model.

Exercise 2.28 *Calculate the average energy of a Gaussian correlated blob. i.e., a blob of size ξ^d - do this by calculating the form of the energy for the Gaussian model and then putting $r = 0$. Study its behaviour for various d and then integrating once with respect to r, estimate the free-energy f_0 of a blob (See Sec. 2.5.2). Justify the violation of hyper-scaling for $d > 4$.*

2.9.2 Cut-off and anomalous dimensions

The above warm-up exercise shows that the cutoff, a relic of the microscopic features of the model, cannot always be ignored even-though the relevant length-scales at which the phenomenon is taking place is much much larger than this. Dimensionality is also important. Dimensional analysis is not expected to yield any unique or useful relation if there are multiple scales in the problem. This is a very important point and we elaborate on this further. In case the cutoff can be ignored, the exponents are the same as predicted by dimensional analysis. This we see directly in the Gaussian model for $d < 4$ - the cut-off just didn't matter.

Exercise 2.29 *Consider a finite Gaussian model with periodic boundary conditions in all directions. Obtain the finite-size scaling behaviour of the specific heat for various d. Note the discrete k sums with no $k = 0$ mode.*

Exercise 2.30 *Do the finite-size scaling analysis for the zero-field susceptibility of the Gaussian model for various d.*

If we go back to the ϕ^4 problem, then we need to treat Eq. (2.50). Question now is what happens for $\xi \to \infty$. If $\mathcal{C}$ goes to a constant, this cutoff can be safely ignored. A general situation would be

$$\mathcal{C}(x,y,z) = z^p \tilde{\mathcal{C}}(xz^{p_1}, yz^{p_2}), \quad \text{as } z \to \infty. \tag{2.52}$$

Setting $\Lambda = 1$,

$$c \sim \xi^{(4-d)+p} \tilde{\mathcal{C}}(u\xi^{(d-4)+p_1}, h\xi^{y_h+p_2}). \tag{2.53}$$

This has the expected scaling form when ξ is replaced by $|t|^{-\nu}$. In fact the same analysis can be done for ξ itself,

$$\xi = r^{-1/2} \, \mathcal{X}(ur^{(4-d)/2}, ..., \Lambda/\sqrt{r}), \tag{2.54}$$

and, in the $r \to 0$ limit, the rhs of Eq.(2.54) might (and would) pick up extra powers of r from the Λ-dependent argument[5]. This

---

[5] There will be a shift in the critical temperature also. We take r as the deviation from the actual critical temperature.

2.9. Models as examples: Gaussian and ϕ^4

would then change the temperature dependent exponent of Eq.(2.53).

Seen from a thermodynamic point of view, these extra powers of ξ are remarkable because these seem to vitiate dimensional analysis. An additional scale is needed, and this comes from the small length-scale $a \sim 1/\Lambda$ whenever there is fluctuations at all length scales. The ultimate exponent one observes are not what dimensional analysis based on ξ as the important length-scale would have predicted, except for Gaussian-type models. (These dimensional-analysis-based exponents are called naive or engineering dimensions.) One needs to understand and explain the origin of the "extra" contribution like the p's, which are to be called *anomalous dimensions*. Once the role of cut-off is recognized, the discrepancy with dimensional analysis (which is infallible in any case) vanishes.

Because of the long range nature of the correlations, any local fluctuation can affect regions away from it. This is the origin of scale-invariance or occurrence of power laws (scaling). However the occurrence of anomalous dimension is something more. In a Gaussian type model, the degrees of freedom can be decomposed into independent modes (*e.g.*, by going over to Fourier modes - "normal coordinates") and then we see the emergence of long range correlation in the long distance limit of $q \to 0$. The modes remain independent so that the fluctuation at one scale determined by q does not affect the fluctuations at other scales. For this reason dimensional analysis gives correct result. In contrast for a ϕ^4 type model, the modes for different q-values are no longer independent (coupled by the ϕ^4 term) and now a fluctuation at a short distance scale (q close to the cut-off) can affect the fluctuations at longer scales even to $q \to 0$.

The mode coupling allows a small fluctuation at some point in space to affect other points and the disturbance seen by any other point is the sum over all the paths that connect the two points in question. For low dimensions, this sum can have fluctuations and this leads to anomalous dimensions. In high enough dimensions, the availability of a large number of paths (phase space volume) helps in averaging out the effects. These two cases are separated by the upper critical dimension.

- Note that u becomes irrelevant for $d > 4$ and it could be a dangerous irrelevant variable. The exponents in such cases would depend on the function also.

2.9.3 Through correlations

It is reasonable to expect that the scaling variable one sees in a given problem should be independent of the actual physical quantity one is looking at. Therefore in the limit of $\Lambda \to \infty$, anomalous dimension like p_2 for h should be the same for all quantities that depend on h. Let us take the example of the correlation function, in zero field, at criticality,

$$g(x) = \frac{1}{x^{d-2}} \mathbf{g}(u\, x^{4-d}, \Lambda x), \qquad (2.55)$$

suppressing the arguments on the left hand side (lhs). Now in the limit $\Lambda x \to \infty$, if $\mathbf{g}(..., z) = z^{-\eta}\tilde{\mathbf{g}}(....)$, where the nature of the other arguments are not so crucial for us right now, we have

$$g(x) = \frac{1}{x^{d-2+\eta}} \tilde{\mathbf{g}}(...) \quad \text{(setting } \Lambda = 1\text{)}. \qquad (2.56)$$

This η changes $y_h = (d+2)/2$ to $y_h = (d+2-\eta)/2$. Identification can therefore be made: $p_2 = -\eta$.

Let us reanalyze the zero field critical correlation function $g(r, \Lambda)$, where the Λ-dependence has been made explicit. Under a rescaling of all lengths $x \to bx$, we see $g(x, t = 0, \Lambda) = b^{-(d-2)} g(bx, 0, \Lambda/b)$. The scale factor for the correlation function picks out the exponent one would expect on dimensional analysis. However if a scale transformation is done that changes the longer length-scales but not the microscopic ones, i.e., $\xi \to b\xi$ but $\Lambda \to \Lambda$, then

$$g(x, 0, \Lambda) = b^{-(d-2+\eta)} g(bx, 0, \Lambda). \qquad (2.57)$$

2.9. Models as examples: Gaussian and ϕ^4

In analogy with the naive dimension, we now define a scaling dimension which is the dimension one observes in the long scale limit keeping the microscopic lengths same.

Denoting the scaling dimension of X by $\mathcal{S}_X$ from how X-X correlation function scales as in Eq.(2.57), and naive dimension by d_X (defined by dimensional analysis), we have $\mathcal{S}_\phi = -(d-2+\eta)/2$ while $d_\phi = -(d-2)/2$. It is easy to check now that

$$\mathcal{S}_h + \mathcal{S}_\phi = -d = d_h + d_\phi. \tag{2.58}$$

This relation for the every pair of conjugate variables is very useful. The scaling behaviour of, say, Eq.(2.53) can then be interpreted as dimensional analysis but *with scaling dimensions*. Renormalization group transformation is a way of doing a transformation that scales the long lengths keeping the short ones same, thereby picking the scaling dimensions.

Generalizing the above discussion, we may define the scaling dimension $\mathcal{S}_\psi$ for any local quantity $\psi(\mathbf{x})$ from the critical autocorrelation function, *i.e.*, the long distance decay of the ψ-ψ correlation. The analogue of dimensional analysis would tell us that for the critical correlation of any combination of local variables $\psi_1, \psi_2, ...$, we should have

$$\langle \psi_1(0)\psi_2(x_2)...\psi_n(x_n)\rangle = |x_2|^{-S_1-S_2..-S_n}\, y\left(\frac{x_3}{x_2}, ..., \frac{x_n}{x_2}\right), \tag{2.59}$$

where S_i is the scaling dimension of ψ_i.

Exercise 2.31 *Why is it that the dimensions add up to $-d$? When can it be something different?*

Exercise 2.32 *Use dimensional analysis to write free-energy $f(r, u, h, \Lambda)$ and magnetization $m(r, u, h, \Lambda)$ as*

$$f = \xi^{-d}\tilde{f}(u\xi^{4-d}, h\xi^{(d+2)/2}, \Lambda\xi), \tag{2.60}$$

and

$$m = \xi^{-(d-2)/2}\tilde{M}(u\xi^{4-d}, h\xi^{(d+2)/2}, \Lambda\xi), \tag{2.61}$$

where $\tilde{f}, \tilde{M}$ are unspecified functions. For $\Lambda\xi \to \infty$ one expects anomalous dimensions:

$$f = \xi^{-d}\hat{f}(u\xi^{4-d+p_u}, h\xi^{(d+2)/2+p_h}), \qquad (2.62)$$

and

$$m = \xi^{-(d-2)/2+p_m}\hat{M}(u\xi^{4-d+p_u}, h\xi^{(d+2)/2+p_h}). \qquad (2.63)$$

Express these new functions in terms of the functions in Eqs. (2.60) and (2.61). Using the relation $m = \partial f/\partial h|_{h=0}$, show that $p_m = p_h \ (= -\eta)$.

2.10 Epilogue

We attempted to give an introduction to critical phenomena especially the idea of scaling, its need and consequences. The emphasis is on diverging length-scales. In case of a diverging length scale, as at a critical point, the correlations become long ranged. Such a point is characterized by (i) the decay exponents of the correlations (which could have anomalous parts like η and η_E of Sec. 2.8.2), and (ii) the number and nature of the relevant variables at that point.

Any problem that does not have any intrinsic or important length-scale would behave like a critical system. This absence of length-scales leads to power law decays of correlations and also power laws for other physical quantities. In low dimensions (less than the upper-critical dimension, which could very well be infinite) fluctuations play an important role near or at the critical point and show up in the anomalous exponents. In such cases (i.e., $d < d_u$), the finite-size effects can be understood in terms of the finite-size scaling.

A distinction needs to be made between scaling and anomalous dimension. Scaling is the rule for criticality, originating from a diverging length scale, while anomalous dimension is seen for fluctuation dominated cases.

Renormalization group provides the proper framework for analyzing such phenomena. We feel it is worthwhile to motivate those

2.10. Epilogue

ideas behind RG at the introductory level. Mean-field theory that ignores fluctuations could then be placed in the same framework as a particular case originating from the special (called dangerous) behaviour of the irrelevant variables.

Acknowledgments

It was a very exciting experience for me to present this set of lectures to the participants of the SERC school at MRI Allahabad. I thank Abhik Basu, Amit K. Chattopadhaya, Harvey Dobbs, Kavita Jain, Parongama Sen, Saugata Bhattacharyya for many comments on the manuscript, and thank Flavio Seno for hospitality at Università di Padova where the final version was completed.

Bibliography

[] * From 90's:

[1] S. M. Bhattacharjee, *"Mean field theories"* in *Models and Techniques of Statistical Physics*, (Narosa, New Delhi, 1997).

[2] H. E. Stanley *Scaling, universality, and renormalization: Three pillars of modern critical phenomena*, Rev. Mod. Phys. **71**, S358 (1999).

[3] M. E. Fisher, *Renormalization group theory: Its basis and formulation in statistical physics*, Rev. Mod. Phys. **70**, 2 (1998).

[4] J. Cardy, *Scaling and Renormalization in Statistical Physics* (Cambridge U Press, 1996).

[5] J. Zinn-Justin, *Quantum field theory and critical phenomena*, 3rd ed (Oxford, 1996).

[6] C. Domb, *The critical point: a historical introduction to the modern theory of critical phenomena* (Taylor and Francis, 1996)

[7] P. M. Chaikin and T. C. Lubensky, *Principles of Condensed Matter Physics* (Cambridge U Press, 1995).

[8] S. V. G. Menon, *Renormaliazation group theory of critical phenomena* (Wiley Eastern, 1995).

[9] J. Yeomans, *Statistical Mechanics of phase transitions*, (Oxford U. Press, 1992)

BIBLIOGRAPHY 115

[10] J. J. Binney, N. J. Dowrick, A. J. Fisher and M. E. J. Newman, *The modern theory of critical phenomena*, (Clarendon Press, 1992).

[11] N. Goldenfeld, *Lectures on phase transitions and the renormalization group*, (Addison Wesley, 1992).

[] ** From 80's:

[12] G. Parisi, *Statistical field theory* (Addison-Wesley, 1988).

[13] D. J. Amit, *Field theory, the renormalization group, and critical phenomena*, 2nd ed. (World Scientific, 1984)

[14] R. J. Baxter, *Exactly solved models in statistical mechanics*, (Academic Press, 1982).

[15] *Critical phenomena*, Ed. by F. J. W. Hahne, Lecture notes in Physics v 186 (Springer, 1982), especially the articles by M. E. Fisher and A. Aharony.

[16] J.-C Toledano and P. Toledano, *The Landau theory of phase transitions: application to structural, incommensurate, magnetic, and liquid crystal systems*, (World Scientific, 1987).

[] *** From 70's:

[17] A. Z. Patashinskii and V. I. Pokrovskii, *Fluctuation theory of phase transitions* (Pergamon, 1979).

[18] G. Toulouse and P. Pfeuty, *Introduction to the renormalization group and to critical phenomena* (Wiley, 1977).

[19] Shang-keng Ma, *Modern theory of critical phenomena* (Benjamin, 1976).

[20] H. E. Stanley, *Introduction to phase transitions and critical phenomena* (Oxford, 1971).

[21] K. G. Wilson and J. Kogut, *The Renormalization Group and the ϵ-Expansion*, Phys. Rep. **12**, 75 (1974).

BIBLIOGRAPHY

[] ****** General texts:**

[22] H. B. Callen, *Thermodynamics and an introduction to thermostatistics* 2nd Ed. (Wiley) 1985.

[23] K. Huang *Statistical Mechanics* 2nd ed. (Wiley) 1987.

[24] L. D. Landau, E. M. Lifshitz and L. P. Pitaevskii , *Statistical Physics*, 3rd Ed. (Pergamon Press) 1980.

[25] Shang-Keng Ma, *Statistical mechanics*, (World Scientific, 1985).

[26] R.. K. Pathria, *Statistical mechanics*, (Oxford, 1972)

[27] L. E. Reichl, *A modern course in statistical physics* 2nd ed. (Wiley, 1998).

[28] F. Reif, *Fundamentals of statistical and thermal physics*, (McGraw-Hill, 1965).

[] ******* Specific topics:**

[29] M. A. Anisimov, *Critical phenomena in liquids and liquid crystals*, (Gordon and Breach, 1991).

[30] B. K. Chakrabarti, A. Dutta and P. Sen, *Quantum Ising phases and transitions in transverse Ising models*, Lecture notes in Physics m 41, (Springer, 1996).

[31] M. F. Collins, *Magnetic critical scattering*, (Oxford, 1989).

[32] P. G. De Gennes, *Scaling concepts in polymer physics*, (Cornell Univ. Press, 1979).

[33] The series on "Phase transitions and critical phenomena" Ed by C. Domb and M. Green (V 1-6), and C. Domb and J. Lebowitz (V 7-).

[34] J. Marro, R. Dickman, *Nonequilibrium phase transitions in lattice models*, (Cambridge, University Press, 1999).

[35] T. Narayanan and Anil Kumar, *Reentrant phase transitions in multicomponent liquid mixtures*, Phys. Rep. **249**, 135 (1994).

[36] S. Sachdev, *Quantum phase transitions*, (Cambridge U. Press, 1999).

[37] D. Stauffer and A. Aharony, *Introduction to percolation theory* (Taylor and Francis, 1992).

[38] C. Vanderzande, *Lattice models of Polymers*, (Cambridge U. Press, 1998).

Chapter 3

Phase Transitions and Critical Phenomena: Renormalisation Group Method

Deepak Kumar

School of Physical Sciences
Jawaharlal Nehru University
New Delhi – 110067

3. Phase Transitions and Critical Phenomena

These lectures describe the physics of phase transitions and critical phenomena. After introducing the thermodynamics of phase transitions, these general considerations are made explicit by presenting statistical mechanical calculations on a lattice gas model in the mean-field approximation. Then the Landau theory of phase transitions is presented. The form of free energy in the Landau theory is motivated by an explicit mean field calculation for the Ising model. This is followed by a discussion of spatial correlations and the breakdown of the mean field theory due to increase of spatial fluctuations near the critical point. This calculation also brings out the role of dimension of the system for the critical behaviour. The basic ideas of Renormalization Group (RG) are then introduced and illustrated by a simple RG calculation for the one-dimensional Ising model. Real-space RG calculations are then applied to the two-dimensional Ising model on the triangular lattice, which, in fact, brings forth several key features of the RG technique. A discussion of more general aspects of the RG formalism, such as scaling for spatial correlation functions, relevant and irrelevant couplings, universality, upper critical dimension, etc. are introduced. Finally momentum-space RG formalism for the Ginzburg-Landau (GL) model is taken up. The lectures end with calculations of critical exponents as expansions in powers of $\epsilon = 4 - d$, where d is the dimension of the system, for the GL model with n-component order parameter.

3.1 Introduction

Phase transitions occur in macroscopic systems when one or more external parameters like temperature, pressure, electrical field, magnetic field, etc are varied. They are rather ubiquitous phenomena which can be observed in a variety of circumstances. For example, the liquid-gas transition can be studied by varying temperature and keeping either pressure or volume fixed. The man-

3.2. Thermodynamic stability

ifestation of transition in the former case is conversion of gas to liquid, at a transition temperature dependent on pressure. In the latter case, it is the conversion of a single gas phase to a two phase gas-liquid coexistence below a fixed temperature, the critical temperature.

Due to this reason, theoretical descriptions of phase transitions require a lot of flexibility, which is very nicely available in thermodynamics through the use of alternative formulations in terms of different thermodynamical potentials like internal energy, Gibbs free energy, etc. Similar flexibility of description also exists in statistical mechanics, in terms of different ensembles, like microcanonical, canonical, etc.

Depending on the convenience of calculation and constraints imposed in experiments, one has the freedom to go from one description to another through the use of Legendre transformations. However, going through the material on phase transitions given in commonly available texts, one finds that the flexibility of theoretical description which is very essential in making contact with observed phenomena is not fully exploited. The purpose of this chapter is to demonstrate the utility of these alternative formulations of thermodynamics and statistical mechanics in the context of phase transitions and show how their use clarifies and unifies the theory of phase transitions.

Another very significant point is that all the diverse phase transitions can be discussed in terms of a few common thermodynamical descriptions. To put it in other words, there are a few generic thermodynamic mechanisms which are operative in all phase transitions. We begin with the basic thermodynamic description of a single component fluid.

3.2 Thermodynamic stability and phase transitions

Let us consider how a phase transition occurs in a single component system like water under the simple situation that its temperature is varied at constant pressure. As temperature increases, liquid passes from liquid to vapour phase at a transition temperature dependent on pressure. At this point, the density of the system changes discontinuously and along with that its internal energy, enthalpy and entropy also change discontinuously as the system absorbs heat at the transition temperature. Thermodynamically, it is most appropriate to discuss the situation in terms of Gibbs potential $G(P,T,N)$ where T, P, N, denote temperature, pressure and number of the molecules. One assigns separate Gibbs potential to the vapour phase $g_v(P,T) = G_v(P,T,N)/N$, and to the liquid phase $g_l(P,T)$ with different variations on P and T, as shown in Figs.3.1a and 3.1b. In these figures, one sees

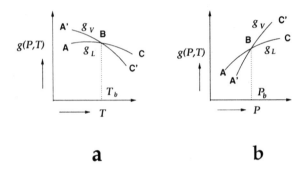

Figure 3.1: Variation of Gibb's potential for vapour and liquid phases with (a) temperature (b) pressure.

that at a particular temperature the two curves cross each other. Below the temperature T_b, $g_l < g_v$ making the liquid phase more stable than the vapour phase, while above above T_b, $g_v < g_l$ and the vapour phase is more stable than the liquid phase. Thus at point B in Fig.3.1a, the thermodynamic system switches from curve AB to curve BC'. Note that there is a discontinuity in the slope at B. Since the entropy per mole $s = -\partial g/\partial T$, this implies a discontinuous change in enthalpy $h = Ts + g$, and hence a latent heat at the transition point. Fig.3.1b shows how the transition occurs as pressure is changed at the fixed temperature by plot-

3.2. Thermodynamic stability

ting g_l and g_v as functions of pressure. This figure illustrates the presence of discontinuity in the pressure derivative of free energy which is related to the volume discontinuity of the system at the transition point.

At this point one might wonder about the origin of two branches of the free energy for a single system. This may be thought of as follows. Consider the system at fixed pressure and temperature with an additional constraint of fixed density ρ. (Note that these considerations apply for densities of other extensive variables like entropy, magnetic moment, etc, as well.). If we now plot g as a function of ρ it has the form as shown in Fig.3.2a. At a given P, T, the density assumed by the unconstrained system is the one with the lowest value of g which corresponds to ρ_v here. But note that the curve shown here is not convex with respect to ρ, and there is another local minima at ρ_l. ρ_v and ρ_l are obviously functions of P and T. The values of g at these two minima are the two curves we plotted in Fig.3.1a. Fig.3.2b schematically shows how the g versus ρ curve looks like at different temperatures. One sees here that when we decrease temperatures from T_5 to T_1, the liquid minima gets lowered compared to the vapour minima. Thermodynamics does not give a way to compute $g(P, T)$ with constrained density as we depict in the figure, but one expects a statistical mechanical calculation to yield curves like this. In a later section, we show how such curves arise for a lattice gas in a mean field approximation.

As a demonstration of our introductory remarks, let us consider the liquid-vapour transition under another set of conditions. This time we consider a fixed quantity of material enclosed in a container of fixed volume, but in contact with a temperature bath. At high enough temperature the system is in the vapour phase, but as the temperature is lowered below a certain temperature the liquid and vapour coexist. The appropriate thermodyamic potential to describe this situation is the Helmholtz potential $F(T, V, N)$ in which the independent variables are V, T, and N. This potential is the Legendre transform of the Gibbs potential $G(T, P, N)$, in which the independent variable P is replaced by V. The relation

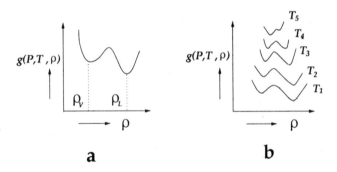

Figure 3.2: (a) Plot of $g(P,T,\rho)$ as function of constrained density ρ. (b) A series of plots for $g(P,T,\rho)$ for different temperatures showing the variation of the relative positions of vapour and liquid minima.

between the two is

$$F(T,V,N) = G(T,P,N) - PV = G - P\frac{\partial G}{\partial P}, \qquad (3.1)$$

$$\text{where } V = \frac{\partial G}{\partial P} \text{ and } P = -\frac{\partial F}{\partial V}. \qquad (3.2)$$

Fig.3.3a plots $f(T,v) = F(T,V,N)/N$, where $v = V/N$, as a function of v at a high temperature where the system is in the vapour phase. Note that $f(v)$ curves must have negative slope, i.e., $P > 0$ and it is a convex function i.e., its tangent lies belows the curve. The latter requirement is a condition of the thermodynamic stability which states that $F(V - \delta V, T, N) + F(V + \delta V, T, N) > 2F(V,T,N)$. We remind the reader that this condition follows from a simple gedanken experiment. Place a movable wall in the container of the system dividing it into two equal portions. If the above condition is violated, the system can lower its free energy by moving the wall one way in a manner that one portion gains the volumes continually at the expanse of the other portion. This is clearly an unstable situation.

The above condition holds as long as the system is in a stable state and in a single phase, the vapour phase in this case. As

3.2. Thermodynamic stability

the temperature of the system is lowered, the vapour phase becomes unstable at a certain point and the system stabilises itself by breaking into two phases. In terms of the Helmholtz free energy, this happens when $f(v,T)$ develops a region of concavity as shown in Fig.3.3b. At this temperature the single phase is stable if the specific volume of the system v is less than v_l or greater than v_v. But if it lies between these values, say v_c, the system can lower its free energy by dividing itself in two phases, one having the specific volume v_l and the other v_v. By this spontaneous division, its free energy is lowered from its value at C to the value at E, which lies on the common tangent of points B and D. Thus between these two volumes, the two phases can coexist. The fraction of the two phases x and $1-x$ follow from the constraint that

$$xv_l + (1-x)v_v = v_c . \tag{3.3}$$

Then

$$x = \frac{v_v - v_c}{v_v - v_l} \; ; \quad 1 - x = \frac{v_c - v_l}{v_v - v_l}, \tag{3.4}$$

which is the well known Lever's rule. Further note that pressure $P = -\partial f/\partial v$, thus the two phases corresponding to values v_l and v_v have the same pressure, as must be true for any intrinsic thermodynamical variable.

From these considerations, one can derive the more familiar picture in terms of $P-V$ isotherms. A visual evaluation of slopes of $f(v)$ in Figs.3.3a and 3.3b yield $P-V$ curves of shapes shown in Figs.3.4a and 3.4b. For Fig.3.4b, one notes that between the volumes v_l and v_v, the pressure remain constant being given by the common tangent. This is the region of phase coexistence. We can derive some more useful information by following the derivative of the original curves. This is shown by the dotted portion, and such isotherms are quite familiar from van-der-Waal's equation of state.

To make this discussion more concrete, we now go over to a statistical mechanical calculation.

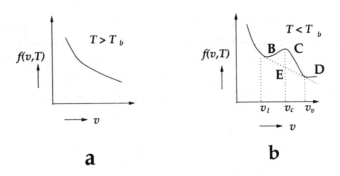

Figure 3.3: Variation of Helmholtz potential $f(v,T)$ with volume per particle v at two temperatures (a) above boiling temperature T_b (b) below boiling temperature T_b.

3.3 Lattice gas : mean field approximation

Consider a gas of molecules in a container of volume V. The interaction between the molecules is described by a spherical symmetric potential $v(r)$, whose schematic form is shown in Fig.3.5. This potential has a repulsive core of radius a and a short-ranged attractive part. Since statistical mechanical calculations with this potential are extremely difficult, it is very useful to consider a lattice model, in which the repulsive part of the potential which is very strong, is treated quite well. The volume of the container is divided into the cells of volume $v_o = a^3$, so that $V = N_o v_o$. Clearly each cell can accomodate only one molecule. So the configuration energy of the molecule can be written in terms of occupancy variable Q_i for each cell, which assumes value 1, if the ith cell is occupied by the molecule, and the value 0, if the ith cell is unoccupied. This simplification ignores the fact that a molecule may occupy two or more cells partially. The attractive part of the interaction is taken care of, by assigning a negative energy to molecules occupying neighbouring cells. Thus the configurational

3.3. Lattice gas : mean field approximation

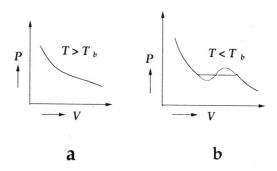

Figure 3.4: Pressure-volume isotherms for (a) $T > T_b$ (b) $T < T_b$.

energy of the system can be written as

$$H_{conf} = -\frac{K}{2}\sum_{(i,j)} Q_i Q_j , \qquad (3.5)$$

where the (i,j) sum is restricted to i and j being the nearest neighbours. The grand canonical partition function of the gas can now be written as

$$Z = \sum_{N=0}^{N_0} {\sum_{Q}}' \exp[\beta(K/2)\Sigma_{i,j}Q_iQ_j + \beta\mu'\Sigma_i Q_i] z_{kin}^N , \qquad (3.6)$$

where z_{kin} is the contribution from the kinetic energy of the molecules, μ' is the chemical potential and the configuration sum involves summing over all Q_i's, with the constraint $\Sigma_i Q_i = N$. Further,

$$\begin{aligned} z_{kin} &= \int d^3 p \exp(-\beta\frac{p^2}{2M}) \\ &= [\frac{2\pi M}{\beta}]^{3/2} . \end{aligned} \qquad (3.7)$$

One absorbs the kinetic contribution by redefining $\mu = \mu' + (3kT/2)\ln(2\pi M/\beta)$. Then one obtains

$$Z = \sum_{\{Q_i=0,1\}} \exp\beta[(K/2)\Sigma_{i,j}Q_iQ_j + \mu\Sigma_i Q_i] . \qquad (3.8)$$

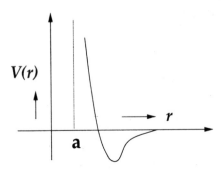

Figure 3.5: Typical form of the intermolecular potential.

We now evaluate Z in the mean field approximation by writing $Q_i = \rho + \delta Q_i$ and neglecting terms quadratic in δQ_i. This yields

$$Z = \exp[-\beta \frac{N_0 z K}{2} \rho^2] \sum_{\{Q_i=0,1\}} \exp \beta \Sigma_i (\mu + \rho K z) Q_i$$

$$= \exp[-\beta \frac{N_0 z K}{2} \rho^2][1 + \exp \beta(\mu + \rho K z)]^{N_0} . \quad (3.9)$$

Here z denotes the number of nearest neighbours. The grand canonical partition function yields pressure through the relation

$$P N_0 v_0 = \frac{1}{\beta} \ln Z \text{ leading to}$$

$$P v_0 = -\frac{z K \rho^2}{2} + kT \ln[1 + \exp \beta(\mu + \rho K z)] . \quad (3.10)$$

Here $\rho = <Q_i>$ is an auxiliary variable to be obtained by minimising the grand canonical free energy whose equation is given by

$$\rho = \frac{\exp \beta(\mu + \rho K z)}{1 + \exp \beta(\mu + \rho K z)} . \quad (3.11)$$

Solving Eq.(3.11) for ρ and substituting the solution in Eq.(3.10) yields the functional relation $P(\mu, T)$, in terms of which all other

3.3. Lattice gas : mean field approximation

thermodynamic information can be derived. This calculation also yields the Gibb's potential $G(P, T, N) = Ng(P, T)$ by noting that $\mu = g$. By inverting Eq.(3.11), we can find μ to be

$$g = \mu = -zK\rho + kT \ln \frac{\rho}{1-\rho}. \quad (3.12)$$

By eliminating μ from Eq.(3.10), one also obtains the isotherm

$$Pv_0 = -\frac{1}{2}zK\rho^2 - kT \ln(1-\rho). \quad (3.13)$$

To calculate $g(P, T)$ one first solves Eq.(3.13) to obtain ρ as a function of P and T, and then substitutes this solution in Eq.(3.12).

The discussion of phase transition is done most easily in terms of the isotherm. To fix ideas qualitatively, we write Eq.(3.13) as

$$Pv_0 = kT\rho + \frac{1}{2}(kT - zK)\rho^2 + kT \sum_{n=3}^{\infty} \frac{(\rho)^n}{n} \quad (3.14)$$

which shows that for $kT > zK$, P is an increasing function of ρ. By differentiating the isotherm it is easily seen that for $kT \geq zK/4$, P is a monotonic function of ρ as shown in Fig.3.6a, while if $kT/zK < 1/4$, two extrema develop at $\rho = (1/2) \pm \sqrt{(1/4) - (kT/zK)}$, so that the $P - \rho$ isotherms look like vander-Waals isotherms as shown in Fig.3.6b.

Now for a range of P, there are three solutions of ρ and corresponding to each of these, there is a branch of g. This is schematically shown in Figs.3.7a and 3.7b. In Fig.3.7b, ABC may be construed as the vapour branch g_v, while DBE can be construed as the liquid branch g_l, and DC is a metastable branch without any physical meaning. As the pressure is increased, the system, staying in the lowest branch moves over from vapour to liquid branch at B. On the isotherm, the point B corresponds to a flat portion where a discontinuous change of volume occurs. The flat portion of the line is drawn by Maxwell's equal area construction, i.e., the line FGH in Fig.3.6b is drawn in such a way that the area of the curve lying above it equals the area lying below it. The logic of

3. Phase Transitions and Critical Phenomena

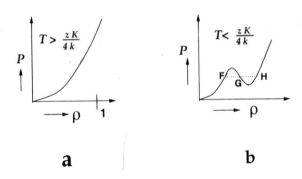

Figure 3.6: P versus ρ plots according to Eq.(3.14) at (a) $kT > \frac{zK}{4}$ (b) $kT < \frac{zK}{4}$.

this construction follows from the relation $dg = vdP - sdT$. So at fixed temperature

$$g(P,T) = g(\rho_0, T) + \int_{P_0,P} vdP \ . \qquad (3.15)$$

To understand this integral, we redraw the Fig.3.6b as Fig. 3.8. We can compute the integral in Eq.(3.15) from this curve starting from the point A corresponding to pressure P_0. Upto point C, we obtain positive additive contribution to g, and this corresponds to the curve of ABC of Fig. 3.7b. From C to D, we get negative contribution so that g decreases as in Fig.3.7b, giving the unphysical branch CD. From D onwards, we again get positive contribution corresponding to the branch DBE. The vertical line BFG is drawn such that the areas BCF and FDG are equal. In this situation it is clear that the integral of the curve from A to B is equal to the integral from A to G, since this implies intersection of g curve with itself. This establishes equal area prescription for drawing the flat portion of the isotherm.

This multiplicity of solutions for density in a certain range of P and T is the essential mathematical mechanism of phase transitions. The phase transition occurs when the system switches from one solution branch to the other, which happens because the Gibbs potential corresponding to the lowest branch crosses that of

3.3. Lattice gas : mean field approximation

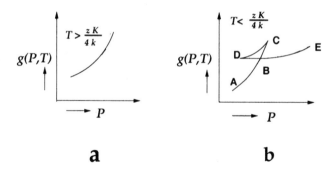

Figure 3.7: Gibb's potential g(P,T) as function of pressure for the lattice gas at (a) $kT > \frac{zK}{4}$ (b) $kT < \frac{zK}{4}$.

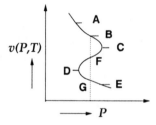

Figure 3.8: Redrawing of the isotherm of Fig.3.6b to compute g(P,T) using Eq.3.15. The vertical line through the curve indicates the Maxwell's equal area construction.

another, thereby changing the branch corresponding to the lowest potential.

The transition pressure at each temperature less than $zK/4k$, is to be obtained from drawing the isotherms and making the equal area construction as in Fig.3.6b. This yields the phase transition line in the $P - T$ plane as shown. The line terminates at the temperature $T_c = zK/4k$, which is the critical temperature. This is shown in Fig.3.9. As the temperature is raised, the densities corresponding to maximum and minimum of $P - \rho$ isotherms come closer to each other and merge at $\rho = 1/2$ at T_c. This gives us the

3. Phase Transitions and Critical Phenomena

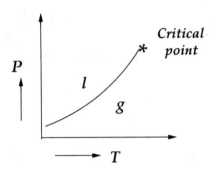

Figure 3.9: The phase transition line in the P-T plane. The line terminates at the critical point.

critical pressure $P_c = (zK/4v_0)[\ln 2 - 0.5]$.

To look at the transition with the constraint of fixed density, let us now obtain the Helmholtz free energy $F(T, V, N) = N_0 f(T, \rho)$ -

$$\begin{aligned} F &= N\mu - PV = N_0[\rho\mu - Pv_0] \\ &= N_0\{\frac{zK\rho^2}{2} + kT\ln(1-\rho) - Kz\rho^2 + kT[\rho\ln\rho - \rho\ln(1-\rho)]\} \\ &= N_0[-\frac{zK\rho^2}{2} + kT(\rho\ln\rho + (1-\rho)\ln(1-\rho))] \ . \end{aligned} \quad (3.16)$$

Fig.3.10a shows f as function of ρ at a temperature larger than $zK/4k$ while Fig.3.10b shows it at a temperature lower than $zK/4k$. This substantiates the discussion given in the previous section. The density discontinuity at the transition can be obtained by following the $f(\rho)$ curve at different temperatures and making the tangent construction. This is shown in Fig.3.11.

The same thing happens with discontinuity in the entropy density which is related to latent heat. To understand the entropy discontinuity, we calculate the entropy density $s(T, P)$ using Eq.(3.16) to be

$$s(T, \rho) = -k[\rho\ln\rho + (1-\rho)\ln(1-\rho)] \ . \quad (3.17)$$

3.3. Lattice gas : mean field approximation

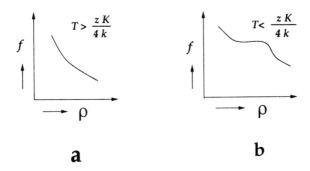

Figure 3.10: The Helmholtz potential for the lattice gas according to Eq.(3.16) at (a) $kT > \frac{zK}{4}$ (b) $kT < \frac{zK}{4}$.

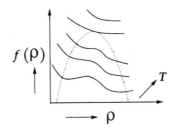

Figure 3.11: Plots of $f(\rho)$ versus ρ at increasing temperatures.

Again, using Eq.(3.13), we can obtain $s(T,P)$ by eliminating ρ. This shows that $s(T,P)$ is a multivalued function in certain ranges of temperature and pressure. In fact if we plot $s(T,P)$ for a fixed P, in this range, one finds curves of the form as shown in Fig.3.12. Further, the Maxwell area construction has to be done to find the entropy density in two coexisting phases. From this construction, one obtains the entropy discontinuity in the same manner as the density discontinuity.

Figs.3.13a and 3.13b show the density discontinuity $\Delta\rho$ and the latent heat $L = T(\rho)\Delta S$ respectively.

We close this section with some remarks about the critical point, which is the point at which the phase transition line comes

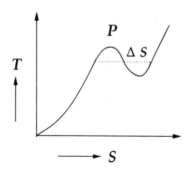

Figure 3.12: Plot of temperature versus entropy at fixed pressure. The plot shows that the transition is accompanied by a discontinuity in entropy as well.

to an end. The transition precisely at this point is characterised by no discontinuities in density (or specific volume) and entropy (or latent heat). For this reason the transition point is termed as second order phase transition. The critical point is further associated with singularities of the higher order derivatives $(\partial V/\partial P)_T$ and $(\partial S/\partial T)_P$. To see this, we plot the $P-V$ and $s-T$ curves respectively in Figs.3.14a and 3.14b around the critical point. From these curves, one notes that at the critical point, $(\partial V/\partial P)_{\rho_c}$ and $(\partial S/\partial T)_{T_c}$ become infinite. This implies that isothermal compressibility and constant pressure specific heat diverges at the critical point. These response functions are in turn related to fluctuations of volume and energy. Thus critical point is marked by large fluctuations in thermodynamic quantities.

3.4 Landau theory

We now present Landau theory which captures the essential and universal concepts of phase transitions in a very transparent manner. For this purpose it is convenient to consider the ferromagnetic transition which is a more familiar and simpler example of phase transitions, due to symmetry. Though it is usual to moti-

3.4. Landau theory

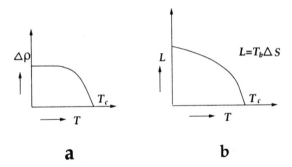

Figure 3.13: Variation of (a) density discontinuity (b) entropy discontinuity with temperature at the phase transition.

vate Landau theory from general considerations of symmetry, it is quite instructive to derive it from a microscopic model.

The simplest theoretical model to describe the ferromagnetic transition is the Ising model, given by the Hamiltonian

$$H = -\sum_{i,j} J_{ij} S_i S_j - h \sum_i S_i \ . \qquad (3.18)$$

As discussed in the thermodynamics section, we shall now determine the Gibbs potential $G(T,h)$ with the additional constraint of fixed magnetization m, which is the analog of density here. We again employ the mean field approximation.

$$e^{-\beta G(T,h,m)} = \sum_{S_i}' \exp[\beta J/2 \Sigma_{i,j} S_i S_j + \beta h \Sigma_i S_i] \qquad (3.19)$$

where the prime on summation implies the constraint that the magnetization $\sum_i S_i = Nm$. Within the mean field approximation this constraint implies that the energy of a configuration can be obtained by replacing S_i by m. Thus

$$e^{-\beta G(T,h,m)} = e^{\beta[\frac{NJzm^2}{2} + Nhm]} W(m) \qquad (3.20)$$

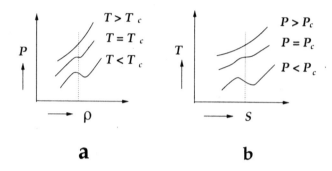

Figure 3.14: (a) P versus ρ curves at three temperatures. Note that $(\frac{\partial \rho}{\partial P})_T$ is infinite at T_c. (b) T versus s curves at three temperatures. Note that $(\frac{\partial s}{\partial T})_P$ is infinite at T_c.

where $W(m)$ is the number of micro-configurations that are consistent with the magnetization constraint. This is given by

$$W(m) = \frac{N!}{(N+Nm)/2!(N-Nm)/2!} \quad (3.21)$$

Using Stirling's approximation, this allows us to write

$$\begin{aligned} g(T,h,m) &= -[\frac{Jzm^2}{2} + hm] + kT[\frac{1+m}{2}\ln(\frac{1+m}{2}) \\ &+ \frac{1-m}{2}\ln(\frac{1-m}{2})] \end{aligned} \quad (3.22)$$

where $g = G(T,h,m)/N$. It is easy to see that this expression is minimized by the value of m determined by the following equation

$$m = \tanh \beta(Jzm + h). \quad (3.23)$$

For high enough temperatures where m is expected to be small, one can expand this g in powers of m to obtain the standard Landau free energy,

$$g(T,h,m) = \frac{1}{2}(kT - Jz)m^2 + \frac{kT}{12}m^4 + O(m^6) + \ldots - mh. \quad (3.24)$$

3.4. Landau theory

This is a typical example of Landau free energy, which is written in the form

$$g(T,h,m) = a(T)m^2 + b(T)m^4 + O(m^6) + \ldots - mh \quad (3.25)$$

The crucial point is the temperature dependence of the coefficient $a(T)$; $a(T) = a_0(T - T_c)$, changes sign when the temperature crosses T_c. From this expression, it is straightforward to derive the critical behaviour at the critical point ($kT_c = Jz, h = 0$).

Minimizing g with respect to m yields the equation for equilibrium magnetization to be

$$k(T - T_c)m + kT\frac{m^3}{3} = h . \quad (3.26)$$

One obtains the spontaneous magnetization by setting $h = 0$.

$$m_s = 0, \quad T > T_c \quad (3.27)$$
$$m_s = [3(T_c - T)/T]^{\frac{1}{2}}, \quad T < T_c \quad (3.28)$$

By differentiating Eq.(3.26) with h, one can find the susceptibility χ to be

$$\chi = \frac{1}{k[T - T_c + 3Tm_s^2]}, \quad (3.29)$$

which yields

$$\chi = \frac{1}{k[T - T_c]} \quad T > T_c \quad (3.30)$$

$$\text{and } \chi = \frac{1}{2k[T_c - T]} \quad T < T_c . \quad (3.31)$$

Finally we obtain the critical isotherm, *i.e.*, the $m - h$ relation at critical temperature to be

$$m = (\frac{3h}{kT})^{\frac{1}{3}} . \quad (3.32)$$

These equations give the usual mean field critical exponents $\beta = 1/2, \gamma = 1$ and $\delta = 3$.

While the above free energy and the critical behaviour is derived for the specific Ising model of ferromagnetic transition, the important point that was emphasized by Landau, is that this form of free energy is generic to all other phase transitions which are characterized by order parameters having the same symmetry and vectorial character. Here we are dealing with a scalar order parameter (OP) m. Further, the symmetry of the problem requires that the thermodynamic potentials do not depend on the sign of m. Even when the two phases coexist, the values of the OP's in the two phases have to be $\pm m_s(T)$ and the free energies of the coexisting phases are equal. A little reflection also shows that due to the symmetry, the transition always occurs at zero magnetic field. The vectorial character along with symmetry determine the kind of terms that are permitted in the Landau expansion. The dependence of the critical behaviour on the vectorial character of the OP will become clearer in the following sections.

Another crucial factor governing the phase transitions and the critical behaviour is the dimensionality of the system. This feature will be brought out in the next section where we consider the spatial correlations of the order parameter.

3.5 Spatial correlations

The mean field approximation assumes that fluctuations in the order parameter like density or magnetisation are small. Furthermore, one assumes that fluctuations at different spatial points are uncorrelated. This approximation becomes grossly inadequate near the critical point, where the fluctuations over increasingly larger spatial regions become correlated. In order to understand quantitatively the shortcomings of the mean-field approximation, it is necessary to analyze the spatial correlations in the system. To this end, the two-spin correlation function is a key quantity that provides considerable insight to the various properties of systems undergoing phase transitions, specially near the critical point, where the spatial and temporal dependences of thermodynamic fluctuations have several important physical manifestations. Such

3.5. Spatial correlations

functions can be directly measured in scattering experiments of light or neutrons from such systems. For example, in a liquid, the light scatters due to spatial and temporal fluctuations of density which, in turn, affect the refractive index of the liquid. If $\mathbf{k}_1$ and $\mathbf{k}_2$ denote the wavevectors of incident and scattered light, then the scattering intensity $I(\theta)$ at an angle θ is proportional to $G(\mathbf{q})$, where $\mathbf{q} = \mathbf{k}_2 - \mathbf{k}_1$, and $G(\mathbf{q})$ is the two-point density correlator defined as

$$G(\mathbf{q}) = \int e^{i\mathbf{q}\cdot(\mathbf{r}-\mathbf{r}')} <\delta\rho(\mathbf{r})\delta\rho(\mathbf{r}')> \qquad (3.33)$$

where $\delta\rho(\mathbf{r})$ denotes the deviation of density from its mean value. The magnitude of $\mathbf{q} \equiv q$ is related to θ through the relation $q = 2k_L \sin(\theta/2)$, for the elastic scattering, where k_L is the magnitude of the wavevector of light.

We now calculate the correlation function $\langle S(0)S(\mathbf{r})\rangle$ for the Ising model. It can also be thought as a conditional average of $S(\mathbf{r})$ when $S(0) = 1$ Thus

$$\langle S(\mathbf{r})S(0)\rangle = \langle S(\mathbf{r})\rangle_{S(0)=1} = \frac{Tr' \, e^{\beta \sum J(\mathbf{r}_1-\mathbf{r}_2)S(\mathbf{r}_1)S(\mathbf{r}_2)} S(\mathbf{r})}{Tr' \, e^{\beta \sum J(\mathbf{r}_1-\mathbf{r}_2)S(\mathbf{r}_1)S(\mathbf{r}_2)}}. \qquad (3.34)$$

Both traces are taken by fixing $S(0) = 1$. This constraint causes a spatial variation of magnetization and other quantities around the origin. We now extend the mean field approximation by assuming a space dependent mean field whose magnitude depends on the distance from the origin. The value of this mean field $h(\mathbf{r})$ is taken to be

$$h(\mathbf{r}) = \sum_{\mathbf{r}'} J(\mathbf{r}-\mathbf{r}')\langle S(\mathbf{r}')\rangle' . \qquad (3.35)$$

We can then write

$$\langle S(\mathbf{r})\rangle' = \tanh\beta \, [\sum_{\mathbf{r}'} J(\mathbf{r}-\mathbf{r}')\langle S(\mathbf{r}')\rangle' \,] . \qquad (3.36)$$

In the paramagnetic phase, when $r = |\mathbf{r}|$ is far from 0, since $\langle S(\mathbf{r})\rangle \approx 0$, we can linearize the above equation to get

$$\langle S(\mathbf{r})\rangle' \approx \beta \, [\sum_{\mathbf{r}'} J(\mathbf{r}-\mathbf{r}')\langle S(\mathbf{r}')\rangle' \,] + corrections . \qquad (3.37)$$

These corrections come in the region of small r, where $S(\mathbf{r}) \simeq S(0)$. This equation is equivalent to the following one for the correlation function,

$$G(\mathbf{r}) = \beta \sum_{\mathbf{r}'} J(\mathbf{r}-\mathbf{r}')G(\mathbf{r}') + corrections \ . \quad (3.38)$$

The small-r corrections are of order unity as $G(0) = 1$. Taking the fourier transform,

$$\begin{aligned} G(\mathbf{k}) &= \beta J(\mathbf{k})G(\mathbf{k}) + C \\ \Rightarrow G(\mathbf{k}) &= \frac{C}{1 - \beta J(\mathbf{k})} \end{aligned} \quad (3.39)$$

with

$$\begin{aligned} J(\mathbf{k}) &= \int J(\mathbf{r})e^{i\mathbf{k}\cdot\mathbf{r}} \ . \\ &= J_0[1 - k^2\xi_0{}^2] \end{aligned} \quad (3.40)$$

where the last expression is valid for small $k = |\mathbf{k}|$; and ξ_0, the bare correlation length, is given by,

$$\xi_0{}^2 = \frac{\int \mathbf{r}^2 J(\mathbf{r})d\mathbf{r}}{\int J(\mathbf{r})d\mathbf{r}} \ . \quad (3.41)$$

This allows us to write the correlation function $G(k)$ as

$$\begin{aligned} G(k) &= \frac{C}{1 - \beta J_0(1 - k^2\xi_0{}^2)} \\ &= \frac{C}{(1 - \frac{T}{T_c}) + \xi_0{}^2 k^2 \beta J_0} \\ &= \frac{C\xi_0{}^{-2}}{k^2 + \xi^{-2}} \ , \end{aligned} \quad (3.42)$$

where we have introduced the temperature-dependent correlation length ξ to be

$$\xi = \xi_0(1 - T_c/T)^{-\frac{1}{2}} \ . \quad (3.43)$$

3.6. Breakdown of mean field theory

Recalling that $I(\theta)$ is proportional to $G(q)$ and $q = 2k_L \sin(\theta/2)$, this expression shows that scattering in the forward direction, i.e., small q, increases very strongly as the critical point T_c is approached. We can also look at the spatial dependence of the two-spin correlation function. Taking the Fourier transform of $G(k)$, one finds that

$$G(r) = C\xi_0^{-2} \frac{e^{-r/\xi}}{r} \quad d = 3 ; \quad (3.44)$$
$$= I_0(r/\xi) \quad d = 2 ; \quad (3.45)$$
$$= e^{-r/\xi} \quad d = 1 . \quad (3.46)$$

This expression shows in a clear fashion how correlations grow in the paramagnetic phase as the temperature is lowered towards T_c. We can now discuss in a more quantitative manner how the mean field approximation breaks down as T_c is approached.

3.6 Breakdown of mean field theory

We are now in a position to make a quantitative criterion regarding the validity of the mean field approximation. The mean field expression for the internal energy per spin is

$$U_{mean} = J_0 \frac{m^2}{2}$$
$$\approx -\frac{J_0}{2}(1 - \beta J_0) \approx -\frac{J_0}{2}(1 - \frac{T}{T_c}) . \quad (3.47)$$

This is good only if the fluctuation contribution to the internal energy is small. The fluctuation contribution, U_f, is estimated in the following manner.

$$U = \sum_{ij} J_{ij}\langle (m + \delta S_i)(m + \delta S_j)\rangle$$
$$\approx U_{mean} + \sum_{ij} J_{ij}\langle \delta S_i \delta S_j \rangle . \quad (3.48)$$

This gives U_f, the second term in the above equation, to be

$$U_f = \sum J(\mathbf{r}_1 - \mathbf{r}_2)G(\mathbf{r}_1 - \mathbf{r}_2) = \sum J_{\mathbf{q}-\mathbf{q}_1}G(\mathbf{q}_1) , \quad (3.49)$$

which for short-ranged interactions can be approximated by

$$U_f = J_0 \sum_q G(0)$$
$$= C\xi_0^{-2} \int \frac{d^d q}{q^2 + \xi^{-2}} = J_0 C' \xi_0^{-2} \xi^{-d+2}. \quad (3.50)$$

Clearly the mean field theory is valid only when $U_{mean} \ll U_f$, i.e., when

$$J_0 C' \xi_0^{-2} \xi^{-d+2} < J_0(1 - \frac{T_c}{T}) = J_0 \frac{\xi_0^2}{\xi^2} \quad (3.51)$$

which reduces to

$$\xi^{4-d} < \xi_0^4. \quad (3.52)$$

For $d < 4$, the left hand side increases, as $\xi \to \infty$ or as $T \to T_c$. Thus for $d < 4$, mean field theory must breakdown when one is sufficiently close to the critical point. In terms of temperature, the criterion becomes

$$(\frac{1}{\Delta T})^{\frac{4-d}{2}} < \xi_0^4$$
$$\text{or } (\Delta T) \leq (\frac{1}{\xi_0})^{\frac{8}{4-d}}. \quad (3.53)$$

The size of the critical region depends on the magnitude of ξ_0, which is physically determined by the range of interaction. The larger the range, the smaller is the critical region.

On the other hand for $d > 4$, the fluctuation contribution to energy is smaller than the mean field contribution. So, for $d > 4$, mean field theory is alright for determining the critical behaviour. Here we see the crucial role of dimension. One notes that the spatial fluctuations are dependent on the dimesion of the system, getting stronger with lowering of the dimension. Above a dimension, known as *Upper Critical Dimension*, which is four here, the role of spatial fluctuations gets asymptotically negligible near the critical point and the mean field critical behaviour is obtained.

3.7 Ginzburg-Landau free energy functional

In order to incorporate the spatial fluctuations, one generalizes the Landau theory to introduce the so-called Ginzburg-Landau free energy functional. This free energy functional $F(H, T, \{m(\mathbf{r})\})$ is like a Helmholtz potential for a system with a fixed but an arbitrary spatially varying magnetisation $\{m(\mathbf{r})\}$. To determine it, one can calculate in principle, $G(H, T, \{m(\mathbf{r})\})$, by computing partition sums where we fix, in addition to H and T, also an arbitrary profile $\{m(\mathbf{r})\}$. This is given as

$$e^{-\beta G(H,T,\{m(\mathbf{r})\})} = \sum_{\{S_i\}} e^{-\beta H(\{S_i\})} \prod_{\mathbf{r}} [\sum_{i \in \mathbf{r}} S_i - m(\mathbf{r})] . \quad (3.54)$$

Here one is envisaging a coarse-graining as the value $m(\mathbf{r})$ is fixed by the sum over S_i's over a small region around the point $\mathbf{r}$. So if the constraints $\{m(\mathbf{r})\}$ are removed the thermodynamic values attained by m is the one that minimizes G. Clearly it is a formidable task to calculate this functional. But keeping in mind the universal features of phase transition, one builds the minimum model that incorporates the basic mathematical mechanism of phase transition - in this case the transition at the critical point.

The typical form for the functional is given by

$$\begin{aligned} F(h, T, M) &= -\frac{1}{2} \int J(\mathbf{r} - \mathbf{r}') m(\mathbf{r}) m(\mathbf{r}') d^3r d^3r' \\ &+ \int [Q(\{m(\mathbf{r})\}) - h m(\mathbf{r})] d^3r \end{aligned} \quad (3.55)$$

where $Q(\{m(\mathbf{r})\})$ is a polynomial function which, in principle, should be obtained from the above calculation. But following Landau, it is chosen to be the simplest polynomial that reflects the symmetry of the problem. This can be done by taking the polynomial that reproduces the mean-field results to the lowest nontrivial order in m when $m(\mathbf{r})$ is taken to be uniform. Since the first term above is the coarse-grained interaction energy, it is plausible to regard $Q(\{m(\mathbf{r})\})$ as the coarse-grained entropy, for which a straightforward generalization is

$$Q(\{m(\mathbf{r})\}) = \int d^3 r kT[\frac{1+m(\mathbf{r})}{2}\ln(\frac{1+m(\mathbf{r})}{2}) \\ + \frac{1-m(\mathbf{r})}{2}\ln(\frac{1-m(\mathbf{r})}{2})] \,. \qquad (3.56)$$

As discussed earlier, near the critical point where $m(\mathbf{r})$ is expected to be small, we need to retain just the lowest order nontrivial terms. For short-ranged interactions, one can also make a local expansion of the energy term. One writes

$$m(\mathbf{r}') = m(\mathbf{r}) + (\mathbf{r}-\mathbf{r}').\nabla m(\mathbf{r}) \\ + \frac{1}{2}\sum_{\alpha,\beta}(\mathbf{r}-\mathbf{r}')_\alpha(\mathbf{r}-\mathbf{r}')_\beta \nabla_\alpha \nabla_\beta m(\mathbf{r}) + \ldots \qquad (3.57)$$

Substituting this expansion and the expansion for $Q(\{m(\mathbf{r})\})$ in the free energy, one obtains

$$F(h,T,\{m(\mathbf{r})\}) \\ = \int d^3r \left(\frac{1}{2}J_0\xi_0{}^2|\nabla m(\mathbf{r})|^2 + \frac{a(T)}{2}m(\mathbf{r})^2 + \frac{b}{4}m(\mathbf{r})^4\right) \qquad (3.58)$$

where $a(T) = kT - J_0$, and its temperature dependence is crucial.

3.8 Renormalisation group (RG)

Any statistical mechanical calculation deals with a macroscopic number of interacting degrees of freedom. However our ability to deal with the dynamics of multiparticle systems is extremely limited. One can not obtain exact results for the dynamics of even three particles. For macroscopic systems, there are some simplifications as physical quantities are related to averages of dynamical quantities. The fluctuations around averages are small by virtue of large numbers. A good example is the mean field approximation (MFA), discussed in the last section, which gives a rather good qualitative account of the basic features of phase transitions. Let us analyse the success of MFA in the context of the Ising model.

3.8. Renormalisation group (RG)

Here one takes account of interactions between spins by assuming that each spin is acted upon by a mean field which is independent of the state of the spin in question, and is taken to be same for all spins. Since the mean field on a given spin is contributed by other spins interacting with it, its fluctuation about its average value can be regarded as small, provided the number of interacting spins is large. However, this is not so with short-ranged interactions as in our nearest-neighbour Ising model, where the number of interacting spins is the coordination number z. Further, when we considered spin correlations, we saw that spins get correlated over increasingly longer ranges as the critical point is approached. This makes the idea of a mean field independent of the state of the spin in question untenable.

A systematic way of improving the MFA, called cluster approximations, involves treating spins within the range of correlation length with greater accuracy by taking into account their correlated states. This kind of approach gets quickly tedious as the correlation length gets bigger, due to the difficulty of enumerating states of even a moderate number of spins like six. Clearly such an approach would fail even qualitatively to describe the physics near the critical point where the correlation length tends to infinity. Indeed, it turns out that these approximations do not lead to a change in the values of the critical exponent, as the analytical structure of the free energy is not altered from its mean-field form. This means that nontrivial changes of critical behaviour arise due to truly long range correlations that are present near the critical point. This is also indicated by the fact that critical behaviour shows such remarkable universality. As one would expect, the long range correlations should be largely independent of the microscopic details of the system.

The renormalization group (RG) method and the associated ideas of scaling arose to tackle problems that involve highly correlated degrees of freedom. In the present context, the scaling approach is based on the premise that the systems with different correlation lengths are similar, except for a change of certain parameters that depend upon the length scale set by the correlation length. These parameters are temperature, magnetic field, and other couplings that occur in the free energy expression and correlation functions. The RG method is a systematic procedure to relate these parameters at different length scales. This is basically done by doing partial traces over short-range degrees of freedom in the partition sum, and writing the remaining terms as a density matrix of a Hamiltonian which has the same form as the original Hamiltonian but with new coupling parameters. This leads to relations between one set of couplings to another set. Since the new Hamiltonian has a larger length scale, these relations are thought of as relations between couplings at different length scales. The physics of the critical phenomena is now derived by examining how various couplings change under the change of length scale, and in particular what happens at very large length scales that are relevant for critical behaviour. This procedure is explained with the help of a few examples below.

3.9 RG for a one dimensional Ising chain

The simplest example to illustrate the basic ideas of RG is the one-dimensional Ising model, whose Hamiltonian is given by,

$$H = -J \sum_{i=1}^{N} S_i S_{i+1} .\tag{3.59}$$

The partition function of the model is given by,

$$Z = Tr e^{\beta J \sum_i S_i S_{i+1}} .\tag{3.60}$$

Here we use periodic boundary conditions and take $S_{N+1} = S_1$. Instead of evaluating the partition function, which is quite straightforward in the present case, here we follow the RG procedure by doing a partial trace. In this spirit, we first sum over all the even spins. After the partial trace, one is left with a function containing the remaining odd spins.

$$Z_{odd}(\{S_i\}) = \sum_{even} e^{K[S_1 S_2 + S_2 S_3 + S_3 S_4 + ...]} \tag{3.61}$$

3.9. RG for a one dimensional Ising chain

where $K = \beta J$. The sum is done by noting that each spin is involved in only two terms as is evident from the above expression. Further note that $S_1 S_2$ can assume only two values ± 1. This allows us to write

$$e^{K S_1 S_2} = \cosh K + S_1 S_2 \sinh K = C_2[1 + x S_1 S_2] \qquad (3.62)$$

where
$$C_2 = \cosh K \quad \text{and} \quad x = \tanh K . \qquad (3.63)$$

Using this result the sum over S_2 can be done in the following manner -

$$\begin{aligned}
Tr_{S_2} e^{K(S_1 S_2 + S_2 S_3)} &= Tr_{S_2} C_2[1 + x S_1 S_2] C_2[1 + x S_2 S_3] \\
&= 2 C_2^2 [1 + x^2 (S_1 S_3)] \\
&= \frac{2 C_2^2 C_3}{C_3}[1 + x' S_1 S_3] \\
&= \frac{2 C_2^2}{C_3} e^{K' S_1 S_3} \qquad (3.64)
\end{aligned}$$

where
$$C_3 = \cosh K' \quad \text{and} \quad x' = x^2 = \tanh K' \qquad (3.65)$$

After the partial trace, the partition sum containing the remaining odd spins can be written in terms of a new Hamiltonian,

$$Z_{odd}(\{S_i\}) = \left(\frac{2 C_2^2}{C_3}\right)^{N/2} e^{K'[S_1 S_3 + S_3 S_5 + S_5 S_7 + \ldots]} \qquad (3.66)$$

with an effective coupling K' given by

$$\tanh K' = \tanh^2 K . \qquad (3.67)$$

This is an identical Hamiltonian to the original one, but with a length scale of $2a$. This enables us to obtain an equation for free energy per spin, $f(K)$, as a function of the coupling constant, which is

$$N f(K) = \frac{N}{2} \ln\left(\frac{2 C_2^2}{C_3}\right) + \frac{N}{2} f(K') \qquad (3.68)$$

or
$$f(K) = A(T) + \frac{1}{2}f(K') \ . \tag{3.69}$$

This procedure can be reiterated by doing the partial sums over the even spins of the new chain, leading to the following RG equations, which connect the couplings between spins at two length scales -

$$\begin{align} \tanh K' &= \tanh^2 K \quad x' = x^2 \\ \tanh K'' &= \tanh^2 K' \quad x'' = x'^2 = x^4 \ . \end{align} \tag{3.70}$$

One now studies how the coupling between spins changes with the length scale. Since x lies in the range (0,1), for any starting nonzero value, its value keeps on decreasing. Thus any coupling other than 1 flows towards zero value as shown in Fig.3.15. Since

```
         x = 1                    x = 0
         T = 0                    T = infinity
      or J = infinity          or J = 0
```

Figure 3.15: Flow of the coupling constant $K = J/kT$ with the length scale.

$K = \beta J$, this corresponds to equivalently zero value of J or infinite temperature. This implies that no matter how low the temperature (K large), the coupling in the system keeps on decreasing as the length scale increases. It asymptotically becomes zero corresponding to the paramagnetic phase.

These RG transformations have two fixed points, which are defined as points which do not change under the transformation. These are $x = 0$ and $x = 1$. All points flow under transformation to the fixed point at $x = 0$, except the point $x = 1$ (or $K = \infty$), which corresponds to zero temperature or infinite coupling. At zero temperature, spins are aligned corresponding to

3.9. RG for a one dimensional Ising chain

the ferromagnetic phase, which occurs at just one temperature in this model. Here we have recovered the well known property of the Ising chain, that it has phase transition only at zero temperature. Even though there is no finite-temperature phase transition in this simple example, the RG equation yields some more interesting information. Let us compare the correlation length of the original system with the one of twice the length scale. Clearly, the following relation should hold -

$$\xi(K) = 2\xi(K')$$
$$\text{and } \xi(x) = 2\xi(x^2) \, . \qquad (3.71)$$

This functional equation has a solution, that can be seen by inspection to be

$$\xi = -\frac{1}{\ln x} = -\frac{1}{\ln \tanh K} \, . \qquad (3.72)$$

This is an exact result. It is interesting to note how the correlation length grows as the temperature is lowered towards zero -

$$\xi \approx -\frac{1}{\ln[1 - 2e^{-K}]} \approx \frac{1}{2} e^{J/k_B T} \, . \qquad (3.73)$$

One also obtains the free energy to be

$$f(K) = \sum_{n=0} \frac{1}{2^n} \ln(\frac{2C_n{}^2}{C_{n+1}}) \qquad (3.74)$$

where C_n's denote the values of C's generated in successive transformations.

Let us verify the result on the correlation length by direct computation.

$$\begin{aligned}
Z &= Tr \cosh^N K \prod [1 + x S_i S_{i+1}] \\
&= Tr \cosh^N K [1 + x(S_1 S_2 + S_2 S_3 + ...) + x^2 (S_i S_j)(S_h S_l)] \\
&= Tr \cosh^N K [1 + x^N S_1 S_2 S_2 S_3 .... S_N S_1] \\
&= 2^N \cosh^N K [1 + \tanh^N K] \, , \qquad (3.75)
\end{aligned}$$

where the last two lines follow from the observation that the sum over any S_i gives zero, so only those terms survive in which each S_i is squared. Now consider the correlation function $G(m)$ -

$$G(m) = \langle S_n S_{n+m} \rangle = \frac{Tr \cosh^N K \prod[1 + xS_iS_{i+1}]S_nS_{n+m}}{2^N \cosh^N K[1 + \tanh^N K]}. \quad (3.76)$$

Now there is only one term that survives in the numerator, which is the product of terms $S_i S_{i+1}$ starting from $i = n$ to $i = n+m-1$, where we are taking $m > n$. This yields in the limit $N \to \infty$,

$$G(m) = x^m = e^{m \ln \tanh K} = e^{-m/\xi(K)} \quad (3.77)$$

which matches our earlier result.

3.10 RG for a two-dimensional Ising model

We next consider a two dimensional Ising model, which has a phase transition at finite temperature. For purposes of implementing the RG coarse graining for connecting coupling constants at different length scales, a two dimensional triangular lattice proves very instructive. We first associate all spins with shaded triangles as shown in Fig.3.16. Now for each shaded triangle we define a single spin out of three in the following manner. The three spins of the ith triangle (s_i^1, s_i^2, s_i^3) are to be replaced by a single superspin S_i, by the following prescription -

$$S_i = sign(s_i^1 + s_i^2 + s_i^3) . \quad (3.78)$$

This implies that the 8 possible spin-configurations of the triangle are to be replaced by just two configurations which are to be labelled by S_i. There are four configurations corresponding to each value of S_i, which are given by

$$\begin{aligned} s_i^1 &= s_i^2 = s_i^3 = S_i \\ s_i^1 &= s_i^2 = -s_i^3 = S_i \\ s_i^1 &= -s_i^2 = s_i^3 = S_i \\ -s_i^1 &= s_i^2 = s_i^3 = S_i . \end{aligned} \quad (3.79)$$

3.10. RG for a two-dimensional Ising model

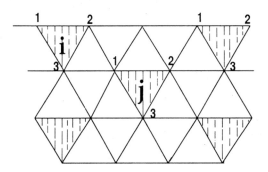

Figure 3.16: The supercell construction used in the RG procedure for a triangular lattice.

We can now obtain an effective Hamiltonian for the variables S_i's by summing over all original spin configurations with the above constraints.

$$Ae^{-\beta H'(S_i)} = \sum_\sigma e^{K \sum_{i,l;j,m} s^l_i s^m_j} \tag{3.80}$$

where the sum over σ denotes the constrained sums. This sum cannot be carried out exactly as the Hamiltonian contains terms that connect spins of different triangles. So one does the sum perturbatively, by dividing the Hamiltonian in two parts - $H = H_0 + V$ where H_0 contains all the couplings within each triangle, while V contains all intertriangle couplings. Thus,

$$H_0 = \sum_i K(s_i^1 s_i^2 + s_i^2 s_i^3 + s_i^3 s_i^1) \ . \tag{3.81}$$

Now we use the standard formalism of perturbation theory of statistical mechanics to get

$$\begin{aligned} Tr' \ e^{-\beta(H_0+V)} &= Tr' e^{-\beta H_0} \frac{Tr' e^{-\beta(H_0+V)}}{Tr' \ e^{-\beta H_0}} \\ &= Z_0 \langle e^{-\beta V} \rangle_0 \end{aligned} \tag{3.82}$$

where the average in the last line is the constrained statistical average with respect to H_0. For evaluating this average we employ the cumulant expansion procedure. The cumulant expansion for a probability distribution $P(x)$ is defined in the following way -

$$\langle e^{\lambda x}\rangle = \int P(x)e^{\lambda x}dx = \exp\left(\sum_n K_n \frac{\lambda^n}{n!}\right). \quad (3.83)$$

K_n's are called cumulants of the distribution $P(x)$. The first few cumulants are given as

$$\begin{aligned} K_1 &= \langle x\rangle \\ K_2 &= \langle x^2\rangle - \langle x\rangle^2 \\ K_3 &= \langle x^3\rangle - 3\langle x^2\rangle\langle x\rangle + \langle x\rangle^3, \end{aligned} \quad (3.84)$$

while the general expression is

$$K_n = \frac{d^n}{d\lambda^n}\ln\langle e^{\lambda x}\rangle|_{\lambda=0}. \quad (3.85)$$

Using this procedure, one obtains the following perturbation expansion for the free energy -

$$F = F_0 - k_B T \sum_n \frac{(-1)^n K_n(\beta V)}{n!}. \quad (3.86)$$

To proceed further, we first obtain Z_0. Since H_0 contains only unconnected triangles,

$$Z_0 = z_0^{N/3} \quad (3.87)$$

where

$$\begin{aligned} z_0 &= Tr' e^{K(s_i{}^1 s_i{}^2 + s_i{}^2 s_i{}^3 + s_i{}^3 s_i{}^1)} \\ &= (e^{3K} + 3e^{-K}). \end{aligned} \quad (3.88)$$

The above line is obtained by giving $\{s_i\}$ the four sets of values in terms of S_i. To obtain the first term in the perturbation series, we note that V involves coupling between spins of neighbouring triangles as shown in Fig.3.16. Between two neighbouring triangles i and j, the interaction is

$$V_{ij} = Js_j{}^1(s_i{}^2 + s_i{}^3). \quad (3.89)$$

3.10. RG for a two-dimensional Ising model

Thus the first term in the series is a sum of terms of the following kind for each pair of neighbouring triangles -

$$\begin{aligned}\langle V_{ij}\rangle &= K\langle s_j{}^1(s_i{}^2 + s_i{}^3)\rangle_0 \\ &= K\langle s_j{}^1\rangle_0\langle s_i{}^2 + s_i{}^3\rangle_0 .\end{aligned} \qquad (3.90)$$

These constrained spin averages are easily evaluated, giving

$$\begin{aligned}\langle s_i{}^3\rangle_0 &= \frac{1}{z_0}Tr'\, s_i{}^3 e^{K(s_i{}^1 s_i{}^2 + s_i{}^2 s_i{}^3 + s_i{}^3 s_i{}^1)} \\ &= \frac{1}{z_0}(S_i e^{3K} - S_i e^{-K} + 2S_i e^{-K}) \\ &= S_i \frac{e^{3K} + e^{-K}}{e^{3K} + 3e^{-K}} \\ &= S_i f_1 .\end{aligned} \qquad (3.91)$$

Further, it is easily seen that

$$\langle s_i{}^3\rangle_0 = \langle s_i{}^2\rangle_0 = \langle s_i{}^1\rangle_0 . \qquad (3.92)$$

This yields

$$\langle V_{ij}\rangle_0 = 2f_1{}^2 K S_i S_j . \qquad (3.93)$$

We can now extract the effective coupling between the superspins,

$$\begin{aligned}e^{H'(K')} &= Tr' e^{H(K)} = z_0{}^{N/3} e^{-\langle \beta V\rangle_0} \\ &= \exp[K' \sum_{\langle ij\rangle} S_i S_j]\end{aligned} \qquad (3.94)$$

with

$$K' = 2f_1{}^2 K = 2K\left(\frac{1 + e^{-4K}}{1 + 3e^{-4K}}\right)^2 . \qquad (3.95)$$

Let us study the flow of couplings implied by this equation. When K is small $K' \approx K/2$, which means than coupling flows to zero, corresponding to a paramagnetic phase. On the other hand when K is large $K' \approx 2K$, which means for a range of large $K's$, the couplings flow to ∞, corresponding to the ferromagnetic phase. Clearly, besides $K = 0$ and $K = \infty$, there must be another fixed point separating these two regimes. This is obtained by finding the nonzero and finite solution of the equation,

$$K^* = 2K^* \left[\frac{1 + e^{-4K^*}}{1 + 3e^{-4K^*}} \right]^2 . \qquad (3.96)$$

The solution is given by

$$K^* = \frac{kT^*}{J} = \frac{1}{4} \ln \left(\frac{3 - \sqrt{2}}{\sqrt{2} - 1} \right) . \qquad (3.97)$$

It will be soon shown that in the neighbourhood of K^*, the coupling flows away from this fixed point. Thus T^* can be identified to be the critical temperature, as for $T < T^*$ ($K > K^*$), the couplings flow to the zero-temperature fixed point corresponding to ferromagnetic phase, whereas for $T > T^*$ the couplings flow to the infinite-temperature fixed point, corresponding to the paramagnetic phase.

One can also write a functional equation for the free energy at this level of approximation, as the Hamiltonian in terms of superspins S_i is identical to the original one except for the value of the coupling. The equation is given by

$$Nf(K) = Nf_0 + \frac{N}{3} f(K')$$
$$\text{with } f(K) = \frac{1}{b^d} f(K') + f_0 , \qquad (3.98)$$

where $b = \sqrt{3}$ and d denotes the dimension of the system.

In order to derive the critical behaviour, we shall consider the behaviour when $K \approx K^*$. The RG recursion relation for small deviations around K^* can be written in a linearized form

$$K' - K^* = \left(\frac{\partial K'}{\partial K} \right)_K^* (K - K^*)$$
$$\text{or } \delta K' = \lambda_b(K^*) \delta K . \qquad (3.99)$$

Note that $\lambda_b(K^*)$ should be of the form $\lambda_b = b^{y_t}$. To see this consider two successive transformations -

$$\delta K' = \lambda_b \delta K$$
$$\delta K'' = \lambda_b \delta K' = \lambda_b^2 \delta K . \qquad (3.100)$$

3.10. RG for a two-dimensional Ising model

But this is also equal to a single transformation of a length scale change of b^2. Thus,

$$\lambda_b{}^2 = \lambda_{b^2} \tag{3.101}$$

which is compatible only with $\lambda_b = b^{y_t}$.

Let us now evaluate the exponent y_t for the above transformation.

$$\begin{aligned} b^{y_t} &= \left(\frac{\partial K'}{\partial K}\right)_{K^*} = 2f_1{}^2 + 4Kf_1f_1' \\ &= 1 + 2K^*\sqrt{2}f_1'(K^*) = 1.634 \ . \end{aligned} \tag{3.102}$$

The physical interpretation of y_t can be obtained by considering the scaling of correlation length -

$$\begin{aligned} \xi(K) &= b\xi(K') \\ \xi(\delta K) &= b\xi(b^{y_t}\delta K) \ . \end{aligned} \tag{3.103}$$

So if we now assume a power-law divergence of the form $\xi(\delta K) = \xi_0(\delta K)^{-\nu}$, we obtain

$$\begin{aligned} \xi_0(\delta K)^{-\nu} &= bb^{-y_t\nu}(\delta K)^{-\nu}\xi_0 \\ \Longrightarrow y_t\nu &= 1 \quad \text{or} \quad \nu = 1/y_t \ . \end{aligned} \tag{3.104}$$

Thus y_t determines the manner in which the correlation length diverges at the critical point.

Another way to see the power-law divergence of the correlation length from the RG scaling relation is to iterate the relation n-times -*i.e.*,

$$\xi(\delta K) = b^n \xi(b^{ny_t}\delta K) = \xi_0 b^n \tag{3.105}$$

where we have chosen n in such a way that $\delta K b^{ny_t} \approx 1$, and at this temperature ξ can be taken to be ξ_0. This gives

$$\begin{aligned} \xi(\delta K) &= \xi_0(\delta K)^{-\frac{1}{y_t}} \\ &= \xi_0 \left(\frac{T - T_c}{T_c}\right)^{-\nu} \ . \end{aligned} \tag{3.106}$$

Next we consider the scaling of free energy. Ignoring the part arising from Z_0, which is nonsingular, we have for the singular part of the free energy,

$$F_s(K) = b^{-d} F_s(K') . \tag{3.107}$$

The singular part of the free energy has a power-law behaviour near the critical point of the form $F_s(\delta K) \approx (\delta K)^s$, $s = 2 - \alpha$. Substituting this in the above relation yields $s = d/y_t = 2 - \alpha$, which also implies the exponent relation $d\nu = 2 - \alpha$.

To evaluate other exponents, we need to consider coupling to the magnetic field. For this purpose we add a term $V_h = h \sum_i s_i$ to the hamiltonian and evaluate the effective field h' coupling to the superspins. This can be done from the first order perturbation term as follows -

$$e^{-\beta H'} = Z_0 e^{\langle V \rangle + \langle V_H \rangle} . \tag{3.108}$$

Further,

$$\begin{aligned}\langle V_H \rangle &= h \sum_i \langle (s_i^1 + s_i^2 + s_i^3) \rangle \\ &= 3h f_1 \sum_i S_i = h' \sum_i S_i . \end{aligned} \tag{3.109}$$

This gives the recursion relation for the magnetic field to be

$$h' = 3 f_1(K) h . \tag{3.110}$$

This, together with the recursion for K in Eq.(3.95), constitute the full RG flow of the Hamiltonian at this level of approximation. The nontrivial fixed point of the transformation is $(K^*, 0)$. To extract the critical behaviour around this point one has to linearize both the relations simultaneously. In the present case the h-relation does not affect our earlier linearized Eq.(3.99), while for the h-field, we have

$$\begin{aligned} h' &= 3 f_1(K^*) h = b^{y_h} h \\ \text{with } y_h &= \frac{\ln 3/\sqrt{2}}{\ln \sqrt{3}} . \end{aligned} \tag{3.111}$$

3.10. RG for a two-dimensional Ising model

Since $y_h > 1$, the effective field increases with the length scale.

To derive further conclusions about the critical behaviour, we again consider the functional relation for the singular part of the free energy as a function of K and h. This is

$$f_s(\delta K, h) = b^{-d} f_s(\delta K b^{y_t}, h b^{y_h}) . \qquad (3.112)$$

Repeating the transformation n times yields

$$f_s(t, h) = b^{-nd} f(t b^{n y_t}, h b^{n y_h}), \qquad (3.113)$$

where we have written $t = (T - T_c)/T_c \simeq \delta K$. As before we choose n such that $t b^{n y_t} \approx 1$, which leads to,

$$\begin{aligned} f_s(t,h) &= t^{d/y_t} f_s\left(1, \frac{h}{t^{y_h/y_t}}\right) \\ &= t^{2-\alpha} g\left(\frac{h}{t^\Delta}\right) . \end{aligned} \qquad (3.114)$$

Further

$$M = \frac{\partial f}{\partial h} = t^{2-\alpha-\Delta} g'\left(\frac{h}{t^\Delta}\right)$$

$$\text{and } \chi = \left(\frac{\partial^2 f}{\partial h^2}\right)_0 = t^{2-\alpha-2\Delta} g''(0) . \qquad (3.115)$$

These relations imply the following exponent relations -

$$\begin{aligned} \beta &= 2 - \alpha - \Delta \\ \text{and} \gamma &= \alpha + 2\Delta - 2 . \end{aligned} \qquad (3.116)$$

In order to calculate δ, the exponent related to the critical isotherm, one differentiates the free energy relation by h to obtain,

$$m(t, h) = b^{-d+y_h} F_s'(t b^{y_t}, h b^{y_h}) . \qquad (3.117)$$

Now by choosing b such that $h b^{y_h} = 1$,

$$\begin{aligned} m(t,h) &= h^{d/y_h - 1} Y\left(\frac{t}{h^{y_t/y_h}}\right) \\ &= h^{1/\delta} Y\left(\frac{t}{h^{1/\Delta}}\right) . \end{aligned} \qquad (3.118)$$

This yields

$$\delta = \frac{y_h}{d - y_h}. \qquad (3.119)$$

Further, note that all exponents are given in terms of y_t and y_h as

$$\begin{aligned}
\alpha &= 2 - \frac{d}{y_t} \\
\beta &= \frac{d}{y_t} - \Delta = \frac{d - y_h}{y_t} \\
\nu &= \frac{1}{y_t} \\
\gamma &= 2\Delta - \frac{d}{y_t} = \frac{2y_h - d}{y_t}.
\end{aligned} \qquad (3.120)$$

This leads to exponent scaling relations which have been known since the early days of critical phenomena studies. Numerical values for the various exponents in this approximation are given by

$$\begin{aligned}
y_t &= 0.88 \\
\nu &= 1.133 \quad \alpha = -0.266 \\
y_h &= \frac{\ln(3/\sqrt{2})}{\ln\sqrt{3}} = 11.37 \\
\Delta &= \frac{y_h}{y_t} = 1.55 \\
\beta &= 0.714 \quad \text{and} \quad \gamma = 0.837.
\end{aligned} \qquad (3.121)$$

The flow of the two couplings are summarized in Fig.3.17. The above numbers can be improved by including higher order terms. However at higher orders one is forced to include more couplings in the Hamiltonian.

3.11 General features of RG

Having worked out explicitly two examples, we are now in a position to state in a general way, those features of renormalization

3.11. General features of RG

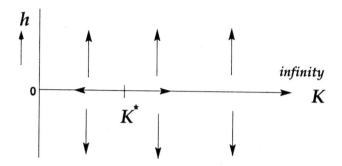

Figure 3.17: RG flows for the couplings (K, h) of the triangular lattice in the first order perturbation theory.

group transformations that are germane to the discussion of critical phenomena. The first step is to do a partial summation over degrees of freedom in the partition function for a given Hamiltonian. By rewriting the remaining terms as a density matrix of a new Hamiltonian, one generates relations between couplings at two length scales of the same system. However the partial summation does not generally lead to Hamiltonians of the same form as the original as a lot of new types of terms, not present in the original Hamiltonian, get generated. In principle, one can start with Hamiltonians that contain all couplings that are possible within the degrees of freedom of the system and write the recursions relation between these. This is not practical in most nontrivial cases, so one has to recognize the important couplings physically, and restrict the parameter space by truncating all others. We shall also see that most of the couplings are not important for the description of phenomena close enough to the critical point.

The recursion relations between the chosen set of couplings are highly non-linear and coupled with each other. They can be schematically written in the form,

$$\{K'_\mu\} = R(b)(\{K_\mu\}) . \tag{3.122}$$

To understand different phases and associated transitions, one studies how these couplings flow in the parameter space under repeated RG transformations. This study is greatly facilitated by determining the fixed points of the transformation and the flows around the fixed points. The flows around the fixed point are easily studied by linearizing the RG equations around the fixed points and determining the eigenvalues and eigenvectors of these linearized RG transformations. The eigenvectors tell us the principle directions along which flows in the neighbourhood of the fixed point occur, and signs of the corresponding eigenvalues determine whether the flow along that direction is towards the fixed point or away from it. Once flows around each fixed point are determined, the global flow can be surmised by connecting these flows in a consistent manner. However for determining the critical behaviour, it is often enough to work out the linearized flow around a particular fixed point.

The fixed points of a transformation are all solutions of the equation,

$$K_\mu{}^* = R_\mu(b)(\{K_\nu{}^*\}) \ . \tag{3.123}$$

Around each fixed point, one linearises the transformation

$$\delta K_i = \sum_j T_{ij} \delta K_j \tag{3.124}$$

where

$$T_{ij} = \left(\frac{\partial R_i}{\partial K_j}(K_1, K_2.....)\right)_{K_i = K_i^*} . \tag{3.125}$$

Note that T_{ij} need not be a symmetric matrix.

We now diagonalize this matrix by constructing its left eigenvectors

$$\sum_i \phi_i{}^{(a)} T_{ij} = \lambda_a \phi_j{}^{(a)} \tag{3.126}$$

where a takes values from 1 to n, if n is the number of couplings in the Hamiltonian. Using these eigenvectors one defines new linear combinations of couplings,

$$u_a = \sum_i \phi_i{}^{(a)} \delta K_i \qquad a = 1, ...n \ . \tag{3.127}$$

3.11. General features of RG

These new variables have rather simple RG transformations by construction, as

$$\begin{aligned} u'(a) = R_b(u_a) &= \sum_i \phi^{(a)}{}_i R_i(\delta K_i) \\ &= \sum_i \phi^{(a)}{}_i T_{ij} \delta K_j \\ &= \lambda_a \sum_i \phi_j{}^{(a)} \delta K_j = \lambda_a u_a \end{aligned} \quad (3.128)$$

The quantities u_a are called scaling fields, and they represent the priciple directions along which the flows occur in the vicinity of the fixed point. Since $\lambda_a = b^{y_a}$, u_a will grow if $y_a > 0$, which means that the couplings will flow away from the fixed point in this direction. On the other hand if $y_a < 0$, u_a will shrink and the flow would be towards the fixed point along that direction. Those fields that grow are called <u>relevant</u> while those which decreases are called <u>irrelevant</u>. Those for which $y_a = 0$ are called <u>marginal</u> fields. For these fields, the linear analysis is not enough to tell whether the field will grow or shrink.

In the neighbourhood of the fixed point, the singular part of the free energy has a rather simple form in terms of the scaling fields. It is given by

$$F_s(u_1, u_2, ... u_n) = b^{-d} F_s(u_1 b^{y_1}, u_2 b^{y_2}, ... u_n b^{y_n}) . \quad (3.129)$$

This is a key equation of the RG analysis. We are now led to the notion of **Critical Surface**. Let $m(m < n)$ scaling fields be relevant, i.e., $y_1, y_2 .... y_m > 0$ and $y_{m+1}, ... y_n < 0$. Then in the space of coupling constants, one defines the Critical Surface to be

$$u_a = \sum_i \phi_i^{(a)} \delta K_i = 0 \quad a = 1, ... m . \quad (3.130)$$

A point on this surface stays on the surface and approaches the fixed point, provided one is close enough to the fixed point so that the linear approximation is valid. This is illustrated for a two-dimensional coupling space (K_1, K_2) in Fig.3.18, where we assume $y_1 > 0$, $y_2 < 0$. A good example of this situation is

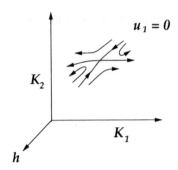

Figure 3.18: The critical surface and flows around the surface for an Ising model with nearest and next-nearest neighbour couplings. The critical surface is $u_1 = 0, h = 0$.

Ising model with nearest neighbour and next-nearest-neighbour couplings. Though the RG transformation generates a lot more couplings, like quadratic couplings of longer range and multi-spin couplings, only one combination of these coupling apart from the magnetic field is relevant. Thus the critical surface is given by

$$u_1 = \sum_i \phi_i^1 (K_i - K_i^*) = 0, \quad h = 0 . \qquad (3.131)$$

We denote the two eigenvalues that are relevant as y_t and y_h. If we do repeated transformations, the free energy relation takes the form

$$F_s(t, h, ...u_n) = b^{-md} F_s(tb^{my_t}, hb^{my_h}, ...u_n b^{my_n}) , \qquad (3.132)$$

where we assume that, due to physical reasons, the two relevant scaling fields are more or less like the temperature difference $t = (T - T_c)/T_c$ and the magnetic field. Now as we increase the number of iterations, all irrelevant scaling fields are driven to zero, leaving us with

$$\begin{aligned} F_s(t, h, ...u_n) &= b^{-dm} F_s(tb^{my_t}, hb^{my_h}, 0, 0...) \\ &= b^{-d} f(tb^{y_t}, hb^{y_h}) \\ &= b^{d/y_t} f(1, h/t^\Delta), \end{aligned} \qquad (3.133)$$

3.11. General features of RG

where, in the second line, we have replaced b^m by b as the total factor for the change of length scale. Thus we can ignore all other variables as irrelevant when one is close to the critical point. The rest of the critical properties follow as discussed in the previous section.

3.11.1 Irrelevant variables

As pointed out above, when one carries out the RG transformations, a lot of couplings that are not present in the original Hamiltonian are generated. But the real utility of the procedure is that most of these are irrelevant. Around the critical point of the Ising model there are just two relevant scaling fields u_t and u_h. u_t is a combination of $K_1, K_2, \ldots$, etc., but can basically be regarded as inverse temperature, whereas u_h is largely h. Now let us examine the role of other couplings $u_3, u_4, \ldots$ which are also some linear combinations of t and h and additional couplings. For these, one can typically write, $u_3 = u_3^0 + at + bh^2 + \ldots$. So near the critical point,

$$F_s(t,h) = b^{-nd} F_s(tb^{ny_t}, hb^{ny_h}, u_3^0 b^{-ny_3}, u_4^0 b^{-ny_4} \ldots). \tag{3.134}$$

Now choosing $1 = tb^{ny_t}$, yields

$$F_s(t,h) = t^{d/y_t} F_s(1, ht^{-\Delta}, u_3^0 t^{|y_3|/y_t}, \ldots). \tag{3.135}$$

Setting $h = 0$ and assuming that F_s can be expanded in a Taylor series in its variables, one gets

$$F_s(t,h) = \left(\frac{t}{t_0}\right)^{d/y_t} [A_0 + A_1 \left(\frac{t}{t_0}\right)^{|y_3|/y_t} + \ldots] \tag{3.136}$$

where t_0 is a temperature deviation of order unity. This shows that irrelevant fields give rise to nonanalytic corrections, as $|y_3|/y_t$, etc, are not integers in general. In real systems, the coefficients $u_3{}^0$ can be quite large, so these irrelevant fields can give rise to numerically important nonanalytical corrections. These are often required if we wish to fit the data to scaling forms.

Implicit in the above discussion is the assumption that the scaling function f is analytic in terms of the irrelevant couplings. But very often, that is not the case. When this happens with regard to an irrelevant variable, such a variable is termed *dangerously irrelevant variable*. We shall see such a situation for $d = 4$ Ginzburg-Landau system.

3.12 RG scaling for correlation functions

So far we have only dealt with scaling properties of free energy, which in turn relied on the fact that the partition function is not affected by the scaling procedure. Now we examine how the correlation functions scale under RG transformations. Let us begin by examining the spin-spin correlation function defined as

$$G(r_1 - r_2, H) = \langle S(r_1)S(r_2)\rangle_H - \langle S(r_1)\rangle_H \langle S(r_2)\rangle_H \quad (3.137)$$

where an average $\langle O \rangle_H$ denotes the usual statistical average with respect to the Hamiltonian H. One can obtain the correlation function in the following manner from an appropriate generating function

$$\begin{aligned} Z[h_1, \ldots, h_n] &= Tr \left[e^{-H + \sum_i h_i S_i} \right] \\ &= \langle e^{\sum_i h_i S_i} \rangle . \end{aligned} \quad (3.138)$$

$\langle e^{\sum_i h_i S_i} \rangle$ is a cumulant generating function, so

$$\frac{\partial \ln Z}{\partial h_m \partial h_n} = \langle S_n S_m \rangle - \langle S_n \rangle \langle S_m \rangle . \quad (3.139)$$

Let us now perform partial tracing in Eq.(3.138). Then,

$$e^{F_0} e^{-H(S') + \sum h'_i S'_i} = Tr_{partial} \left[e^{-H + \sum h_i S_i} \right] . \quad (3.140)$$

Now summing over the remaining variables S' yields the identity

$$\begin{aligned} Z'(h'_i) &= Tr_{S'} \left[e^{-H' + \sum h'_i S'_i} \right] \\ &= Tr_{S'} \, Tr_{partial} \left[e^{-H + \sum S_i h_i} = Z(h_i) \right] . \end{aligned} \quad (3.141)$$

3.12. RG scaling for correlation functions

Using this we can write,

$$\frac{\partial \ln Z'}{\partial h'(r_1) \partial h'(r_2)} = \frac{\partial \ln Z}{\partial h'(r_1) \partial h'(r_2)} . \quad (3.142)$$

Now using the scaling form $h'(r_1) = b^{y_h} h(r_1)$, one obtains the following relation for the correlation function,

$$\begin{aligned}
G(\frac{r_1 - r_2}{b}, H') &= b^{-2y_h} \frac{\partial^2 \ln Z}{\partial h'(r_1) \partial h'(r_2)} \\
&= b^{-2y_h} \langle (s^1{}_{r_1} + s^2{}_{r_1} + ...)(s^1{}_{r_2} + s^1{}_{r_2} + ...) \rangle \\
&= b^{2d - 2y_h} G(r_1 - r_2, H) , \quad (3.143)
\end{aligned}$$

where the last two lines follow from the fact that $h'(r_1)$ couples to to the superspin $S'(r_1)$ which is arising from a block of b^d original spins. Thus we arrive at the relation,

$$G(r_1 - r_2, u_t, u_h, ...) = b^{-2d + 2y_h} G(\frac{r_1 - r_2}{b}, u_t b^{y_t}, u_h b^{y_h}, ...) . \quad (3.144)$$

We consider the situation for $h = 0$ first. In this case,

$$G(r_1 - r_2, t) = b^{-2d + 2y_h} G(\frac{r_1 - r_2}{b}, t b^{y_t}) . \quad (3.145)$$

Again taking $t b^{y_t} = 1$, we get

$$\begin{aligned}
G(r, t) &= t^{\frac{2(d - y_h)}{y_t}} g\left(\frac{r}{t^{-1/y_t}}\right) \\
&= t^{\frac{2(d - y_h)}{y_t}} g\left(\frac{r}{\xi(t)}\right) . \quad (3.146)
\end{aligned}$$

Another useful scaling form is to choose b such that $r/b = a$, then

$$\begin{aligned}
G(r, t) &= \left(\frac{a}{r}\right)^{2(d - y_h)} G(a, t(\frac{r}{a})^{1/\nu}) \\
&= \frac{1}{r^{2(d - y_h)}} g\left(\frac{r}{\xi}\right) . \quad (3.147)
\end{aligned}$$

Exactly at the critical point, the scale in the system $\xi \to \infty$. So there is a complete scale invariance.

$$G(r, T_c) = \frac{g(0)}{r^{2(d-y_h)}} \propto \frac{1}{r^{d-2+\eta}}, \qquad (3.148)$$

where η is a new exponent defined by this equation. $x_h = d - y_h$ is known as the scaling dimension of the operator S_i. Clearly $\eta = d + 2 - 2y_h$.

Let us now calculate the correlation function of other local operators. We can define, for example, a local energy operator in the Ising model, $E(r) = \sum_\delta S_r S_{r+\delta}$, where the sum is over a small neighbourhood of r. One would now like to calculate the correlation function

$$G_{EE}(r_1 - r_2) = \langle E(r_1) E(r_2) \rangle - \langle E(r_1) \rangle \langle E(r_2) \rangle. \qquad (3.149)$$

This can again be done with the help of a generating function defined as

$$Z(\mu(r)) = e^{-H + \sum_r \mu(r) E(r)}. \qquad (3.150)$$

We can now repeat the exercise that was done for the spin-spin correlation function to obtain

$$\frac{\partial \ln Z'}{\partial \mu'(r_1) \partial \mu'(r_2)} = \frac{\partial \ln Z}{\partial \mu'(r_1) \partial \mu'(r_2)} \qquad (3.151)$$

where $\mu'(r)$ is the coupling to local energy in the coarsened system. Just as for the magnetic field h, we introduce a relation connecting $\mu'(r)$ to $\mu(r)$, which is $\mu'(r) = b^{y_E} \mu(r)$. This leads to the relation,

$$G\left(\frac{r_1 - r_2}{b}, H'\right) = b^{2(d-y_E)} G(r_1 - r_2, H) \qquad (3.152)$$

or

$$G(r, t, h) = b^{-2(d-y_E)} G\left(\frac{r}{b}, tb^{y_t}, hb^{y_h}\right). \qquad (3.153)$$

At the critical point, we again have,

$$G_{EE}(r) = \langle E(r) E(0) \rangle \propto \frac{1}{r^{2(d-y_E)}}. \qquad (3.154)$$

3.13. RG for Ginzburg-Landau model

One can also discuss cross correlations. For example, at the critical point, due to scale invariance, the energy-spin correlator assumes the following form -

$$G_{ES}(r) = \langle E(r)S(0) \rangle \propto \frac{1}{r^{(d-y_h)}} \frac{1}{r^{(d-y_E)}} \propto \frac{1}{r^{x_h+x_E}}. \quad (3.155)$$

Clearly, at the critical point every local operator is associated with a scaling dimension which governs the spatial decay of correlators of the fields involved. The above formula can be generalized to obtain the power law correlations between any combination of local operators at the critical point. The power law exponent is given by the sum of the scaling dimensions of all the operators in the correlator.

3.13 RG for Ginzburg-Landau model

We shall now apply RG ideas to a simple form of Ginzburg-Landau free energy functional. This calculation introduces the momentum-space RG technique. Through this technique a number of important results in critical phenomena have been obtained. In this calculation one also sees the important roles of dimension and symmetry.

We begin with the free energy functional given in Eq.(3.58). We can write it in momentum-space variables by using the definitions,

$$\phi(\mathbf{q}) = \frac{1}{L^{d/2}} \int m(\mathbf{r}) e^{i\mathbf{q}\cdot\mathbf{r}} d^d r$$

$$\text{and } m(\mathbf{r}) = \frac{1}{L^{d/2}} \int_0^{\Lambda} \phi(\mathbf{q}) e^{-i\mathbf{q}\cdot\mathbf{r}} \frac{d^d q}{(2\pi)^d} \quad (3.156)$$

where the upper limit $\Lambda = 2\pi/a$ reflects the underlying lattice structure in the problem. Then we find,

$$H = \frac{1}{2} \int_q^{\Lambda} r(\mathbf{q})\phi(\mathbf{q})\phi(-\mathbf{q})$$
$$+ u \int_{\mathbf{q}_1}^{\Lambda} \int_{\mathbf{q}_2}^{\Lambda} \int_{\mathbf{q}_3}^{\Lambda} \phi(\mathbf{q}_1)\phi(\mathbf{q}_2)\phi(\mathbf{q}_3)\phi(-\mathbf{q}_1-\mathbf{q}_2-\mathbf{q}_3) \quad (3.157)$$

where we use a shorter notation, $\int_q$ for the momentum integral of Eq.(3.156) i.e., $\int_q^\Lambda = \int_0^\Lambda d^d q$, with $q \equiv |\mathbf{q}|$ and

$$r(\mathbf{q}) = J_0[(T-T_c)/T_c + \xi_0^2 q^2] = r_0 + \xi_0^2 q^2 . \qquad (3.158)$$

The partition function to be evaluated is

$$Z = \int D\phi(\mathbf{q}) e^{-G\{\phi(\mathbf{q})\}} \qquad (3.159)$$

where $D\phi(\mathbf{q})$ denotes integration over all $\phi(\mathbf{q})$ variables. To follow the RG procedure we shall again do a partial tracing. The first step is to integrate over a partial set of $\phi(\mathbf{q}) \equiv \phi_q$'s with q's in the range $\Lambda > q > \Lambda/b$, as shown in Fig.3.19. For this purpose we write $H = H_< + H_>$, with

$$H_< = \frac{1}{2} \int_q^{\Lambda/b} r(\mathbf{q}) \phi_q \phi_{-q} + u \int_{q_1}^{\Lambda/b} \int_{q_2}^{\Lambda/b} \int_{q_3}^{\Lambda/b} \phi_{q_1} \phi_{q_2} \phi_{q_3} \phi_{-q_1-q_2-q_3} . \qquad (3.160)$$

$H_>$ represents all the remaining terms which involve $\phi(\mathbf{q})$'s of both ranges. For convenience, we denote the variables as $\phi_q^>$ if

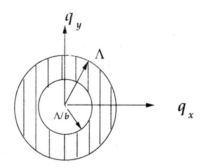

Figure 3.19: The momentum space of the integration variables. The shaded shell indicates the portion from which the variables are integrated in the RG procedure.

$q > \Lambda/b$, and as $\phi_q^<$ if $q < \Lambda/b$. One now has to integrate over all $\phi^>$ variables. Since the Hamiltonian contains quartic terms, the

3.13. RG for Ginzburg-Landau model

partial tracing will have to be done perturbatively with respect to the quadratic part of the Hamiltonian. So we write $H_>$ as,

$$H_> = \frac{1}{2}\int_{\Lambda/b}^{\Lambda} r(\mathbf{q})\phi_q\phi_{-q}\frac{d^dq}{(2\pi)^d} + H_4 = H_0 + H_4 \qquad (3.161)$$

and the partial trace is written as a cumulant expansion with averages with respect to H_0, in the following manner -

$$\begin{aligned}
Tr'e^{-H} &= e^{-H_<}\int D\phi_q^> e^{-H_0-H_4} \\
&= e^{-H_<}\int D\phi_q^> \frac{\int D\phi_q^> e^{-H_0-H_4}}{\int D\phi_q^> e^{-H_0}} e^{-H_0} \\
&= e^{-H_<} K_G \langle e^{-H_4}\rangle_0 \qquad (3.162)
\end{aligned}$$

with

$$K_G = \int D\phi_q^> e^{-\sum r_q \phi_q \phi_{-q}} = \prod_{q>\Lambda/b} \frac{2\pi}{r_q}. \qquad (3.163)$$

H_1 contains all the quartic terms involving $\phi^>$ variables. These come from the u term, which can be written as

$$u \int\int\int [\phi_{q_1}^< + \phi_{q_1}^>][\phi_{q_2}^< + \phi_{q_2}^>][\phi_{q_3}^< + \phi_{q_3}^>][\phi_{q_4}^< + \phi_{q_4}^>] \qquad (3.164)$$

where $q_4 = -q_1 - q_2 - q_3$. To keep track of these terms it is most convenient to represent them diagrammatically in the following way. We draw a dotted line for u, a full line for $\phi_q^<$ and a wiggly line for $\phi_q^>$. This is shown in Fig.3.20, where we also give the number of times a particular type of term occurs in the product. The perturbation expansion is again done by making the following cumulant expansion,

$$\langle e^{-H_4}\rangle_0 = \exp[-\langle H_4\rangle_0 + \frac{1}{2}[\langle H_4^2\rangle_0 - \langle H_4\rangle_0^2] + \ldots] \qquad (3.165)$$

This yields

$$Tr'e^{-H} = \exp[-(H_< + \langle H_4\rangle_0 - \frac{1}{2}[\langle H_4^2\rangle_0 - \langle H_4\rangle_0^2] + \ldots)]. \qquad (3.166)$$

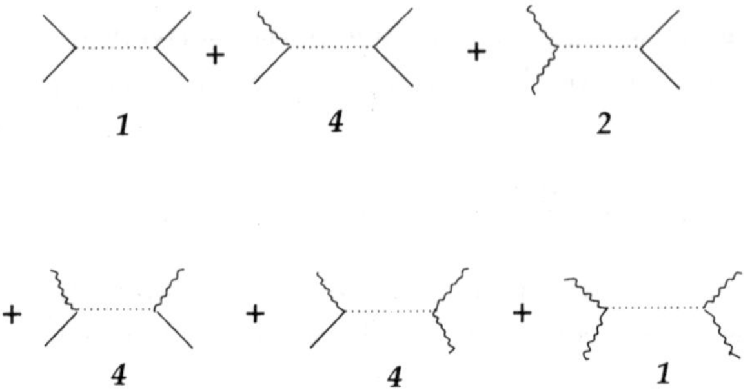

Figure 3.20: The various terms of the quartic interaction as separated in Eq.(3.164). The full line denotes $\phi^<$, the curly line denotes $\phi^>$ and the dotted line represents the interaction u. The number below each figure gives the number of times that particular type of term occurs in the product.

3.13.1 Tree-level approximation

Before we examine the perturbation terms any further, let us carry out some more steps of the RG transformation at zeroth order which is also called the "tree level". At this level, the new Hamiltonian is just $H_<$. Thus

$$H' = \int^{\Lambda/b} (r_0 + cq^2)\phi_q \phi_{-q} + u \int_{q_1}^{\Lambda/b} \int_{q_2}^{\Lambda/b} \int_{q_3}^{\Lambda/b} \phi_{q_1} \phi_{q_2} \phi_{q_3} \phi_{-q_1-q_2-q_3} \quad (3.167)$$

Since we want to express it in the same form as the original Hamiltonian, we convert the integration range to the same as that before, by relabelling q variable as $q = q'/b$. This gives for the quadratic term

$$\int^{\Lambda/b} d^d q (r_0 + cq^2)\phi_q \phi_{-q} \to \int^{\Lambda} \frac{d^d q'}{b^d}\left(r_0 + c\frac{q'^2}{b^2}\right)\phi_{\frac{q'}{b}} \phi_{-\frac{q'}{b}}. \quad (3.168)$$

We now do another transformation, which comes from the freedom that ϕ_q is an integration variable, and hence a constant multiplying factor will affect the partition function only trivially. This

3.13. RG for Ginzburg-Landau model

freedom can be exploited to keep one coupling the same as before. So here we rescale ϕ so that the q^2 term does not change; the transformation is

$$\phi_{q'/b} = \psi_{q'} b^{\frac{d+2}{2}} . \tag{3.169}$$

This transformation reduces the quadratic term to

$$\int^\Lambda \frac{d^d q'}{b^{d+2}} (r_0 b^2 + cq'^2) \phi_{\frac{q'}{b}} \phi_{-\frac{q'}{b}} = \int^\Lambda (r_0 b^2 + cq'^2) \psi_{q'} \psi_{-q'} . \tag{3.170}$$

Now we carry out the same step on u-term,

$$u \int^{\Lambda/b} d^d q_1 d^d q_2 d^d q_3 \phi_{q_1} \phi_{q_2} \phi_{q_3} \phi_{-q_1-q_2-q_3} =$$

$$ub^{-3d} \int^\Lambda d^d q_1 d^d q_2 d^d q_3 \phi_{q_1/b} \phi_{q_2/b} \phi_{q_3/b} \phi_{(-q_1-q_2-q_3)/b} =$$

$$ub^{4-d} \int^\Lambda \psi_{q_1} \psi_{-q_2} \psi_{q_3} \psi_{-q_1-q_2-q_3} . \tag{3.171}$$

Finally we look at magnetic field term, which for uniform field goes as,

$$h\phi_0 \to h\psi_0 b^{(d+2)/2} \to h'\psi_0 . \tag{3.172}$$

This allows us to identify the parameters of the new Hamiltonian. These are given by

$$\begin{aligned} r'_0 &= b^2 r_0, \quad c' = c, \\ u' &= ub^{(4-d)}, \text{ and } h' = hb^{(d+2)/2} . \end{aligned} \tag{3.173}$$

These transformations have a fixed point, $h = 0, r_0 = 0, u = 0$. This is called the Gaussian fixed point (GFP). The eigenvalues of the linearized relations around GFP are trivially found to be

$$y_t = 2, \quad y_h = (d+2)/2, \text{ and } y_u = 4 - d = \epsilon . \tag{3.174}$$

For $d > 4$, r_0 and h are relevant, but u which represents interactions is irrelevant.

One may think that the above conclusions are particular to the simple ϕ^4 Ginzburg-Landau (GL) functional. In order to appreciate the generality of these considerations, we shall examine the

$d > 4$ situation more thoroughly. We can consider more general GL functionals, by adding more terms and see their relevance at the tree-level. The scaling for the 6th order term goes as

$$u_6 \int^{\Lambda/b} \phi_{q_1}\phi_{q_2}\phi_{q_3}\phi_{q_4}\phi_{q_5}\phi_{q_6} \to u_6 b^{-5d} b^{3d+6} \int^{\Lambda} \psi_{q'_1}\psi_{q'_2}\cdots \quad (3.175)$$

This yields $u'_6 = u_6 b^{6-2d}$, which is irrelevant for $d > 4$. It is a simple matter to check that the 8th order coupling would scale like, $u'_8 = u_8 b^{8-3d}$, which is again irrelevant. This argument also shows that the higher order polynomial terms are all irrelevant.

Next we consider the more general wavevector dependences of couplings. The higher order terms in $r(q)$ transform as follows,

$$\int^{\Lambda/b} (r_0 + cq^2 + fq^4)\phi_q\phi_{-q}$$
$$\to \int^{\Lambda} (r_0 + c\frac{q'^2}{b^2} + \frac{fq'^4}{b^4})\phi_{(q'/b)}\phi_{(-q'/b)} \frac{d^d q'}{b^d} \quad (3.176)$$
$$= \int^{\Lambda} (r_0 b^2 + cq'^2 + \frac{fq'^4}{b^2})\psi_{q'}\psi_{-q'} \quad (3.177)$$

This shows that $f' = f/b^2$, which makes it irrelevant. Still higher order terms in momenta are even more irrelevant. Suppose u_4 had momentum dependences so that we can expand it as

$$u_4(q_1, q_2, q_3) = u_4 + (q_1^2 + q_2^2 + q_3^2)v_4 + \ldots \quad (3.178)$$

From the above it is obvious that the each power of the wavevector gives a factor of b^{-1} to the coupling. This gives $v'_4 = v_4 b^{\epsilon-2}$, which makes v_4 and other momentum-dependent terms irrelevant. Thus our simple GL model omits largely irrelevant terms, and should therefore describe quite general critical phenomena.

3.13.2 Critical exponents for $d > 4$

Let us now examine the critical exponents above dimension 4. From the values of y_t and y_h given above, one calculates all other

3.13. RG for Ginzburg-Landau model

exponents. These come out to be,

$$\Delta = (d+2)/4, \quad \alpha = (4-d)/4, \quad \beta = (d-2)/4,$$
$$\gamma = 2\Delta - d/2 = 1, \quad \nu = 1/2, \quad \eta = 0,$$
$$\text{and} \quad \delta = \frac{d+2}{d-2}. \qquad (3.179)$$

From these expressions it seems that the exponents continuously depend on the dimension for $d > 4$. It should be noted that at $d = 4$ the critical exponents obtained from the above eigenvalues are the same as the mean field values. This shows that the mean field approximation gives a correct account of the critical behaviour at dimension 4. Due to irrelevance of the u-couplings, one expects the mean field behaviour to be correct for all dimensions above 4. This is indeed true. To show this, one has to keep in mind that though u is irrelevant it is a dangerous parameter, as thermodynamic functions are not analytic in u.

The scaling expression for the free energy is given by

$$f(r_0, u, h) = b^{-d} f\left(r_0 b^2, u b^{-\epsilon}, h b^{\frac{d+2}{2}}\right)$$
$$= t^{\frac{d}{2}} f(-1, u|t|^{-\epsilon/2}, h|t|^{\frac{d+2}{4}}), \qquad (3.180)$$

where we have set $b = (-r_0)^{\frac{1}{2}}$ and $r_0 \propto t$. Differentiating this with h, we obtain the magnetization,

$$m(t, u, 0) = |t|^{\frac{d-2}{4}} f'(-1, u|t|^{-\epsilon/2}). \qquad (3.181)$$

For $f'(-1, u|t|^{-\epsilon/2})$, being away from the critical point, one can use the mean field expression, which is $m(t, u) = \sqrt{[-t/u]}$. Substituting this yields,

$$m(t, u) = |t|^{\frac{d-2}{4}} (u|t|^{-\epsilon/2})^{-1/2} = u|t|^{1/2}. \qquad (3.182)$$

This shows that $\beta = 1/2$ for all $d > 4$. The important point to note here is that f' is not analytic in variable u, which leads to a constant exponent for all dimensions greater than 4. To get other exponents, we look at the critical isotherm. Setting $b = (h/h_0)^{-\frac{2}{d+2}}$, one gets,

$$m(0, u, h) = (\frac{h}{h_0})^{\frac{d-2}{d+2}} M(0, u(\frac{h}{h_0})^{-\frac{2\epsilon}{d+2}}, h_0) . \tag{3.183}$$

Again being away from the critical point, one is allowed to use the mean field form of the isotherm $M(0, u, h) = (h/u)^{1/3}$, which is also a nonanalytic expression. This yields,

$$\begin{aligned} m(0, u, h) &= (\frac{h}{h_0})^{\frac{d-2}{d+2} - \frac{2(d-4)}{3(d+2)}} \\ &= (\frac{h}{h_0})^{\frac{1}{3}} , \end{aligned} \tag{3.184}$$

which gives $\delta = 3$ for all $d > 4$. Thus above $d = 4$, all critical exponents are independent of dimension and are equal to the one obtained in the mean-field approximation. For this reason, $d = 4$, is called the upper critical dimension.

Another way to see why $d = 4$ is an upper critical dimension, is the following. Since

$$H = \int d^d r [\frac{1}{2}(\nabla \phi)^2 + \frac{1}{2}r_0\phi^2 + u\phi^4] = \beta G \tag{3.185}$$

is dimensionless, we can calculate the dimensions of all the quantities in terms of length -

$$\begin{aligned} \frac{L^d}{L^2}\phi^2 &= 1 \Rightarrow [\phi] = L^{1-d/2} \\ r_0 L^d L^{2(1-d/2)} &= 1 \Rightarrow [r_0] = L^{-2} \\ u L^d L^{4(1-d/2)} &= 1 \Rightarrow [u] = L^{d-4} . \end{aligned} \tag{3.186}$$

By using the length unit to be $r_0^{-1/2} = \xi(T)$ and

$$\begin{aligned} \psi &= \frac{\phi}{\xi^{1-d/2}}, \\ r &= r'\xi(T), \end{aligned} \tag{3.187}$$

3.13. RG for Ginzburg-Landau model

one can now rewrite the hamiltonian as

$$H = \int d^d r' [\frac{1}{2}(\nabla \psi)^2 + \frac{1}{2}\psi^2 + u'\psi^4] \qquad (3.188)$$

where $u' = u/\xi^{d-4}$. In this expression, all the temperature dependence of the problem is in $u' = u t^{(d-4)/2}$. So if $d > 4$, as $t \to 0$, $u' \to 0$, and one can do the perturbation theory in u' and the Gaussian theory becomes increasingly accurate as $t \to 0$. On the other hand if $d < 4$, u' becomes increasingly large as $t \to 0$.

3.13.3 Anomalous dimensions

Now let us look at the dimension of the correlation function -

$$\begin{aligned} [G(k)] &= [\int e^{ik \cdot (r-r')} G(r,r') d^3 r'] \\ &= L^d [\phi]^2 = L^2 \: . \end{aligned} \qquad (3.189)$$

This matches with the dimension of the Ornstein - Zernike form which is given by

$$G(k) = \frac{1}{k^2 + \xi^{-2}} \: . \qquad (3.190)$$

But at $T = T_c$,

$$G(r) \approx \frac{1}{r^{d-2+\eta}}; \qquad G(k) \approx \frac{1}{k^{2-\eta}} \: . \qquad (3.191)$$

Dimensionally, we always require $G(k) \propto 1/k^2$. This means that the operator ϕ must change its dimension from the value $L^{1-d/2}$. Now the question arises as to why the above dimensional analysis should breakdown.

One can also pose the problem of critical phenomena in this manner. The deviation of critical exponents from the mean field values is related to operators like ϕ acquiring anomalous dimensions, which are different from what is dictated by the above dimensional analysis. One way to understand the problem is to recognise that $\xi = r_0^{-1/2}$ is not only the length scale in the problem. There are other equally important length scales which need to be analysed.

3.14 Perturbation series for $d < 4$

We now return to the consideration of higher order terms of the perturbation expansion in H_4. Clearly these terms are important for $d < 4$, as the coupling u is relevant now. To evaluate these terms, we need averages over Gaussian distribution, which obey an analog of Wick's theorem. We briefly recall this result, noting the identity,

$$\begin{aligned} Z(\{h_i\}) &= \int e^{-\frac{1}{2}\sum_{ij}\phi_i A^{-1}{}_{ij}\phi_j + \sum_i h_i \phi_i} \prod_i d\phi_i \\ &= \frac{(2\pi)^{N/2}}{(\det A)^{1/2}} \exp\left(-\frac{1}{2}\sum_{i,j} h_i A^{-1}{}_{ij} h_j\right). \end{aligned} \quad (3.192)$$

Using this identity, it is straightforward to derive the following results -

$$\begin{aligned} \frac{1}{Z_0}\int e^{-\frac{1}{2}\sum_{ij}\phi_i A^{-1}{}_{ij}\phi_j} \phi_k \phi_l &= \frac{1}{Z}\left(\frac{\partial^2 Z}{\partial h_k \partial h_l}\right)_{\{h_i=0\}} \\ &= A^{-1}{}_{ij} = G_{ij}. \end{aligned} \quad (3.193)$$

Similarly,

$$\frac{1}{Z_0}\int e^{-\frac{1}{2}\sum_{ij}\phi_i A^{-1}{}_{ij}\phi_j} \phi_k \phi_l \phi_m \phi_n = [G_{kl}G_{mn} + G_{km}G_{ln} + G_{kn}G_{ml}]$$
$$= [\langle\phi_k\phi_l\rangle\langle\phi_m\phi_n\rangle + \langle\phi_k\phi_m\rangle\langle\phi_l\phi_n\rangle + \langle\phi_k\phi_n\rangle\langle\phi_l\phi_m\rangle]. \quad (3.194)$$

For the present case, we have

$$\begin{aligned} H_0 &= \frac{1}{2}\int_{\Lambda/b}^{\Lambda} r(q)\phi_q\phi_{-q} \\ \langle\phi^>_{q_1}\phi^>_{q_2}\rangle &= \delta_{q_1+q_2,0}\, G(q_1) \\ \text{and}\ G(q) &= \frac{1}{r(q)} = \frac{1}{r_0 + cq^2}. \end{aligned} \quad (3.195)$$

Now we evaluate the first term of the series which is $\langle H_4\rangle$. It is easiest to do it with the help of diagrams. For this purpose we

3.14. Perturbation series for d < 4

consider one of the terms of H_4 given by

$$u \int \langle \phi_{q_1}^< \phi_{q_2}^< \phi_{q_3}^> \phi_{q_4}^> \rangle = u\phi_{q_1}^< \phi_{q_2}^< \int_{\Lambda/b}^{\Lambda} \delta_{q_3+q_4=0} G(q_3)$$

$$= u\phi_{q_1}^< \phi_{-q_1}^< \delta_{q_1+q_2=0} \int_{\Lambda/b}^{\Lambda} G(q_3) \ . \ (3.196)$$

Here we have integrated over only $\phi^>$ variables. Diagrammatically we represent this by joining the wiggly lines. Thus a closed wiggly line denotes $G(q)$, and there is integration over the momenta of these lines in the range Λ/b to Λ. In fact it is easily verified that all averages can be just obtained by closing wiggly lines in all possible ways. The first order terms are shown diagrammatically in Fig.3.21. Fig.3.21a just gives a constant contribution, while

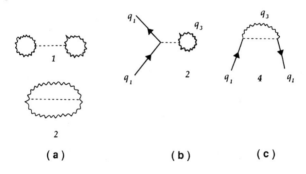

Figure 3.21: The diagrammatic representation of the first order perturbation terms. (a) constant terms contributing to the free energy (b) and (c) contributions to the quadratic coupling. The numbers below the figures give the degeneracy of the diagram.

Figs.3.21b and 3.21c, which are left with two external legs each give contributions proportional to $\phi_q^< \phi_{-q}^<$. These contributions modify the coupling of such terms in the new Hamiltonian. Due to momentum conservation and the fact that the interaction u can be taken to be momentum independent, both sets of diagrams contribute equally. Thus,

$$\langle H_4 \rangle = A_0 + 6u \int_{\Lambda/b}^{\Lambda} G_0(q) d^d q \sum_q \phi_q^< \phi_{-q}^< . \qquad (3.197)$$

Let us evaluate the integral in dimension close to 4.

$$\begin{aligned}
\int_{\Lambda/b}^{\Lambda} G_0(q) d^d q &= S_d \int \frac{1}{r_0 + q^2} q^{d-1} dq \\
&= S_d \left[\frac{\Lambda^{d-2}}{d-2} \left(1 - \frac{1}{b^{d-2}}\right) - r_0 \frac{\Lambda^{-\epsilon}}{\epsilon} \left(1 - \frac{1}{b^{\epsilon}}\right) \right]
\end{aligned} \qquad (3.198)$$

where

$$S_d = \frac{2\pi^{d/2}}{\Gamma(d/2)} \frac{1}{(2\pi)^d} . \qquad (3.199)$$

From this we obtain for $d \approx 4$,

$$\langle H_4 \rangle = A_0 + 3u S_d [\, \Lambda^2 \{1 - \frac{1}{b^2}\} - 2r_0 \ln b \,] \int^{\Lambda/b} \phi_q^< \phi_{-q}^< . \qquad (3.200)$$

Combining this with similar terms in $H^<$, changing the integration range and rescaling the ϕ-variables one finds that

$$r' = b^2 [\, r + 6u S_d \Lambda^2 (1 - \frac{1}{b^2}) - 12u S_d r_0 \ln b \,] . \qquad (3.201)$$

We now consider the second order terms. This contribution can again be most easily determined using diagrams, which we show in Fig.3.22. These diagrams are drawn in the following manner. One first draws two dotted lines each with four legs corresponding to u-interaction in all possible configurations for each as shown in Fig.3.22. Now the averaging corresponds to joining wiggly lines in all possible ways. Depending on external single lines that are left, we shall get contributions to r_0, u, etc. Note that the r_0 terms, which are shown in Fig.3.23, contain two loops, which give higher order corrections to the leading order terms that we had obtained earlier, whereas for u, we get diagrams which have

3.14. Perturbation series for $d < 4$

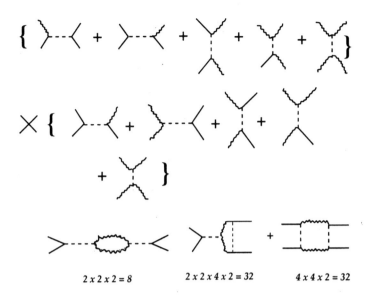

Figure 3.22: Diagrams representing the second order perturbation contributions to the u-coupling. Again the figures below each diagram gives the number of times the particular diagram occurs.

just one loop. These are the leading order terms in the recursion relations for u. The idea is to keep contributions from diagrams, that contain only one loop. It will become clear later that these terms enable us to obtain the first order corrections in the dimensional parameter ϵ. The contribution of a diagram is obtained by writing a factor of G for each double line and u for each interaction line. The momentum associated with each G is determined from momentum conservation at each vertex. All the internal momenta are then integrated between Λ/b and Λ. All the second order terms contributing to u along with their occurence number are shown in Fig.3.22. Since one is ignoring the q-dependence of u, all these 72 terms contribute equally. A typical contribution to the ϕ^4 term is

$$u^2 \int^{\Lambda/b} d^d q\, G(q) G(-q - q_1 - q_2) \approx u^2 \int^{\Lambda/b} d^d q \frac{1}{(r_0 + q^2)^2}$$

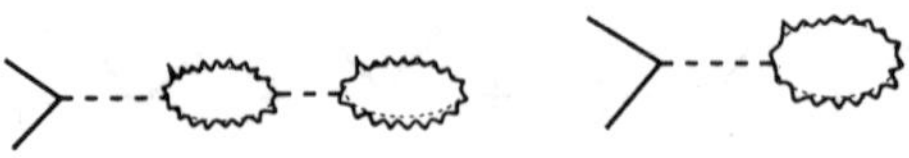

2 x 2 x 2 x 2 = 16   2 x 2 x 2 x 4 = 32

Figure 3.23: Diagrams representing the second order perturbation contributions to the r-coupling. Again the figures below each diagram gives the number of times the particular diagram occurs.

$$\approx u^2 \int^{\Lambda/b} \frac{q^{d-1}}{q^4}$$
$$\approx u^2 S_d \ln b , \qquad (3.202)$$

where we have set external momenta to zero as we are ignoring the momentum-dependence of the coupling. Further, we have evaluated the integral to zeroth order in r_0 and for $d = 4$. Again combining this with the similar term in $H^<$, and rescaling momenta and fields, one obtains the following recursion relation -

$$u' = b^\epsilon [\, u - 36u^2 S_4 \ln b \,] . \qquad (3.203)$$

Now we can write the recursion relations for both the parameters using the one-loop calculation, which also turns out to be correct to first order in the dimensional parameter ϵ. These are

$$r'_0 = b^2 [\, r_0 + 6u S_d \Lambda^2 (1 - \frac{1}{b^2}) - 12 u S_d r_0 \ln b \,]$$
$$\text{and} \quad u' = b^\epsilon [\, u - 36u^2 S_d \ln b \,] . \qquad (3.204)$$

We now study the fixed points of these relations. Clearly one fixed point is $r_G^* = 0, u_G^* = 0$, the Gaussian fixed point which we studied earlier. This fixed point is unstable in both variables, with eigenvalues $y_t = 2, y_u = \epsilon$. So we look for other fixed points. Writing the u-relation as

$$u^* = (u^* - 36u^{*2} S_d \ln b)(1 + \epsilon \ln b) ,$$
$$\text{we get} \quad u^* S_4 = \epsilon/36 \qquad (3.205)$$

3.14. Perturbation series for $d < 4$

to leading order in ϵ. Further for the r-relation one gets

$$r_0{}^*\left(\frac{1}{b^2} - 1\right) = 6u^*S_d\Lambda^2(1 - \frac{1}{b^2}) - 12u^*S_d^*r_0{}^*\ln b \ . \tag{3.206}$$

Clearly the second term on rhs is of order ϵ^2, and can be ignored giving us

$$r_0{}^* = -\frac{\Lambda^2\epsilon}{6} \ . \tag{3.207}$$

This is a new fixed point known as Wilson-Fisher fixed point (WFP). It is instructive to write these relations as differential relations by taking the scaling parameter b close to unity. So we write $b = 1 + \delta l$ or $\ln b = \delta l$. Then

$$u' = u(1 + \epsilon\delta l)[u - 36S_d u^2 \delta l]$$
$$\Rightarrow \frac{du}{dl} = \frac{u' - u}{\delta l} = \epsilon u - 36S_d u^2 \ . \tag{3.208}$$

Similarly

$$r'_0 = [1 + 2\delta l][r_0 + 6u\Lambda^2\{1 - (\frac{1}{1+\delta})^2\} - 12ur_0\delta l]$$
$$\Rightarrow r'_0 - r_0 = \delta l[2r_0 + 12uS_d\Lambda^2 - 12uS_d r_0]$$
$$\Rightarrow \frac{dr_0}{dl} = 2r_0 + 12uS_d\Lambda^2 - 12uS_d r_0 \ . \tag{3.209}$$

The fixed points of these equations to first order in ϵ are obviously the same as the ones found from Eq.(3.204).

It is easier to find eigenvalues at the WFP with the differential relations. Linearising about this fixed point, one gets

$$\frac{d}{dl}\begin{vmatrix} \delta u \\ \delta r \end{vmatrix} = \begin{vmatrix} (2 - \epsilon/3) & B(1 + \epsilon/6) \\ 0 & -\epsilon \end{vmatrix} \begin{vmatrix} \delta u \\ \delta r \end{vmatrix} \tag{3.210}$$

where $B = 12S_d\Lambda^2$. The eigenvalues of the matrix are clearly $y_t = 2 - \epsilon/3$ and $y_u = -\epsilon$. Now we must consider the flows in the $u - r_0$ plane with regard to these two fixed points. All flows are outwards around the GFP, whereas around the WFP there is

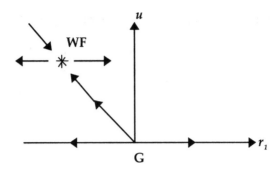

Figure 3.24: Fixed points and the flows around them for the Ginzburg-Landau model in the epsilon expansion.

one inward direction corresponding to u and one temperature-like outwards direction. The schematic flows are shown in Fig.3.24.

Near the WFP, the nature of the flows can be made more precise by considering the eigenvectors of the linearized relations. Left eigen-vectors are seen to be $(1, \beta(1 + \epsilon/6)/(2 + \epsilon/3))$ and $(0, 1)$. These eigenvectors and linearized flows are also shown in Fig.3.24. This allows us to obtain the scaling fields to be

$$u_a = (r - r^*) + \frac{B(1 + \epsilon/6)}{2(1 + \epsilon/3)}(u - u^*)$$
$$\text{and} \quad u_b = u - u^* . \tag{3.211}$$

The critical surface, *i.e.*, the line along which the flow is inwards to the WFP, is obtained by setting $u_a = 0$, which yields

$$r_c = r^* - \frac{\beta(1 + \epsilon/6)}{2(1 + \epsilon/3)}(u - \frac{\epsilon}{36 S_d}) . \tag{3.212}$$

This equation gives us the dependence of the transition temperature $T_c(u)$ on u as

$$T_c(u) = T_c^{MF}[1 + \frac{\epsilon \Lambda^2}{6} + \frac{\beta(1 + \epsilon/6)}{2(1 + \epsilon/3)}(u - \frac{\epsilon}{36 S_d})]^{-1} . \tag{3.213}$$

Further note that the temperature exponent y_t has acquired a nonclassical correction which depends on the dimesionality. Now

3.15. *Generalisation to a n-component model* 183

we can list all the critical exponents. They are

$$
\begin{aligned}
y_t &= 2 - \epsilon/3; \quad \nu = 1/y_t = \frac{1}{2}(1 + \epsilon/6); \\
2 - \alpha &= \frac{1}{2}(1 + \epsilon/6)(4 - \epsilon); \quad \alpha = \epsilon/6; \\
y_h &= \frac{d+2}{2} = 3 - \epsilon/2; \\
\Delta &= \frac{y_h}{y_t} = \frac{3(1 - \epsilon/6)}{2(1 - \epsilon/6)} \approx 3/2; \\
\gamma &= 3 - 2(1 - \epsilon/6) = 1 + \epsilon/6; \\
\delta &= \frac{y_h}{d - y_h} = \frac{3 - \epsilon/2}{4 - \epsilon - (3 - \epsilon/2)} = 3 + \epsilon \,. \quad (3.214)
\end{aligned}
$$

This show how the various exponents get modified as the dimension is lowered below four.

3.15 Generalisation to a n-component model

In the above calculation we have seen how critical behaviour depends on the dimensionality of the system. Next we want to demonstrate how the vectorial character of the order parameter characterizing the transition affects critical exponents. For this purpose, we consider the Ginzburg-Landau model of an order parameter with n components. We shall first consider the situation in which there is isotropy with respect to components. The appropriate effective Hamiltonian is given by

$$ H = \frac{1}{2}\int dr [r_0 \sum_{\alpha=1}^{n} \phi_\alpha^2(\mathbf{r})] + c[\sum_\alpha (\nabla\phi_\alpha).(\nabla\phi_\alpha)] + u[\sum_\alpha \phi_\alpha(\mathbf{r})^2]^2 \,. \quad (3.215) $$

This Hamiltonian has the O_n symmetry, *i.e.*, its symmetry group consists of all transformations that do not change the vector magnitude $\phi_1^2 + \phi_2^2 + ... + \phi_n^2 = \vec{\phi}^2$.

We can again do the momentum shell RG, by doing the partial trace over field variables with momenta in the range $\Lambda/b < q < \Lambda$.

3. Phase Transitions and Critical Phenomena

As before, one obtains the renormailzed Hamiltonian perturbatively using the equation

$$Tr_{partial} \exp(-H) = \exp(-H_< - [\langle H_4 \rangle_0 - \frac{1}{2}(\langle H_4^2 \rangle_0 - \langle H_4 \rangle_0^2)]) ,$$
(3.216)

where

$$H_< = \int^{\Lambda/b} (r + cq^2) \sum_{\alpha=1}^{n} \phi_\alpha(q) \phi_\alpha(-q)$$
$$+ u \sum_{\alpha\beta} \int^{\Lambda/b} \int^{\Lambda/b} \int^{\Lambda/b} \phi_\alpha(q_1) \phi_\alpha(q_2) \phi_\beta(q_3) \phi_\beta(-q_1 - q_2 - q_3) .$$
(3.217)

H_4 is diagrammatically similar to the diagram shown in Fig.3.21, with the difference that each line is now labelled with both momentum and component index.

The first order term is given by

$$\langle H_4 \rangle_0 = u \sum_\beta \int_0^{\frac{\Lambda}{b}} \phi_\beta(q) \phi_\beta(-q) \left[2 \int_{\frac{\Lambda}{b}}^{\Lambda} \sum_\alpha G_\alpha(k) + 4 \int_{\frac{\Lambda}{b}}^{\Lambda} G_\beta(k) \right]$$
$$= \sum_\beta (2n + 4) u \int_0^{\frac{\Lambda}{b}} \phi_\beta(q) \phi_\beta(-q) \int_{\frac{\Lambda}{b}}^{\Lambda} \frac{1}{r_0 + ck^2} . \quad (3.218)$$

As shown in Fig.3.25, the factor of n comes from summing over component labels of the internal lines. Following the earlier rescaling procedure this leads to the recursions given by

$$r' = b^2[r - 2(2n + 4)u I_1] \quad (3.219)$$

with

$$I_1 = S_4[\frac{\Lambda^2}{2}(1 - \frac{1}{b^2}) - r_0 \ln b] . \quad (3.220)$$

In second order perturbation theory, the one-loop u-terms are shown along with their occurence factors are shown in Fig.3.25. As the momentum dependence of these terms is ignored, the entire contribution to the coefficient of a term like

$$\phi_a^<(q_1) \phi_a^<(q_2) \phi_b^<(q_3) \phi_b^<(q_4)$$

3.15. Generalisation to a n-component model

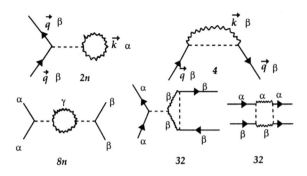

Figure 3.25: The first and second order perturbation diagrams for the n-component GL model.

can be written as

$$(n+8)u^2 \int_{\frac{\Lambda}{b}}^{\Lambda} G(q_3) G(-q_1 - q_2 - q_3) dq_3$$
$$= 8(n+8)u^2 \int_{\frac{\Lambda}{b}}^{\Lambda} \left(\frac{1}{r_0 + cq_3^2}\right)^2$$
$$= 8(n+8)u^2 S_d \ln b . \qquad (3.221)$$

Including this term and rescaling leads to the recursion relation,

$$u' = b^\epsilon [\, u - 4(n+8)u^2 S_d \ln b \,] . \qquad (3.222)$$

Now we may write the differential form of these recursions -

$$\frac{dr}{dl} = 2r + 2(2n+4)uS_4\Lambda^2 - 2(2n+4)(uS_4)r,$$
$$\text{and} \quad \frac{du}{dl} = \epsilon u - 4(n+8)u^2 S_4 . \qquad (3.223)$$

The Wilson-Fisher fixed point is given by

$$u^* S_d = \frac{\epsilon}{4(n+8)}$$
$$\text{and} \quad r^* = -(2n+4)\Lambda^2 u^* S_d = -\frac{(n+2)\Lambda^2 \epsilon}{2(n+8)} . \qquad (3.224)$$

It is a straightforward exercise to linearize the recursion relations around the fixed point and find the eigenvalues. These are found to be

$$y_t = 2 - \frac{(n+2)\epsilon}{(n+8)}; \quad y_h = 3 - \epsilon/2;$$

$$\alpha = \frac{(4-n)\epsilon}{2(n+8)}; \quad \beta = \frac{1}{2} - \frac{3\epsilon}{2(n+8)};$$

$$\gamma = 1 + \frac{(n+2)\epsilon}{2(n+8)}; \quad \delta = 3 + \epsilon;$$

$$\nu = \frac{1}{2} + \frac{(n+2)\epsilon}{4(n+8)}; \quad \text{and} \quad \eta = 0 \ . \qquad (3.225)$$

Now one can see the explicit dependence of exponents on the dimensionality as well as on the number of components of the order parameter. Other parameters like the various higher order polynomial couplings and the momentum dependences of the couplings have been shown to be irrelevant at least to the lowest order in $\epsilon = 4 - d$. This clearly leads to rather wide universality classes for systems, for which we have have common critical behaviour.

Bibliography

[] *The literature on phase transitions and critical phenomena is vast. Here we mention just those books and articles which were used in preparing this chapter. For thermodynamics of phase transitions and critical phenomena, two excellent texts are:*

[1] *Thermodynamics and an Introduction to Thermostatics (Second Edition)*, H.B. Callen (John Wiley, New York).

[2] *Introduction to Phase Transitions and Critical Phenomena*, H.E. Stanley (Oxford University Press, Oxford).

[] ** *For an introduction to critical phenomena and scaling ideas, the following books and articles are suggested.*

[3] *General Scaling Theory for Critical points*, M.E. Fisher in *Collective Properties of Physical Systems* ed. by B. Lundqvist and S. Lundqvist, Nobel Symposium 24 (Academic Press, London).

[4] *Modern Theory of Critical Phenomena* by S.K. Ma (Benjamin, New York).

[5] L.P. Kadanoff et. al., Rev. Mod. Phys. 39, 395 (1967).

[6] *Scaling and Renormalization in Statistical Physics* by J. Cardy (Cambridge University Press, Cambridge).

[] ******* *For advanced material on Renormalization Group, readers may consult the following books and articles, in addition to the ones mentioned above.*

[7] K.G. Wilson and J.B. Kogut, Phys. Reports 12C, 75(1974).

[8] *Phase transitions and Critical Phenomena Vol.6*, ed. C. Domb and M.S. Green (Academic Press, London).

[9] *Lectures on Phase Transitions and the Renormalization Group* by N. Goldenfeld (Addison and Wesley, New York).

[10] *Principles of Condensed Matter Physics* by P.M. Chaiken and T.C. Lubensky (Cambridge University Press, Cambridge).

Chapter 4

Topological Defects in Condensed Matter Physics

Ajit M. Srivastava

Institute of Physics
Sachivalaya Marg
Bhubaneswar – 751 005

4. Topological Defects

These lectures present an introduction to the theory of topological defects. Basic concepts relating to topological defects are explained using examples in condensed matter systems. Correspondence with topological defects in particle physics is also discussed.

Topological concepts play an important role in a wide range of physical phenomenon. Topological defects are one such example where ideas of topology have been crucial in understanding the subject. Many ideas relating to topological considerations find direct physical manifestation in terms of topological defect. These lectures review general theory of topological defects. The topics covered are, concept of order parameter and order parameter space, notion of topological invariants, characterization of defects in terms of free homotopy of maps, based homotopy, homotopy group and its relation to defect classification, association with phase transitions and spontaneous symmetry breaking, and representation of order parameter space as a coset space. Defects in nematic liquid crystals are discussed as an illustration of these concepts. Theories of formation of defects, and experimental results in liquid crystal systems, will be briefly discussed. Similarities and differences between defects in condensed matter systems and in other systems (such as in particle physics) will be discussed.

For a general theory of topological defects, references [1, 2, 3, 4, 5] can be consulted. The most comprehensive and clear article is by Mermin [1] and I strongly recommend a thorough reading of this article to anyone seriously interested in understanding theory of defects. For relevant concepts in topology, see refs.[6]. During the lectures, notions of topology will be introduced as and when needed. For detailed discussions of topological defects in liquid crystals, see refs.[7].

An important point needs to be emphasized in the beginning. The theory of topological defects is a highly interdisciplinary field.

Thus, although the following discussion will typically be in the language of condensed matter physics, almost all the things apply to theories in particle physics as well but with an appropriate translation of the terminology. [For example, the order parameter space is called the vacuum manifold in particle physics, and the order parameter field is typically called a Higgs field]

4.1 The subject of topological defect

Topological defects occur in many different branches of physics, and are of interest to the researchers in these fields for varied reasons. In condensed matter physics, topological defects play a crucial role in understanding properties of superfluids, superconductors, liquid crystals, and crystals. Presumably the earliest reference to a topological defect is in terms of disclination lines in crystals. A much more sophisticated example of a topological defect came with the study of vortex filaments in superfluid helium and subsequently with flux tubes in superconductors. Topological defects in liquid crystals have been known for a very long time. Due to ease in observation of defects in liquid crystals using optical microscopy, their properties have been investigated in great detail.

In the last two decades or so, certain developments have taken place in the fields of particle physics as well as in condensed matter physics, which have made it possible for the study of topological defects in one field to directly relate to the developments in the other field. An important example of this is provided by the study of the processes of formation of topological defects and the evolution of defect networks. The first serious attempts in this direction were made in particle physics models of the early universe [8, 4, 5] where the possibility of occurrence of topological defects provided the long sought source of density fluctuations which could lead to the formation of structure in the universe, *i.e.*, galaxies, clusters of galaxies, etc. It was predicted that during various stage of unification of forces, phase transitions would have occurred in the universe, as an early super hot universe expanded and cooled. Present observations imply that the matter distribution was roughly uniform at the beginning. Small density fluctuations in this otherwise uniform background grew later and lead to clumping of matter in the form of galaxies, clusters of galaxies, superclusters, etc. However, there was no natural source of this required small density fluctuation. The possibility that superheavy topological defects could form during the phase transitions in the early universe gave the first candidates for the source of density fluctuations. The rest of the matter could then be assumed to be uniform, which would eventually clump around the topological defects. Of course, by now there are many different possibilities for creating density fluctuations, such as inflation, fluctuations due to quantum gravity effects, etc. In fact, the most natural models of structure formation based on the formation of topological line defects (called cosmic strings) seem not to be in agreement with data on pulsar timings, etc. Still, the possibility of structure formation due to topological defects is a real one. (Recent, more accurate simulations have raised the possibility that cosmic string models may be consistent with pulsar data.)

The entire picture of the formation of these defects in phase transitions in the early universe is very similar to what has been experimentally observed in condensed matter physics for long time. However, the issue of details of the processes of formation of these defects and their evolution had not attracted much attention. Later we will discuss in detail, how not only the general picture of topological defects in both fields is similar, but even the theoretical framework for discussing these defects is almost identical. Thus, the processes of defect formation which were proposed for the early universe, are completely valid for describing the formation of topological defects during a phase transition in a condensed matter system. Recognition of this has made it possible to carry out direct experimental investigation of such aspects of theories of topological defects using suitable condensed matter systems [9, 10, 11, 12, 13]. Needless to say, such experimental verification would be out of question for particle physics since these defects are supposed to form only at extreme densities and temperatures, which is not present anywhere in the present universe (and not conceivable in laboratory experiments in any foreseeable future).

To summarize this discussion, there are exciting new developments happening in the field of topological defects. Exchange of ideas from different branches of physics have made it possible to see those features of these systems which were lying unexplored.

4.2 What is a topological defect?

Topological defects are defects of topological origin in a configuration of the order parameter. This basically is just rephrasing the word topological defect. In the following sections, we will explain each new term in the above statement. To begin with, let us start by having a simple physical picture of a topological defect. What is most important to emphasize, at the start, is the word topological. This is what distinguishes these defects from other defects, like impurities, in condensed matter systems.

In fact, the word defect has its origin in condensed matter systems where, in an otherwise uniform system, occasionally there are localized regions representing some type of non-uniformity. A simple example is a steam bubble immersed in water. In order not to get confused with co-existence of phases in this case, consider the simple situation when most of water is at a temperature much lower than water-steam transition temperature of 100 ^{0}C. For example suppose water temperature is 50 ^{0}C almost everywhere in the sample, except in a very tiny region where the water is locally heated to above 100 ^{0}C. This will lead to a steam bubble embedded in (relatively) cold water, which naturally justifies the term defect for such a localized region of steam in an otherwise homogeneous medium of water.

However, it is also clear that this particular defect is really not stable. All that one needs to do is to let that localized region of steam cool to a temperature below 100 ^{0}C, and we will have water everywhere, with no (steam) defect. In other words, one has been able to make this defect disappear by simple modification in the local properties of the medium.

This is where topological defects differ from the above example. They still appear as defect like regions in otherwise uniform systems. However, it is just not possible to make them disappear by simple modifications of the local properties of the medium. The origin of the term *topological* lies here. Very qualitatively, things which are topological, cannot be altered by simple (meaning smooth) local modifications of the system. In order to make a topological defect disappear, one needs to make modifications in the entire system, *i.e.*, modifications should be global, like heating (or cooling) of the entire system and not just a local region.

Another simple example of a defect, which is not topological, is a vacancy at a lattice site in a crystal. It clearly represents a defect in crystal structure. However, by simply substituting an atom at the vacant site (which clearly is a local process) one can make this defect disappear. In contrast, disclination lines in crystals are topological defects which cannot be removed by local modifications of crystal sites.

Now we are all set to start appreciating the non-trivial aspects of a topological defect. For this it is convenient to go back to the example of water and steam bubble and start generalizing from there. It turns out that the steam-water system is not rich enough to allow for any topological defect. So we will need to consider some different systems. Still, there is one similarity between systems allowing for topological defects and the water-steam system. All these systems can exist in several (at least two) phases. Like water when heated up converts to a different phase - *i.e.*, steam. These different phases have to be distinguished by some means, for example, water phase and steam phase can be distinguished by the density. This brings us to the concept of an order parameter.

4.2.1 Meaning of order parameter

Order parameter is introduced to describe, what is called, an ordered medium in condensed matter physics. By ordered medium we mean a region of space described by a function $f(\mathbf{r})$ (*i.e.*, a

4.2. What is a topological defect?

field) that assigns to every point of the region a value of the *order parameter*.

Order parameter characterizes the system under consideration. Later we will see that an order parameter typically characterizes a certain phase transition happening in a system. In one phase, say the high temperature phase, the order parameter can be zero, while after the phase transition, in the low temperature phase, the order parameter will assume some non-zero value which characterizes the low temperature phase. For the example of water-steam system, one may think of the density as an order parameter, which assumes very small values in the steam phase (and can be set to be equal to zero by subtracting a constant) and becomes large in the water phase.

A simple example of order parameter which, as we will show later, allows for topological defects, is the magnetization **M**. Above some critical temperature T_c (the Curie temperature), **M** = 0 while at lower temperatures, **M** is non zero.

4.2.2 Order parameter and spontaneous symmetry breakdown

Order in a system determines symmetry properties of the system which typically change across a phase transition. Lowering temperature through the transition temperature leads (often) to a state with a higher degree of order and hence of lower symmetry. This is what happens when a symmetry is *spontaneously broken*. We discuss a simple example to illustrate the basic idea.

Consider a system whose free energy is plotted in Fig.4.1 as a function of the order parameter ϕ which is a real scalar field. Thus the ordered medium corresponds here to prescribing a real number at each point of space. The dashed curve in Fig.4.1 shows the plot of the free energy at high temperature, while the solid curve shows the plot at low temperature.

This type of plot of free energy can arise with the following expression for the free energy density

196   4. Topological Defects

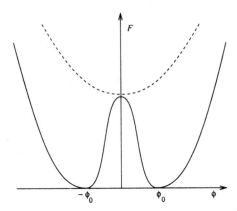

Figure 4.1: Plot of free energy F as a function of the real scalar order parameter ϕ. The dashed curve and the solid curve shows the plot at high temperature and at low temperature respectively.

$$F = k(\nabla\phi)^2 - \frac{1}{2}(m^2 - T^2)\phi^2 + \frac{\lambda}{4}\phi^4 \qquad (4.1)$$

Note that this free energy is invariant under the transformation $\phi \to -\phi$. At high temperatures, with $T \geq m$, the lowest energy configuration will have $\nabla\phi = 0$ and $\phi = 0$. This ground state configuration (the lowest free energy configuration) is invariant under $\phi \to -\phi$ and hence reflects the symmetry of the free energy F. The lowest energy configuration for low temperatures, with $T < m$, will have $\nabla\phi = 0$ and $\phi = \pm\phi_0$ where $\phi_0 = \sqrt{(m^2 - T^2)/\lambda}$. However, now, only one of the values, $+\phi_0$, or $-\phi_0$, has to be chosen. Clearly, either of the choices, $\phi = \phi_0$ or $\phi = -\phi_0$, for the ground state does not respect the symmetry of the free energy F under $\phi \to -\phi$. This is called *spontaneous symmetry breaking* - i.e., the situation when a symmetry of the free energy is not respected by the ground state.

4.2. What is a topological defect?

4.2.3 Spontaneous symmetry breaking in particle physics

An exactly similar situation arises in the context of relativistic field theory models in particle physics. There, plots like that in Fig.4.1 represent the effective potential $V(\phi)$ as a function of an order parameter (OP) field ϕ. This order parameter field is typically called *the Higgs field*. [Strictly speaking, in Fig.4.1, the order parameter characterizing the phase transition is the vacuum expectation value of the Higgs field.] The effective potential is the analog of free energy here.

So, in particle physics also, at low temperature, the physical space represents an ordered medium with the value of the order parameter (the Higgs field) prescribed at each point of the physical space.

4.2.4 Order parameter space

The order parameter space S consists of all possible values of the order parameter at a given temperature (*i.e.*, all possible values of the order parameter field in the lowest free energy configuration). For example, for the plot corresponding to high temperature in Fig.4.1, the order parameter space consists of one point, *i.e.*, $S = \{0\}$, whereas for the plot corresponding to low temperature, S consists of two points, *i.e.*, $S = \{\phi_0, -\phi_0\}$. [Note that S here is just a point set, and not a smooth manifold. This will be important later when we discuss the need for generalized concepts of continuity requiring notion of topological spaces, etc.]

For the case of magnetization, we note that at some fixed (low) temperature, the magnitude of magnetization **M** is fixed. However, at a given point in the sample, the direction of **M** is completely arbitrary as the free energy here is independent of the direction of **M** (in the absence of any external fields). [This is just like the example shown in Fig.4.1 where $V(\phi)$ was independent of the sign of ϕ.] Thus all orientations are allowed for the order parameter and the order parameter space S consists of all such configurations. S then consists of all possible orientations of a vector which has fixed magnitude, namely the magnetization vector **M**. If **M** is a 3-dimensional vector, then it will span the surface of a 2-sphere S^2. Thus the order parameter space S is S^2 in this case.[1] Note here that S is a smooth manifold, in contrast to the case of Fig.4.1 where S was a point set.

4.3 Simple example of topological defect: the domain wall

The typical situation of a topological defect is that, given an ordered medium in which the order parameter is prescribed at each point, some times, there are small regions where the order parameter cannot take values in the order parameter space. If there are such regions then they will be like defects in an order parameter configuration.

In Fig.4.2(a) we have reproduced the free energy plot corresponding to the low temperature case of Fig.4.1. Fig.4.2(b) shows a certain configuration of the order parameter ϕ. To minimize the free energy, ϕ must take values in S in a region. That is $\phi = \phi_0$ or $\phi = -\phi_0$. In Fig.4.2(b), ϕ is assumed to take value ϕ_0 in the region to the left of the thick solid curve, while $\phi = -\phi_0$ in the region to the right of the solid curve.

We mention here that this specific configuration has been chosen to illustrate the structure of a topological defect. Still, it may seem bothersome how such a configuration can arise to begin with, in which ϕ does not assume a single value throughout the space. This is actually the issue of the formation of defects and will be dealt with in detail later. To make the reader somewhat comfortable for now, we give some physical arguments to show why a configuration as in Fig.4.2(b) is natural to arise during a phase

---

[1]This example of magnetization illustrates one important point that the ideas of topology must be supplemented with physical considerations in order to reach proper conclusions. For example, a reader familiar with topological defects will conclude (and as we will explain later) that order parameter space being S^2 implies existence of monopoles (in 3 space dimensions). This conclusion is wrong because we know that ordinary magnetization does not allow for any monopoles.

4.3. The domain wall

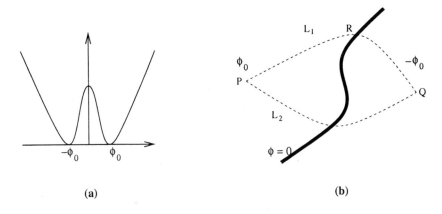

Figure 4.2: (a) Plot of free energy F is reproduced from Fig.4.1. (b) ϕ takes value ϕ_0 at a point P in the region left of the thick solid curve, while $\phi = -\phi_0$ at the point Q in the region right of it.

transition. Note that the plot in Fig.4.2(a) represents a situation when a phase transition has occurred in a system, say by lowering the temperature so that the free energy plot changes from the dashed curve in Fig.4.1 to the solid curve in Fig.4.1 (and in Fig.4.2(a)). For the high temperature case ϕ takes value equal to zero. When the phase transition occurs, ϕ has to assume one of the values ϕ_0 or $-\phi_0$, and this should happen typically in the time scale of the phase transition. Now, for a sufficiently large system, in which phase transition is carried out rather fast, it will not be possible to make the same choice for ϕ at two far separated points. This is certainly the case when the transition is so fast that the two points are causally disconnected (meaning that no signal can travel during the transition time between the two points; signal may have speed of light for particle physics case, or the speed of sound for condensed matter case). Thus, for far separated points, the choice for ϕ in S will have to be made locally and there is no way to avoid spatial variation of ϕ, with ϕ taking different values in S in far separated regions. These arguments are somewhat simplistic, but they illustrate how a configuration like the one shown in Fig.4.2(b) can arise naturally during a phase transition.

Now consider a path L_1 joining points P and Q. As $\phi = \phi_0$ at point P and $\phi = -\phi_0$ at point Q, it is clear that somewhere on the path L_1 there must be at least one point (say, R) where $\phi = 0$. [Note that ϕ must vary continuously, otherwise the gradient term in the expression for the free energy will become infinite.] Clearly, this argument does not depend on the choice of path connecting points P and Q. Thus even on path L_2 there should be at least one point where $\phi = 0$. If we vary path L_1, say, by a very small amount, then continuity of ϕ will imply that the point R will also vary by a very small amount. In other words, continuity of ϕ implies that point R should trace a wall (in 3-dimensions) such that $\phi = 0$ on the wall. This is the domain wall (kink in 1 dimension) defect. This is the simplest example of a topological defect.

Note from the plot of free energy that $\phi = \pm\phi_0$ have lowest energy (at $T < T_c$) and that at such low temperatures, $\phi = 0$ costs lot of free energy, with free energy density equal to the height of the central bump at $\phi = 0$ in Fig.4.1. This domain wall thus costs energy.

> **Exercise 4.1** *Show that, in 3-space dimensions the domain wall must either form a closed surface (like a 2-sphere), or it should end at the boundary of the system. (In 2-space dimensions it forms a closed curve, or ends at the boundary).*

4.3.1 Why defect?

As we mentioned earlier, the origin of the word defect lies in the fact that these objects represent regions of certain shapes (*e.g.*, wall in the above example) in which the order parameter cannot remain in the order parameter space S whereas the remaining region has well defined order parameter taking values in the order parameter space S. Thus it is natural to call these localized regions (note that the region may be localized in only one direction and extended in the other directions, such as domain wall) as *defects*.

4.3. The domain wall

Historically, the term *defect* was used because of the work of Voltera on crystal dislocations. Later works also focused on the plastic deformations of solids. These were called defects as they were nuisances to the perfect ordering of the crystal. This terminology was kept in condensed matter physics even for more complex systems. On the other hand particle physicists and cosmologists developed independent terminology guided by different motivations. Thus point defects, line defects and planar defects of condensed matter physics are typically called as monopoles, strings, and domain walls respectively. Though, now the term topological defect is used quite uniformly, in fact even for the cases where it is not quite appropriate, like Skyrmions and textures.

4.3.2 Why topological?

Now we come to the *topological* part of the topological defect. What is meant by the term *topological*? A careful definition will be given later, but for now we will try to give an intuitive understanding.

A given property of a system is said to be of topological nature if smooth deformations do not lead to any change in that property. Consider, for example, two closed surfaces, a torus and a 2-sphere. Among other differences between these surfaces, such as the rotational symmetry, etc., there is one important difference which is the presence of a hole at the center in the torus and its absence in the sphere.

If we think of both these surfaces to be made up of rubber sheets, then smooth deformation of these surfaces will mean any type of stretching, shrinking, etc., but not tearing (that will not be smooth). Note that many of the properties of the torus and the sphere change under such deformations. For example the spherical symmetry of the sphere can be lost by such deformations. Similarly cylindrical symmetry of the torus can be lost. One thing which does not change under these deformations is the fact that there is still one hole in a (deformed) torus, while there is no such hole in a (deformed) sphere. The presence or absence of such a hole is then an example of a topological property which is unaltered by smooth deformations (or, continuous changes).

With this intuitive understanding of the meaning of the word *topological*, let us go back to topological defects. Recall that these defects exist in an order parameter configuration where in any given region of space, the order parameter takes values in the order parameter space S. This defect will be a topological defect if any smooth deformation (continuous changes) in the order parameter configuration (within S) does not affect its existence.

Consider again the case of a real scalar order parameter with $S = \{\phi_0, -\phi_0\}$. The resulting topological defect is the domain wall. Smooth deformations cannot turn $-\phi_0$ to ϕ_0 (for non-zero ϕ_0) within S. This shows that the conclusion that region 1 and region 2 are separated by a topological planar defect is unaffected by smooth deformations, hence the topological nature of the defect.

In fact, the considerations are somewhat more general. For the planar defect example, even if we allow all smooth variations of the order parameter (even outside S), one still cannot remove this planar defect by making changes only in a localized region. So, either one has to take the wall to the infinity, or change the order parameter to zero value (*i.e.*, set $\phi = 0$) everywhere costing an infinite amount of energy.

It should be clear by now that the topological nature of these defects is coming from the non-trivial variation in the order parameter from one region of space to another with the net variation having a topological nature, which will be elaborated upon later. So, if the order parameter was zero (as in the high temperature phase in Fig.4.1), there would be no chance of getting a topological defect. One may consider a localized region of some non-zero ϕ, embedded in the background of $\phi = 0$, but clearly one can shrink this region of non-zero ϕ completely to get $\phi = 0$ everywhere.

4.3.3 Energy considerations in topological arguments

It is important to note that almost all these topological notions are strongly tied with the considerations of energetics of the problem. What is topological for a given energy scale may not be so at a higher energy scale. For example, in Fig.4.1, as long as the energetics of the problem (thermal fluctuations, etc.) is on an energy scale much smaller than V_0, domain walls will be well defined and will be topological defects. At higher scales (say, higher temperatures) fluctuations in energy density with magnitude V_0 may become too frequent. Then the topological classification breaks down. Actually, this is typically the scale at which $\phi \to -\phi$ symmetry is restored.

As will be discussed in detail later, it is the topology of S which determines what kind of topological defects can exist. However, the topology of S depends on the energy scale relevant to the problem. In Fig.4.1, at low temperatures ($T \ll T_c$), $S = \{\phi_0, -\phi_0\}$, while at high temperatures ($T \gg T_c$), $S = \{0\}$. The topology of S changes here so for intermediate energy scales (with $T \sim T_c$), topological classification is not particularly meaningful.

4.4 Examples of topological defects

It should be clear by now that the origin of topological defects crucially relies on the nature of the order parameter and order parameter space. Let us first see what sorts of order parameter spaces typically arise in physical systems.

• **Examples of order parameter spaces in condensed matter physics**

We first discuss some examples of order parameter spaces in condensed matter systems. Intuition obtained from these examples will be useful in discussing various cases in particle physics where the mathematical structure of the order parameter, etc., is the same, although the actual physical meaning may be somewhat less obvious.

Planar Spins

Consider a plane R^2 such that at each point of R^2, a vector of fixed magnitude is prescribed which lies in the plane as shown in Fig.4.3(a). One can think of these vectors as corresponding to spins lying in the plane. The order parameter is thus a vector of unit magnitude which lies in a plane and the order parameter space S consists of an angle θ denoting the orientation of this vector. θ lies between 0 and 2π, so the order parameter space S is a circle S^1, see Fig.4.3(b).

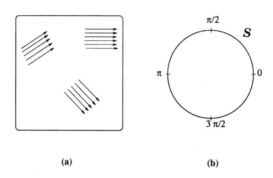

(a)   (b)

Figure 4.3: (a) A system of planar spins. The order parameter is a vector of unit length which lies in a plane. (b) The order parameter space S which is a circle S^1.

Superfluid ^{4}He

Interestingly, the same order parameter space (S^1) also describes superfluid ^{4}He where the order parameter is a complex scalar field of fixed magnitude ψ_0 but arbitrary phase:

$$\psi(r) = \psi_0 e^{i\phi(r)} \tag{4.2}$$

$\psi(r)$ is essentially the wave function of the superfluid condensate.

For ^{4}He there is a phase transition from the normal phase to the superfluid phase. This is a second order transition, with the critical temperature $T_c = 2.18\ ^0K$. In the Ginzburg-Landau (GL) theory of second order transitions, specific free energy of the system is given by,

4.4. Examples of topological defects

$$F(\psi) = \frac{\hbar^2}{2m}|\nabla \psi|^2 + \alpha|\psi|^2 + \beta|\psi|^4 \qquad (4.3)$$

$$\alpha = \alpha'\frac{(T-T_c)}{T_c} \; ; \; \alpha', \beta > 0$$

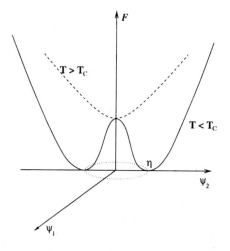

Figure 4.4: Free energy for superfluid ^{4}He. Order parameter is a complex scalar field with real components ψ_1 and ψ_2.

Fig.4.4 shows a plot of $F(\psi)$ for spatially constant ψ. We see that for $T < T_c$, the minimum of F occurs at

$$|\psi| = \sqrt{-\frac{\alpha}{2\beta}} \equiv \eta \qquad (4.4)$$

This only determines the magnitude of ψ, leaving its phase arbitrary. The order parameter space consists of variations of this phase and, hence, is a circle S^1.

Superconductors

It is remarkable that the same order parameter space S^1 describes yet another very important condensed matter system, that is, a superconductor. Here again, the order parameter is a complex scalar field $\psi(r)$ which is directly related to the wave function of the Cooper pairs which condense, giving rise to the superconducting state. The only additional thing here, compared to the case of superfluid ^{4}He, is the inclusion of gauge fields. However, that only affects the energetics of the problem and has no effect on the topological properties of the order parameter space.

Superfluid ^{3}He

We just mention here that superfluid ^{3}He has many phases which have a whole variety of topological defects. For example in one of the phases, the order parameter space is SO(3) while in another it is $S^2 \times SO(3)/Z_2$ (which is a five dimensional manifold). For details see ref. [14].

Ordinary Spins in Three Dimensions

The order parameter here is a unit vector which can point in any direction in the 3-dimensional space. The order parameter space is thus the surface of a unit sphere S^2. From the example of magnetization **M** discussed earlier, note that S^2 is also the order parameter space for that case. Spins in 3-d can be taken as a macroscopic continuum model of a ferromagnetic crystal.

Nematic Liquid Crystals

Here the molecules have rod like (sometimes disk like) shape and their local ordering gives an order parameter which is a vector of fixed length but without any arrowhead. That means that the opposite orientations of the arrow are indistinguishable. We can say that the order parameter space is the same S^2 as for 3-d spins, but now diametrically opposite points are identified. The resulting order parameter space S is the projective plane $RP^2 = S^2/Z_2$.

Other interesting examples, like the biaxial nematics, will be discussed later.

4.4. Examples of topological defects

• **Examples of order parameter spaces in particle/nuclear physics**

The only topological defect in relativistic field theories which are accessible to direct laboratory investigations are the Skyrmions in nuclear/particle physics which are supposed to represent the nucleons. In particle physics, the standard context in which other topological defects arise is the early universe. Starting with extremely high temperatures ($T > 10^{19}$ GeV), the universe cools and, in the process, goes through several phase transitions.

Grand Unified Theory Phase Transitions

The phase transition at the highest temperature is supposed to be the Grand Unified Theory (GUT) transition with $T_c \sim 10^{16}$ GeV. (Sometimes one talks about transitions at the Planck scale $\sim 10^{19}$ GeV). Above GUT transition temperature, all forces (except gravity) are unified, while at lower temperatures, spontaneous symmetry breaking occurs leading to differences between strong and electroweak forces.

The order parameter space here depends on the specific model under consideration. One point which is important is that in most GUT models, the order parameter space is such that it implies the existence of monopole defects (*i.e.*, it has non-trivial 2nd homotopy group). An example, would be SU(5) breaking to SU(3) × SU(2) × U(1).

Electroweak Phase Transition

This phase transition occurs when the universe cools below the critical temperature $T_c \simeq 100 GeV$. The transition corresponds to the spontaneous breaking of SU(2) × U(1) electroweak symmetry to the U(1) of electromagnetism. The order parameter space is a three sphere S^3.

Models with Discrete Symmetry

Many particle physics models have spontaneous breaking of discrete symmetries (like $\phi \to -\phi$). These lead to order parameter spaces which are point sets. For example $S = \{\phi_0, -\phi_0\}$ as discussed earlier. The Lagrangian density for this model can be written as

$$L = \frac{1}{2}\partial_\mu \phi \partial^\mu \phi - V(\phi) \; ; \; V(\phi) = \frac{\lambda}{4}(\phi^2 - \eta^2)^2. \quad (4.5)$$

ϕ is a real scalar field and $V(\phi)$ is the effective potential. This has the same interpretation as the free energy.

S^1, S^3/Z_2 etc.

The abelian Higgs model has the order parameter space S^1, just as for the case of superconductors. The effective potential for this case will be the same as given in Eq.(4.5) except that ϕ will be replaced by $|\psi|$, where ψ is a complex scalar field. S^1 order parameter space is frequently discussed as a simple model for cosmic strings. There are other order parameter spaces which arise in specific contexts, such as S^3/Z_2 which also allows for cosmic strings.

Sigma Model

Chiral sigma models in nuclear/particle physics provide the only example where topological defects are discussed in the context separate from the early universe. For example, in the Skyrme model, the nucleon is identified with a topological object (similar to defect) in the O(4) sigma model, called as the Skyrmion, with the order parameter space being S^3.

- **Examples of topological defects**

A detailed discussion of various kinds of defects will be given later. For now, we mention a broad characterization of topological defects in terms of their shapes.

Domain Walls

A domain wall is the simplest example of a topological defect. This results when the order parameter space is disconnected (*e.g.*

4.5. Condensed matter versus particle physics

a point set), as has been described in detail earlier. The resulting structure of defect is that of a membrane with a certain surface tension.

Vortices, Strings

For planar spins, the order parameter space $S = S^1$. This leads to point like topological defects called vortices in two space dimensions and line like defects called string defects (with some string tension) in three space dimensions. As superfluid ^{4}He and superconductors have the same order parameter space one gets string defects for these systems as well. String defects arise in certain particle physics models of the early universe and they are usually called cosmic strings. String defects also arise when $S = S^2/Z_2$, as in nematic liquid crystals.

Monopoles

Monopoles, or point defects, arise when $S = S^2$, as for ordinary spins in 3-d. Most of GUT models in particle physics lead to existence of these monopole defects. Monopoles also arise with $S = S^2/Z_2$, as for nematic liquid crystals.

Textures, Skyrmions

This is a class of topological defects which differs in some essential manner from the other topological defects. In some sense, these are not defects because the system throughout remains in the same, symmetry broken, phase. Still, the rest of the characterization of these objects is identical to other defects. The simplest example of these defects in 3-d is with $S = S^3$ as for Skyrmions and textures in nuclear/particle physics. Another example is nematic liquid crystal with $S = S^2/Z_2$ which also supports textures.

4.5 Qualitative difference between topological defects in condensed matter and particle physics

From the above discussion it should be clear that topological characterization of defects only requires knowledge of the order parameter space. Later we will see that one only needs to know certain topological properties of S. Thus, the fact that $S = S^1$ for the case of superfluid ^{4}He as well as (say) in the Abelian Higgs model, is enough to conclude that in both cases one will have string defects. It does not matter here that in one case, one is dealing with a non-relativistic system, while in the other case, one is dealing with a relativistic field theory. This difference is relevant only when one discusses the energetics and dynamics of these defects. As far as the topological characterization of defects is concerned, the entire discussion applies to both cases equally well.

One essential difference between topological defects in condensed matter physics and in particle physics is the following. In condensed matter physics, the free energy does not include the rest mass of the particles constituting the system. Thus, a plot like the one in Fig.4.1 means that for $|\phi| = \phi_0$, the system is in the state of the lowest free energy, which, for superconductors, will mean that the system is in the superconducting phase. For $\phi = 0$ the system will be in the normal phase. Both phases still have the same number of nucleons and electrons, hence the same rest mass. For example, the energy needed to break a Cooper pair will be in milli-electron volts. Compare this to the rest mass energy of proton ~ 1 GeV, or even for the electron ~ 0.5 million electron volts. What this means is that the energy of a topological defect is higher than the background configuration only in terms of the free energy which constitutes a negligible part of the total mass-energy density. These defects are therefore *embedded* in the system and their dynamics can be completely controlled by the background. For example, in superfluid ^{4}He, strings move along with the flow of the superfluid just as eddy currents move in flowing water.

In contrast, in relativistic field theories, as in particle physics, one only talks about the total energy. Thus, now, a plot like the one in Fig.4.1 means that for $|\phi| = \phi_0$, energy is zero (so, no cosmological constant), and this is the vacuum of the theory (just as you expect in outer space where there is supposed to be nothing). So, there is no background medium here. Now the non-zero energy density of defects makes them into completely

4.5. Condensed matter versus particle physics

separate entities. For example, magnetic monopoles in particle physics models are expected to have a mass $\sim 10^{16}$ GeV, that is, 10^{16} times heavier than a proton, while having almost point like structure. Its dynamics is that of a point mass moving freely in empty space. (Of course, due to the fact that it has a magnetic charge, it will be subject to forces if there are electromagnetic fields present). Similarly, cosmic strings behave as strings with certain string tension, moving in empty space. These can be as heavy as having mass density of 10^{16} metric tons per centimeter, again having almost zero thickness.

- **A brief summary of implications of topological defects**

Most of the important implications of topological defects arise due to their core structure. Sometimes, the long range variation of order parameter also plays an important role (like in generating superflow for vortices in superfluid helium). Let us recall the example of a domain wall defect. Inside the wall $\phi = 0$, so $\phi \to -\phi$ symmetry is restored, while outside the wall, this symmetry is spontaneously broken as $\phi_0 \to -\phi_0$ does not leave the configuration invariant.

Thus, we conclude that, inside the wall, one has a disordered phase (*i.e.*, the symmetric phase) while outside one has the ordered phase. Note here that, strictly speaking, we have only argued earlier that ϕ should vanish *at least at a point* on any curve joining a region with $\phi = \phi_0$ to a region with $\phi = -\phi_0$. This will imply that the region inside the wall which is in the symmetric phase, has zero thickness. Non-zero thickness of this symmetric region in the wall arises because ϕ is close to zero over a region. Certainly, there is a broad region around the $\phi = 0$ point in which ϕ deviates significantly from the equilibrium value $|\phi| = \pm \phi_0$, and the free energy density is higher than the equilibrium value in all this region. The thickness of this region (hence the thickness of the wall) is given by an appropriate correlation length. Note that, due to the topological nature of the defect, both these phases are co-existing at all values of temperatures below the critical temperature. In the absence of topological defects this is just not possible when the phase transition is a second order transition, and even for a first order transition, coexistence of phases is possible only at the critical temperature. If the domain wall happens to move inside the system then it corresponds to the motion of the boundary of phase-coexistence.

Why does a wall move? We know that the wall corresponds to a region of higher free energy (compared to the situation when there is no wall and the system is in uniform, symmetry broken phase). The system will try to lower its free energy as it evolves towards equilibrium. It cannot get rid of the wall due to its topological nature, but it can certainly reduce the total free energy by decreasing the area of the wall. So the wall will try to become as planar as possible [2]. This implies that a wall with arbitrary shape will evolve to become a planar wall.

Thus, the wall behaves exactly like an elastic membrane. It has a surface tension which tries to minimize the curvature (bending, etc.) of the wall. One can thus conclude that a domain wall which forms a closed surface will shrink to become spherical and eventually collapse away. Exactly the same thing happens for string defects. The $\psi = 0$ region corresponds to higher free energy. Again, to minimize the free energy the string tries to become straight, *i.e.*, it has string tension. Thus a closed string loop of arbitrary shape collapses to smaller sizes, and at the same time becomes more circular.

Exercise 4.2 *Give arguments to show that the topological nature of the domain wall does not forbid collapsing away of a closed wall (use continuity arguments, similar to the one which showed the existence of the domain wall in the first place).*

---

[2] It is important to clearly distinguish between the energy and the free energy. As the wall evolves, the energy of the wall is simply transferred to other modes of excitations of the system. However, the free energy of the total system keeps decreasing as the system evolves towards equilibrium basically due to the fact that the system evolves to maximize its entropy. So, in other words, the movement of the domain wall is due to fluctuations in the system which drive the system towards the state of maximum entropy.

4.6 Detailed understanding of a topological defect

Defects and Free Homotopy of Maps

We will use the example of strings in superfluid ^{4}He to explain various mathematical concepts involved in classifying topological defects. As mentioned earlier, the order parameter which describes the superfluid phase is a complex scalar field ψ. The phase transition from the normal to the superfluid phase is of second order and the free energy density F near the critical temperature T_c is given in Eq.(4.3). The plot of F for a spatially uniform configuration of ψ is shown in Fig.4.4 for different values of T.

We see that for $T > T_c$, F is minimized for $\psi = 0$ while in the superfluid phase ($T < T_c$) F is minimized for $|\psi| = \eta$. Clearly this does not fix the phase of ψ. By writing $\psi = |\psi|e^{i\theta}$ we conclude that in the superfluid phase the magnitude of the order parameter ψ is fixed to be equal to η but its phase θ can vary spatially. This remaining degree of freedom spans the order parameter space S, which is a circle S^1 in this case. Of course any spatial variation of ψ will cost energy due to the gradient term in the expression for F.

Consider now a region of the superfluid with the distribution of θ on (and nearby) a closed path L in physical space as shown in Fig.4.5. We assume that $|\psi| \simeq \eta$ for the regions where arrows have been drawn. Orientation of the arrow from positive x axis denotes the value of θ at that point. We note that θ changes by 2π as we go around the loop L. Another way to say it, is that as we go around the loop L in physical space, ψ traces a closed path going around the valley of the minima of F in Fig.4.4. It is now easy to see that as we shrink the loop L in physical space down to a point, the corresponding loop in the plot of F vs. ψ must also shrink to a point due to the fact that F is single valued. Clearly this can only be done by taking the loop (in the $\psi_1 - \psi_2$ plane) through the origin where $\psi = 0$. This means that as the original loop L is shrunk in the physical space, ψ cannot lie in the order parameter space S always and in fact L must enclose at least one point where ψ vanishes identically. As L can be shrunk on any surface whose boundary is L, we conclude that L encloses a line like region where $\psi = 0$. Since vanishing of ψ implies normal phase of ^4He, one obtains a string like region of normal phase embedded in the superfluid phase of ^4He. This is the vortex in superfluid ^4He.

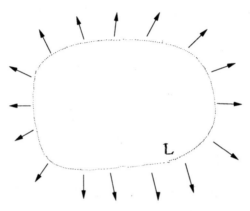

Figure 4.5: A non-trivial winding of θ in space.

Exercise 4.3 *Just as for the domain wall case, show that the string cannot end in the middle of the superfluid sample. That is, the string should either form a closed loop, or it should end at the boundary of the sample. Also, argue that for the present case of superfluid strings, a closed string loop has no topological stability; it can collapse away completely. (This latter part may not be true for strings in some other systems. For example, in nematic liquid crystals, a string loop can carry monopole charge. When such a loop collapses, a monopole defect is left behind.)*

Note that in this example, the existence of the string was related to the fact that there was a closed curve in the order param-

4.6. Detailed understanding of a topological defect

eter space S (which was S^1 in the above example), corresponding to the distribution of θ along the closed path L in the physical space, which could not be shrunk to a point within S. In fact, it is also easy to see that there are other kinds of loops in S here corresponding to net change in θ around L being $2n\pi$ ($n = \pm 1, \pm 2..$) which cannot be deformed to a point in S and hence correspond to string defects. Furthermore, a loop in S corresponding to one value of n cannot be continuously deformed to another with a different value of n, but can always be deformed to any other loop with same value of n. This leads to topologically distinct strings, each characterized by a different value of n usually called the winding number of the string, such that one string cannot be continuously deformed into the other unless both correspond to same n.

These arguments can be extended to other models with different order parameter spaces. Topologically distinct string defects are characterized by equivalence classes of loops in the order parameter space S. Two loops in S are considered equivalent if and only if they can be continuously deformed to each other. The trivial class consists of all loops which can be shrunk to a point and this class corresponds to no defect.

Mathematically speaking, one is considering continuous maps of a circle into the order parameter space and finding out when one map can be continuously deformed into another map. The circle can be thought of as representing a loop L in the physical space and its image under the map in S as the distribution of θ around L. A given map ϕ corresponds to a specific order parameter configuration. When two maps can be continuously deformed to each other, then they are said to be homotopic to each other, otherwise they are homotopically inequivalent. At this point one can appreciate the importance of using the notions of topology. In the above discussion of a string defect, it was nowhere required that the loop L should be a circle, or even that it should be a smooth manifold. For example, if the loop L had sharp edges, or kinks, it does not affect anything in the above arguments. Thus for classifying topological defects one must be able to define maps (continuous order parameter configurations) from general spaces into the order parameter space. The most general space which allows for the notion of continuity is a topological space. For definition of a topological space, see refs. [6]. Thus, for all the discussion below, relevant spaces are to be taken as topological spaces.

4.6.1 Free homotopy of maps

We now give the general definition of free homotopy of maps. Let $\phi_1 : M \to S$ and $\phi_2 : M \to S$ be two (continuous) maps from a topological space M to a topological space S. (In the above example, M was the loop L in the physical space and S was the order parameter space.) ϕ_1 and ϕ_2 are said to be (freely) homotopic if there exists a (continuous) family of maps $\Phi_s : M \to S$ with $0 \leq s \leq 1$ such that $\Phi_0(r) = \phi_1(r)$ and $\Phi_1(r) = \phi_2(r)$.

Note that $\Phi_s(r)$ defines a map from $M \times I$ to S where I denotes the unit real interval [0,1]. Thus, ϕ_1 and ϕ_2 are homotopic if there exists a continuous map

$$\Phi : M \times I \to S \tag{4.6}$$

such that $\Phi(r, 0) = \phi_1(r)$ and $\Phi(r, 1) = \phi_2(r)$. Here we are denoting $\Phi_s(r)$ by $\Phi(r, s)$. The map Φ is called a homotopy from ϕ_1 to ϕ_2.

The homotopy we have been discussing is called free homotopy. Here two loops in a space (actually maps of a circle into S whose images are these loops) are called freely homotopic if one loop can be continuously deformed to the other. We have just argued that string defects are classified by the equivalence classes of freely homotopic loops.

However, this classification is not very convenient because in this way we cannot give a simple scheme for combining two defects. For example suppose loops L_1 and L_2 (Fig.4.6) correspond to two different defects. If we know all non trivial loops in the order parameter space S, we would like to ask whether there is a simple classification of the combined defect composed of 1 and 2; *i.e.*, how to characterize loop L encircling both defects in terms of L_1

4.6. Detailed understanding of a topological defect

and L_2. This is a relevant question because the two defects can be brought together and merged to form a single defect. The problem here is that it is not clear how to compose two loops such as L_1 and L_2 for the case of free homotopy.

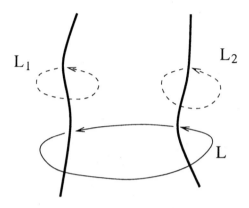

Figure 4.6: Combining two defects.

4.6.2 Based homotopy and the fundamental group

The above problem can be solved by defining a more restricted notion of homotopy called based homotopy. Here one considers maps of a circle (*i.e.*, a space which is topologically equivalent to a circle) into S which take a given point of the circle to a fixed point $x_0 \in S$. The images of these maps consist of all loops passing through a given point, x_0, in S. Then two loops f and g (again, maps whose images are these loops) are called homotopic if there is a continuous family of loops, all passing through x_0 such that f and g are members of that family. In other words, f and g should both pass through x_0 and should be continuously deformable to each other, without leaving point x_0 from the loops.

Equivalently, one can consider maps from the unit interval I into S,

$$\phi : I \to S \tag{4.7}$$

with the condition that $\phi(0) = \phi(1) = x_0$. This compactifies the unit interval I to a circle. With this, two maps $\phi_1 : I \to S$ and $\phi_2 : I \to S$ are called (based) homotopic if there exists a continuous map

$$\Phi : I \times I \to S \tag{4.8}$$

such that

$$\Phi(0, s) = x_0 = \Phi(1, s) \text{ for all } s \in I \tag{4.9}$$

$$\Phi(t, 0) = \phi_1(t) \quad \& \quad \Phi(t, 1) = \phi_2(t) \tag{4.10}$$

Recall that the definition of free homotopy required only the condition in Eq.(4.10). The extra condition (Eq.(4.9)) is required here so that all homotopies are based homotopies. In Fig.4.7, (based) loops are drawn for a region of R^2 with a hole. Here f is homotopic to g, h is homotopic to k but h is not homotopic to f.

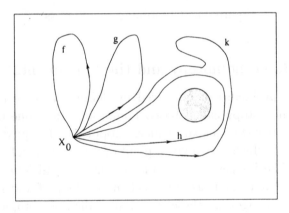

Figure 4.7: Based homotopy of loops.

For such based loops (maps) it is possible to define product of two loops. So $f_o g$ means traverse loop f first and then continue to traverse loop g. More precisely, in terms of maps, let

$$\phi_1, \phi_2 : I \to S$$

be two maps such that

$$\phi_{1,2}(0) = \phi_{1,2}(1) = x_0$$

Then the product $\phi_1 o \phi_2$ of the two maps ϕ_1 and ϕ_2 is defined as,

$$\phi_1 o \phi_2 : I \to S$$

such that

$$\phi_1 o \phi_2(t) = \phi_1(2t) \text{ for } 0 \leq t \leq 1/2$$
$$= \phi_2(2t-1) \text{ for } 1/2 \leq t \leq 1$$

Note that $\phi_1 o \phi_2$ is also based at x_0. One can show that this composition law becomes associative if one considers (based) homotopy equivalence classes of maps. If f and g are homotopically equivalent (so they belong to same homotopy class) then we denote it as f $\sim$ g. One can give a group structure to homotopy classes of loops by defining product of two loops (as described above) as the composition law for group elements. Group identity consists of the equivalence class of maps which are homotopic to the constant map. Inverse of a map (whose image is loop f) is the map which traces loop f in reverse order. (Note that we have been using the term loop to represent the map as well as its image.) One can show that these groups (denoted as $\pi_1(S, x_0)$) for different choices of x_0 are all isomorphic for a connected space so one can just call it $\pi_1(S)$. $\pi_1(S)$ is called the fundamental group, or the first homotopy group of S.

4.7 Classification of defects using homotopy groups

We now describe how homotopy groups can be used to classify topological defects. As mentioned earlier, classification of string defects is done by equivalence classes of freely homotopic loops and *not* in terms of based homotopy. However, it is possible to use the structure of the fundamental group to classify the defects.

• Characterization of string defects: free homotopy

Consider three loops f, g and h belonging to $\pi_1(S)$ based at x_0. Now suppose that f and g are homotopic (under based homotopy). Then clearly f and g are also freely homotopic. As it is the free homotopy which actually classifies topologically distinct defects, f and g correspond to the same defect.

Now consider the case when f and g are not homotopic to each other. So they correspond to different elements of $\pi_1(S)$. We need to check whether f and g are freely homotopic in order to decide if they actually correspond to same defect or not. To do this we should be able to take loop f any where we please (by releasing it from the base point x_0) and then try to deform it to loop g. If in this manner we can ultimately deform f to g then f and g will be freely homotopic and hence will correspond to the same defect. Clearly in this process we start with loop f based at x_0, take it any where and then bring it back so that finally one of its points is again at x_0, in trying to deform it to loop g (since loop g is based at x_0). See Fig.4.8(a).

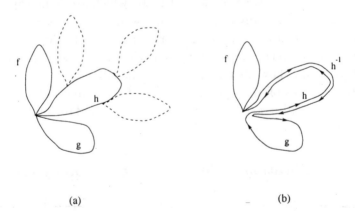

Figure 4.8: (a) Loop f is freely homotopic to loop g. (b) If f can be deformed to g by taking f along loop h, then $f \sim h\, g\, h^{-1}$.

To do this it is convenient to mark one point on loop f which is

4.7. Classification of defects using homotopy groups

at x_0 initially and finally. This point on loop f then moves around some other loop, say h, when loop f is brought back to x_0 in trying to deform it to loop g. If the final loop (*i.e.*, loop f taken along loop h and brought back to point x_0) is f_1 (not shown in Fig.8), then $f \sim h\, f_1\, h^{-1}$. Now, f and g will be freely homotopic if f_1 is homotopic (under based homotopy) to g, *i.e.*, f_1 and g correspond to the same element in $\pi_1(S)$.

Thus we see that two loops f and g are freely homotopic if there is a third loop h such that $f \sim h\, g\, h^{-1}$. Taking these loops as elements of group $\pi_1(S)$ (where now they represent equivalence classes under based homotopy) we say that f and g are conjugate to each other. *We conclude that topologically distinct string defects are classified by the conjugacy classes of the elements of* $\pi_1(S)$.

If π_1 is Abelian then $h\, g\, h^{-1} = g$, so conjugacy classes consist of single group elements. Thus "for Abelian fundamental group, topologically distinct line defects are classified by distinct elements of the fundamental group of the order parameter space."

- **Characterization of point defects, second homotopy group**

Consider the order parameter space S to be a two sphere S^2 (as for ordinary spins in 3-dimensions). Now consider a closed surface A (topologically equivalent to 2-sphere S^2) in the physical space. The order parameter takes values in S which can be taken to be unit vectors along the radial direction. At each point of A there is a value of the order parameter prescribed. This amounts to giving a map from $A \to S^2$. The image of this surface A will be a closed surface A' in S^2. As A is shrunk in the physical space down to a point, one would like to know whether the corresponding image A' can also be completely shrunk in S^2. If this can be done then A does not enclose any point defect. However, if A' cannot be shrunk completely in S^2 then there must be at least one point inside A where the order parameter cannot lie in S. This will be a point defect.

Thus we come to the conclusion that point defects are classified by closed surfaces (images of S^2 under continuous maps) which cannot be shrunk to a point in S. Clearly, one has to consider the equivalence classes of maps where one equivalence class consists of maps of spheres which are *freely homotopic* to each other.

Again, for combining defects, etc., it is convenient to introduce the concept of based homotopy where in constructing homotopy between two maps of S^2, the value of order parameter is kept fixed at one point of S^2. More precisely, consider the images of spheres in S as being given by mappings f of a unit square into S with the restriction that f takes the entire circumference of the square onto the single point x_0 in S. The sphere is thus represented by closing up of the borders of the square. This allows for the product of two such maps to be defined in the following manner. One simply joins the two squares and maps the entire boundary of the rectangle onto $x_0 \in S$. As we are interested in homotopy classes of maps, we can re-scale the rectangle so that one finally gets a map from a square into S.

We say that a map f is homotopically equivalent to another map g (f $\sim$ g) if f can be continuously deformed to g while keeping the restriction that the boundary of the square always goes to $x_0 \in S$. Equivalence classes of these maps under this based homotopy form the second homotopy group $\pi_2(S, x_0)$.

One can show that these maps commute with each other. One can combine two squares from different directions, still leading to same map (homotopically) as the entire boundary goes to x_0 (note that for homotopy classes of maps the square can be suitable deformed.) So, slowly, one can interchange the order of two squares being combined leading to the results that $f_1 f_2 = f_2 f_1$. We thus conclude that $\pi_2(S, x_0)$ is always Abelian. Further, $\pi_2(S, x_0)$ is isomorphic to $\pi_2(S, x_1)$ for a connected space (just as for the first homotopy group). It is easy to show this by realizing that a sphere at x_0 can be considered to be a sphere at x_1 by drawing a string (or a thin tube) from x_0 to x_1. One can, therefore, associate a single group $\pi_2(S)$ denoting the second homotopy group of S.

We know that we need free homotopy to classify point defects. Consider now two spheres S_1 and S_2 at $x_0 \in S$ (which are im-

4.7. Classification of defects using homotopy groups

ages of a square under two maps with the boundary of the square mapped to x_0). Just as for string defects, if S_1 and S_2 belong to two different elements (more appropriately the two maps whose images are S_1 and S_2) in $\pi_2(S)$ then, in order to know whether they correspond to topologically distinct defects, we need to know whether S_1 and S_2 can be continuously deformed to each other using *free homotopy*. The arguments here again are similar to the ones used for the string case. Starting from S_1 and S_2, both based at x_0, we take S_1 anywhere we please and then try to see if we can deform it to S_2. Clearly, we have to take S_1 (or, more precisely, a given point on S_1) along some arbitrary loop L in order to see if it is freely homotopic to S_2. If it is possible then clearly S_1 should be in the same homotopy class (*i.e.*, corresponds to the same element in $\pi_2(S)$) as S_2 with a loop L attached to it at the base point. Attaching loops corresponds to an action of $\pi_1(S)$ on $\pi_2(S)$ and leads to an automorphism of $\pi_2(S)$.

We conclude that point defects are classified by distinct automorphism classes of $\pi_2(S)$ under the action of $\pi_1(S)$.

- **Characterization of domain walls**

One can also associate a zeroth homotopy group $\pi_0(S)$ which characterizes disconnected components of S. If $\pi_0(S)$ is non trivial (S consists of many disconnected pieces) then one gets domain wall defects. These are similar to the kink solutions in 1 space dimension. For example, if the order parameter is a real scalar with free energy plot as shown in Fig.4.1 then we have seen that a wall like defect must separate two regions of space, which correspond to the order parameter belonging to two disconnected pieces of S (*i.e.*, $+\phi_0$ and $-\phi_0$). This is because one cannot go from one such region to the other continuously, along any trajectory, while keeping the order parameter belonging to S all along. All such trajectories must intersect some region where the order parameter has to get out of S, thus leading to a domain wall. We mention here that $\pi_0(S)$ is not a group in general (unless S is a group).

• Higher homotopy groups

We have seen that different homotopy groups of the order parameter space S lead to different types of topological defects. [More accurately, certain automorphism classes of different homotopy groups characterize topologically distinct defects.] For example, nontrivial $\pi_0(S)$ corresponds to domain walls, nontrivial $\pi_1(S)$ leads to string defects while nontrivial $\pi_2(S)$ leads to point defects.

One can extend this correspondence of different homotopy groups with different types of defects to the third homotopy group $\pi_3(S)$ (for 3 space dimensions; for D space dimensions one can associate topological objects to all homotopy groups up to $\pi_D(S)$). In 3-dimensions, when $\pi_3(S)$ is non trivial, then the associated topological object is called a texture. A texture is not a defect in the conventional sense as there is no region where the order parameter has to get out of the order parameter space. Still these are topological structures and play an important role in various theories in physics. For example, in particle physics, certain effective field theory models lead to the description of baryons in terms of such textures (which are called Skyrmions). Textures also exist in liquid crystals, though none has been seen experimentally.

• n^{th} homotopy group $\pi_n(S)$

Thus for the general classification of defects, we need to study the nth homotopy group for various values of n. $\pi_n(S)$ is defined in a manner analogous to the homotopy groups discussed earlier, say $\pi_2(S)$. Thus $\pi_n(S)$ consists of equivalence classes of maps of the n-dimensional unit cube into S such that all the faces of the cube are mapped to a fixed point $x_0 \in S$. [Clearly this leads to a mapping of a n-sphere S^n into S with one point kept fixed.] The composition of two maps is just as for π_2, i.e., by putting two cubes side by side and mapping the entire surface to x_0 (with appropriate rescaling, one again gets a map from a cube to S). Again, as for π_2, two cubes can be joined from any direction and

4.7. Classification of defects using homotopy groups

hence their order can be reversed using a continuous family of maps. Thus all higher homotopy groups (n $\geq$ 2) are Abelian.

Again from a topological point of view, an n-sphere S_1 at point x_0 can be considered as an n-sphere S_2 at any other point x_1 for path connected space, where S_2 is just S_1 with a string attached to it. This string is any path connecting x_1 to x_0. This way one can show that $\pi_n(S, x_0)$ is isomorphic to $\pi_n(S, x_1)$.

- **General scheme for classifying topological defects in terms of homotopy groups**

Defects will be classified by the equivalence classes of freely homotopic maps. [There are some subtle points for textures which, as we mentioned earlier, are not defects in the conventional sense.] Arguments used for the case of point defects can be extended for the general case and we see that topologically distinct defects will be classified by the automorphism classes of $\pi_n(S)$ under the action of $\pi_1(S)$.

For the classification of defects one needs to calculate various homotopy groups of a given order parameter space. This is done using exact sequences, which requires the concept of a relative homotopy group. We will not be discussing these here. Ref. [1] contains a detailed discussion of these techniques. One important, and very useful concept in this context is to represent the order parameter space as a coset space. We discuss this below.

- **Order parameter space as a coset space**

In most situations the ordered state represents spontaneous breakdown of a symmetry of the system. The concept of spontaneous symmetry breaking has been described above. A typical situation is that in one (disordered) phase of the system, (say at high temperatures), the order parameter describing the system is invariant under the action of a Lie group G, while in the ordered phase, the order parameter is invariant under a smaller group H which is a subgroup of G.

The order parameter space S consists of distinct values of the order parameter in the ordered state. As the action of H leaves the order parameter unchanged, all those values of the order parameter which are related to each other by the action of any element of H belong to same point in S and hence should be considered equivalent. Action of any other element of G will change the order parameter and hence will correspond to a different point in S. We see then that in the ordered state, the points in S are in one to one correspondence with the space of cosets G/H. This correspondence can be shown rigorously [1] and we conclude that the order parameter space S can be described as the coset space G/H.

- **Examples of order parameter spaces as coset spaces**

Magnetization

The order parameter here is the magnetization vector **M**. Above a critical temperature **M** vanishes, so the symmetry group G can be taken to be SO(3). In the ordered phase, below critical temperature, **M** is non-zero and only those elements of G leave **M** invariant which correspond to rotation about the axis in the direction of **M**. Thus H should be identified with SO(2). The order parameter space S is then the coset space SO(3)/SO(2), which is known to be the 2-sphere S^2. Since $\pi_1(S^2) = 1$ (*i.e.*, trivial) while $\pi_2(S^2) = Z$, we conclude that there are no string defects, though there are point defects characterized by the set of integers.

Superfluid ^4He

The order parameter is a complex scalar ψ. In the normal phase ψ vanishes and the symmetry group is U(1). In the superfluid phase ψ is non-zero and only the identity element of U(1) leaves it invariant. S in this case is then U(1) itself. Only the first homotopy group of U(1) ($\equiv S^1$) is non trivial; $\pi_1(S^1) = Z$, so there are string defects. These are the superfluid vortices discussed earlier. As discussed earlier, a string defect is characterized by vanishing of ψ along a line like region. The vanishing of ψ corresponds to the normal phase of ^4He. We thus conclude that a

4.8. Defect structure in liquid crystals

vortex in superfluid ^4He has normal ^4He in its core.

Another system which has the same order parameter (and order parameter space) is a superconductor. In fact the only difference between the superfluid case and the superconductor case is the presence of a gauge field in the latter case. It does not affect the topological classification of defects but affects its energetics. For example, in the absence of gauge fields strings have a logarithmically divergent energy per unit length which comes from the variation in the orientation of the order parameter around the string. However, for the case of superconductors, the orientation of the order parameter is a gauge degree of freedom and its spatial variation around the string does not contribute to the energy, leading to finite energy per unit length. For superconductors, the vanishing of ψ corresponds to the normal phase of the metal. For the case of type II superconductors, strings correspond to line like regions of normal phase where the magnetic field penetrates. These are also called flux tubes.

4.8 Defect structure in liquid crystals

Liquid crystals are orientationally ordered liquids formed by elongated (rod like) or flat (disk like) molecules. We briefly mention some of the liquid crystalline orders.

Nematic

Nematic liquid crystals (NLC) show purely orientational order. Molecules align locally but at large distances, the average orientation can change continuously.

Smectic

These have layered structures with molecules in each layer having uniform orientation. There are many variations of this phase depending on the orientation of these molecules from the normal to each layer.

Cholesteric

These may be looked upon as spontaneously twisted nematic layers where each layer has nematic order while the orientation of the molecules twists in the direction normal to the layers.

Biaxial Nematic

The molecules here have the symmetry of a rectangular box. There are three mutually perpendicular twofold axes (along x,y and z). Looking down any of the axis, the biaxial nematic will have the symmetry of a rectangle.

There are many other variations but for our purpose this classification will suffice as we will be only focusing nematic liquid crystals.

4.8.1 Defects in nematics

Here the order parameter describes the local axis of orientation of the molecules. The magnitude of the order parameter is related to the strength of ordering and is fixed by the temperature. (See, ref.[7] for details.) In the high temperature isotropic phase, molecules are randomly oriented, so the magnitude of the order parameter vanishes. In the low temperature nematic phase, the orientation of the order parameter can be represented by a unit vector without the arrowhead. The order parameter space is then the surface of a unit sphere with diametrically opposite points identified.

Thus $S = S^2/Z_2 \equiv RP^2$, where RP^2 is the projective plane. We can write it as a coset space by realizing that the isotropic phase (where molecules are randomly oriented so there is no orientational order even locally) is invariant under all spatial rotations, *i.e.*, the symmetry group G = SO(3). In the nematic phase with orientational order, the symmetry group of the order parameter is H = D_∞ consisting of rotations about the local axis of orientation and 180^0 rotations about axes perpendicular to the axis of orientation. The order parameter space is then

4.8. Defect structure in liquid crystals

$$S = G/H = SO(3)/D_\infty = RP^2 \qquad (4.11)$$

For classifying defects in nematics it is convenient to take S as S^2/Z_2. One can then use exact sequence to classify defects (even though S^2 is not a group, since exact sequence is valid for any general fibration). We will only give the results for various homotopy groups.

Domain walls
$\pi_0(S^2/Z_2) = 1$, so there are no domain walls.

Strings
$\pi_1(S^2/Z_2) = Z_2$. Thus, there are Z_2 string defects in nematics. As this is an Abelian group, strings are classified by Z_2 itself. This means that a string is the same (topologically) as an antistring. The string here corresponds to the nontrivial loop in S^2/Z_2 which corresponds to a path joining two diametrically opposite points on S^2. *i.e.*, the director rotates by 180^0 as we circle around the string as shown in Fig.4.9. For this reason the string here is called a strength $s = 1/2$ defect. Remember that due to the Z_2 nature of the strings, $s = 1/2$ and $s = -1/2$ line defects are topologically the same, so one can be continuously deformed into the other.

Monopoles
$\pi_2(S^2/Z_2) = Z$. One may think that there are monopoles characterized by a set of integers Z. However, it turns out that in this case there is a non trivial action of $\pi_1(S^2/Z_2) = Z_2$ on this Z. This action transforms a point defect with $n \in Z$ to the one with $-n \in Z$. (A monopole, taken around a $s = 1/2$ string defect in a closed loop, transforms into an antimonopole.) Thus n and $-n$ correspond to the same automorphism class of π_2 and correspond, topologically, to the same defect. Monopoles in nematics, therefore, are classified by the set of positive integers only.

The nontrivial aspect of monopole classification in this case can be anticipated from the following. Recall that for the case

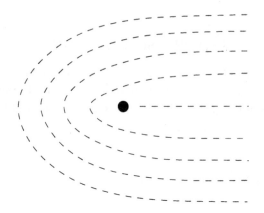

Figure 4.9: Director variation around a strength 1/2 defect.

when the order parameter space S is S^2, a monopole configuration with winding number one corresponds to radially outward vectors on S^2, while an anti-monopole corresponds to S^2 with radially inward going vectors. The only difference for nematics is that there are no arrowheads for nematics. However, without an arrowhead, there seems to be no difference in monopole and antimonopole configurations as described above.

4.8.2 Non abelian π_1 - biaxial nematics

As we mentioned, the order parameter here has the symmetry of a rectangular box. Thus if the symmetry group of the disordered state is taken to be SO(3), then H is D_2, the four element group consisting of the identity and 180^0 rotations about three mutually perpendicular axes. The order parameter space S is then $S = \mathrm{SO}(3)/D_2$. It is convenient to take G to be the universal cover of SO(3) which is SU(2). Then H is the lift of D_2 in SU(2) which is the quaternion group Q. So,

$$S = \mathrm{SU}(2)/Q \qquad (4.12)$$

where $Q == \{\pm 1, \pm i\sigma_x, \pm i\sigma_y, \pm i\sigma_z\}$, σ_i being Pauli matrices.

As SU(2) is simply connected, one can easily show that $\pi_1(S) = Q$ while $\pi_2(S)$ is trivial (since $\pi_0(H) = Q$ but $\pi_1(H)$ is trivial). This is an example of a non Abelian fundamental group and in this case strings are characterized by the conjugacy classes of Q.

4.9 Formation of topological defects

In the end, we briefly discuss how topological defects can form during a phase transition. Conventionally, defect production has been studied in two different situations. In certain condensed matter systems, it is possible to produce defects by external influence, such as in an external magnetic field (which leads to formation of flux tubes in type II superconductors), or due to moving boundaries, such as production of vortices in a rotating vessel containing superfluid helium, (or nucleation of vortex rings in superflow through a small orifice). The other method of producing topological defects is due to thermal fluctuations at temperatures close to the transition temperature (with defect density suppressed by the Boltzmann factor).

It was proposed by Kibble [8], in the context of early universe, that defect production can happen due to formation of a kind of domain structure during a phase transition with defects forming at the junctions of these domains. This is commonly referred to as *the Kibble mechanism*. Even though the Kibble mechanism was originally proposed for defect formation in the early universe, the mechanism as such has complete general applicability and, in fact, has been recently verified by studying defect formation in liquid crystal systems [9, 10]; see also [11]. Fig.4.10 shows a sequence of pictures where strings are seen to arise during isotropic to nematic phase transition in a liquid crystal system.

In the Kibble mechanism, the domain like structure arises from the fact that during phase transition, the orientation of the order parameter field can only be correlated within a finite region. The order parameter can be taken to be roughly uniform within a correlation region (domain), while varying randomly from one domain to the other. This situation is very natural to expect in a first order phase transition where the transition to the spontaneous symmetry broken phase happens via nucleation of bubbles. Inside a bubble the orientation of the order parameter (say, phase θ of ψ in Eq.(4.3)) will be uniform, while θ will vary randomly from one bubble to another. Eventually bubbles grow and coalesce, leaving a region of space where θ varies randomly at a distance scale of the inter-bubble separation, thereby leading to a domain like structure. The same situation happens for a second order transition where the orientation of the order parameter (OP) field is correlated only within a region of the size of the correlation length. This again results in a domain like structure, with domains being the correlation volumes. There are non-trivial issues in this case in determining the appropriate correlation length for determining the initial defect density. This is due to the effects of critical slowing down of the dynamics of the OP field near the transition temperature. For a discussion of these issues, see ref.[12] and references therein.

In order to determine whether a topological defect (vortex) has formed in a region, one needs the information about θ at every point on a closed path in that region. For this, one needs to know how θ will vary in between any two domains, once θ is known inside the two domains. An important ingredient in the Kibble mechanism is the assumption that in between any two adjacent domains, the OP field is supposed to vary with least gradient. This is usually called the *geodesic rule* and arises naturally from the consideration of minimizing the gradient energy (at least for the case without any gauge fields).

One can take the example of a superfluid phase transition in ^{4}He corresponding to spontaneous breaking of U(1) symmetry and characterized by a complex order parameter. GL free energy for this is given in Eq.(4.3). In the superfluid phase, magnitude of the order parameter ψ gives the degree of superfluidity, while its phase θ takes values in the order parameter space S^1, and can vary spatially over distances larger than the correlation length. In this case, string defects (superfluid vortices) arise at the junctions

4.9. Formation of topological defects

of domains, if θ winds non trivially around a closed path going through adjacent domains. Consider a junction of three domains, with the values of θ in the three domains being θ_1, θ_2 and θ_3. One can show that with the use of the geodesic rule (*i.e.*, the variation of θ in between any two domains is along the shortest path on the order parameter space S^1), a non-trivial winding of θ along a closed path, encircling the junction and going through the three domains, will arise only when θ_3 lies in the (shorter) arc between $\theta_1 + \pi$ and $\theta_2 + \pi$. Maximum and minimum values of the angular span of this arc are π and 0, with average angular span being $\pi/2$. Since θ_3 can lie anywhere in the circle, the probability that it lies in the required range is $p = (\pi/2)/(2\pi) = 1/4$. Thus, we conclude that the probability of vortex formation at a junction of three domains (in 2-space dimensions) is equal to 1/4.

Exercise 4.4 *Complete the arguments in the above estimate of probability of string formation.*

It is important to mention that in the above argument, no use is made of the field equations. Thus, whether the system is a relativistic one, or a non-relativistic one appropriate for condensed matter physics with the dynamics of the order parameter being given by time dependent Ginzburg Landau (TDGL) equations, there is no difference in the defect production (per domain). It is this universality of the prediction of defect density (number of defects per domain) in the Kibble mechanism which has been utilized to experimentally verify this prediction (which was originally given for cosmic defect production) in liquid crystal systems [9]. Note that defect density as such is not universal since it depends on the domain size. For a second order transition, domains are not easily identified (as in the case of superfluid ^{4}He [13]), making it difficult to utilize the universality of defect density (defects per domain) for a clean experimental verification of the theoretical prediction. For a first order transition proceeding via bubble nucleation, bubbles are the domains, which can be easily identified facilitating the measurement of defect density (defects per domain), as in the case of liquid crystals, see [9]. See, also, [10] for the experimental verification of another universal prediction of the Kibble mechanism relating to the defect-antidefect correlations.

Acknowledgments I would like to thank the organizers for a great school. Special thanks to all the students for their tremendous enthusiasm and truly interactive participation. I also thank Somen Bhattacharjee and Supratim Sengupta for useful comments and suggestions on this writeup.

4.9. Formation of topological defects

Figure 4.10: These pictures show how strings arise in a first order isotropic to nematic phase transition in a liquid crystal sample. Bubbles of nematic phase nucleate in the background of isotropic phase. Bubbles coalesce, leading to a dense network of strings, which subsequently coarsens due to shrinking of strings.

Bibliography

[1] N.D. Mermin, Rev. Mod. Phys. **51**, 591 (1979).

[2] "Physics of Defects", Proceedings, Les Houches Summer School in Theoretical Physics, session 35 (1980), Editors, R. Balian, M. Kleman, and J.P. Poirier, (North-Holland, 1981)

[3] Louis Michel, Rev. Mod. Phys., **52**, 617 (1980); V. Poinaru and G. Toulouse, J. Phys. (Paris), **8**, 887 (1977); H. -R Trebin, Advances in Physics, **31**, 195 (1982).

[4] "Formation and interaction of topological defects", Proceedings of the Advanced Study Institute at the Newton Institute for Mathematical Sciences, Cambridge, England (1994), Editors, A.C. Davis and R. Brandenberger.

[5] A.Vilenkin and E.P.S.Shellard, "Cosmic Strings and Other Topological Defects", (Cambridge University Press, Cambridge, 1994).

[6] C. Nash and S. Sen, "Topology and Geometry for Physicists", (Academic Press, 1983); N. Steenrod, "The Topology of Fibre bundles", (Princeton University Press, 1951); P.J. Hilton, "An Introduction to Homotopy Theory", (Cambridge University Press, 1966).

[7] P. -G deGennes, "The Physics of Liquid Crystals" (Clarendon, Oxford, 1974); S. Chandrasekhar and G.S. Ranganath, Advances in Physics, **35**, 507 (1986).

BIBLIOGRAPHY 237

[8] T.W.B. Kibble, J. Phys. A **9**, 1387 (1976).

[9] M.J. Bowick, L. Chandar, E.A. Schiff and A.M. Srivastava, Science **263**, 943 (1994).

[10] S. Digal, R. Ray, and A.M. Srivastava, Phys. Rev. Lett. **83**, 5030 (1999).

[11] I. Chuang, R. Durrer, N. Turok and B. Yurke, Science **251**, 1336 (1991).

[12] W.H. Zurek, Phys. Rep. **276**, 177 (1996), see also A. Yates and W.H. Zurek, Phys. Rev. Lett. **80**, 5477 (1998).

[13] P.C. Hendry et al., J. Low. Temp. Phys. **93**, 1059 (1993); G.E. Volovik, Czech. J. Phys. **46**, 3048 (1996) Suppl. S6; M.E. Dodd et al., Phys. Rev. Lett. **81**, 3703 (1998); G. Karra and R.J. Rivers, Phys. Rev. Lett. **81**, 3707 (1998).

[14] Salomaa and Volovik, Rev. Mod. Phys. **59**, 533 (1987).

Chapter 5

An Introduction to Bosonization and Some of its Applications

Sumathi Rao

    Harish-Chandra Research Institute
    Chhatnag Road, Jhusi
    Allahabad – 211 019

Diptiman Sen

    Centre for Theoretical Studies
    Indian Institute of Science
    Bangalore – 560 012

We discuss the technique of bosonization for studying systems of interacting fermions in one dimension. After briefly reviewing the low-energy properties of Fermi and Luttinger liquids, we present some of the relations between bosonic and fermionic operators in one dimension. We use these relations to calculate the correlation functions and the renormalization group properties of various operators for a system of spinless fermions. We then apply the methods of bosonization to study the Heisenberg antiferromagnetic spin 1/2 chain, the Hubbard model in one dimension and transport in clean quantum wires and in the presence of isolated impurities.

5.1 Fermi and Luttinger liquids

In two and three dimensions, many systems of interacting fermions at low temperatures are described by the Fermi liquid theory developed by Landau (see Ref. 1 for a brief review). According to this theory, at zero temperature, the ground state of each species of fermions has a Fermi surface in momentum space located at an energy called the Fermi energy E_F, such that all the states within that surface (*i.e.*, with energies less than E_F) are occupied while all the states outside it are unoccupied. An elementary low-energy excitation is one in which a particle is added (annihilated) in a state just outside (inside) the Fermi surface; these are called particle and hole excitations respectively. In an interacting system, these one-particle excitations are accompanied by a cloud of particle-hole pairs, and they are more commonly called quasiparticles; these carry the same charge as a single particle (or hole). If the particle number is held fixed, the low-energy excitations of the system consist of particle-hole pairs in which a certain number of particles are excited from states within the Fermi surface to states outside it. A few of these excitations have both low wave numbers and low energies with the energy being proportional to

5.1. Fermi and Luttinger liquids

the wave number; such excitations can be thought of as sound waves. But most of the particle-hole excitations do not have such a linear relationship between energy and wave number; in fact, for most such excitations, a given energy can correspond to many possible momenta. Another interesting property of a Fermi liquid

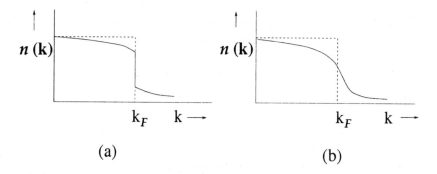

Figure 5.1: One-particle momentum distribution function. (a) shows the finite discontinuity at the Fermi momentum $\mathbf{k}_F$ for a system of interacting fermions in more than one dimension. (b) shows the absence of a discontinuity in an interacting system in one dimension.

in two and three dimensions is that the one-particle momentum distribution function $n(\mathbf{k})$, obtained by Fourier transforming the one-particle equal-time correlation function, has a finite discontinuity at the Fermi surface as shown in Fig.5.1(a). This discontinuity is called the quasiparticle renormalization factor $z_\mathbf{k}$; it is also equal to the residue of the pole in the one-particle propagator. For non-interacting fermions, $z_\mathbf{k} = 1$; but for interacting fermions, $0 < z_\mathbf{k} < 1$ because a quasiparticle is a superposition of many states, only some of which are one-particle excitations. To compute $z_\mathbf{k}$, we consider the one-particle Green's function $G(\mathbf{x}, t)$ defined as the expectation value of the time-ordered product of the fermion operator $\psi(\mathbf{x}, t)$ in the ground state $|0\rangle$, namely,

$$G(\mathbf{x}, t) = \langle 0| T\psi(\mathbf{x}, t)\psi^\dagger(\mathbf{0}, 0) |0\rangle \ . \qquad (5.1)$$

(We will ignore the spin label here). The Fourier transform of this function can be written as

$$\mathcal{G}(\mathbf{k},\omega) = \frac{i}{\omega - \epsilon_\mathbf{k} - \Sigma(\mathbf{k},\omega)}, \qquad (5.2)$$

where $\epsilon_\mathbf{k}$ is the dispersion relation for the non-interacting theory; we absorb the chemical potential μ in the definition of $\epsilon_\mathbf{k}$ so that $\epsilon_\mathbf{k} = 0$ for $\mathbf{k}$ lying on the Fermi surface. (We will set $\hbar = 1$). The self-energy $\Sigma(\mathbf{k},\omega)$ contains the effects of all the interactions as well as any prescription necessary to shift the pole slightly off the real axis in ω. For a Fermi liquid, $\mathcal{G}(\mathbf{k},\omega)$ has a pole near the real axis of ω for any value of $\mathbf{k}$ on the Fermi surface. In addition, Σ is sufficiently analytic at all such points so that the derivative $\partial \Sigma/\partial \omega$ has a finite value. The quasiparticle renormalization factor $z_\mathbf{k}$ is then given by the residue at the pole, *i.e.*,

$$z_\mathbf{k} = \left(1 - \frac{\partial \Sigma}{\partial \omega}\right)^{-1}. \qquad (5.3)$$

This gives the discontinuity in $n(\mathbf{k})$ at the Fermi surface.

Finally, in a Fermi liquid, the various correlation functions decay asymptotically at long distances as power laws, with the exponents being independent of the strength of the interactions. Thus non-interacting and interacting systems have the same exponents and there is a universality.

The discussion above does not apply if the ground state of the system spontaneously breaks some symmetry, for instance, if it is superconducting, or forms a crystal or develops charge or spin density ordering.

In contrast to a Fermi liquid, interacting fermion systems in one dimension behave quite differently [2, 1, 3]; we will assume again that the ground state breaks no symmetry. Such systems are called Luttinger liquids and they have the following general properties. First of all, there are no single particle or quasiparticle excitations. Thus *all* the low-energy excitations can be thought of as particle-hole excitations; further, all of these take the form of sound waves with a linear dispersion relation. (As we will see

5.1. Fermi and Luttinger liquids

below, there are also excitations of another kind possible which correspond to adding a small number of particles N_R and N_L to the right and left Fermi points. However, these correspond to only two oscillator degrees of freedom, and therefore do not contribute to thermodynamic properties like the specific heat). Secondly, there is no discontinuity in the momentum distribution function at the Fermi momentum, as indicated in Fig.5.1(b). Rather, there is a cusp there whose form is determined by a certain exponent. Finally, this exponent depends on the strength of the interactions in a non-universal manner, and it also governs the power-law fall-offs of the correlation functions at large space-time distances [4].

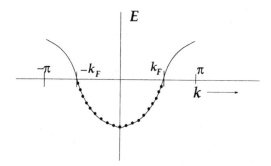

Figure 5.2: Picture of the Fermi sea of a lattice model; the momentum lies in the range $[-\pi, \pi]$. The occupied states (filled circles) below the Fermi energy $E_F = 0$ and the two Fermi points at momenta $\pm k_F$ are shown.

Let us be more specific about the nature of the low-energy excitations in a one-dimensional system of fermions. Assume that we have a system of length L with a boundary condition to be specified later. The translation invariance and the finite length make the one-particle momenta discrete. Suppose that the system has N_0 particles with a ground state energy $E_0(N_0)$ and a ground state momentum $P_0 = 0$; we are assuming that the system conserves parity. We will be interested in the thermodynamic limit $N_0, L \to \infty$ keeping the particle density $\rho_0 = N_0/L$ fixed.

Let us first consider a single species of non-interacting fermions which have two possible directions of motion, right-moving with $d\epsilon_k/dk = v_F$ and left-moving with $d\epsilon_k/dk = -v_F$. Here ϵ_k is the energy of a low-lying one-particle excitation, k is its momentum measured with respect to a right Fermi momentum k_F and a left Fermi momentum $-k_F$ respectively, and v_F is called the Fermi velocity. (See Fig.5.2) for a typical picture of the momentum states of a lattice model). The values of k_F and v_F are defined for the non-interacting system; hence they depend on the density ρ_0 but not on the strength of the interaction. Then a low-lying excitation consists of two pieces [2],

(i) a set of bosonic excitations each of which can have either positive momentum q or negative momentum $-q$ with an energy $\epsilon_q = v_F q$, where $0 < q << k_F$, and

(ii) a certain number of particles N_R and N_L added to the right and left Fermi points respectively, where $N_R, N_L << N_0$. (Note that N_R and N_L can be positive, negative or zero. It is convenient to assume that $N_R \pm N_L$ are even integers; then the total number of particles $N_0 + N_R + N_L$ is always even or always odd. We can choose the boundary condition (periodic or antiperiodic) to ensure that the ground state is always non-degenerate).

It turns out that the Hamiltonian and momentum operators for a one-dimensional system (which may have interactions) have the general form

$$H = E_0(N_0) + \sum_{q>0} vq \, [\, \tilde{b}^\dagger_{R,q}\tilde{b}_{R,q} + \tilde{b}^\dagger_{L,q}\tilde{b}_{L,q} \,]$$

$$+ \mu(N_R + N_L) + \frac{\pi v}{2LK}(N_R + N_L)^2 + \frac{\pi v K}{2L}(N_R - N_L)^2,$$

$$P = \sum_{q>0} q \, [\, \tilde{b}^\dagger_{R,q}\tilde{b}_{R,q} - \tilde{b}^\dagger_{L,q}\tilde{b}_{L,q} \,]$$

$$+ \left[\, k_F + \frac{\pi}{L}(N_R + N_L) \,\right](N_R - N_L), \qquad (5.4)$$

where v is the sound velocity, q is the momentum of the low-energy bosonic excitations created and annihilated by $\tilde{b}^\dagger_q$ and $\tilde{b}_q$, K is a positive dimensionless number, and μ is the chemical potential of

5.1. Fermi and Luttinger liquids

the system. We will see later that v and K are the two important parameters which determine all the low-energy properties of a system. Their values generally depend on both the strength of the interactions and the density. If the fermions are non-interacting, we have

$$v = v_F \quad \text{and} \quad K = 1 \; . \tag{5.5}$$

Note that one can numerically find the values of v and K by studying the $1/L$ dependence of the low-energy excitations of finite size systems.

It is interesting that the expression for the momentum operator in Eq. (5.4) is independent of the interaction strength. We can understand the last term in the momentum as follows. For a continuum system, the Fermi momentum $k_F(N)$ is related to the density by the relation

$$L \int_{-k_F(N)}^{k_F(N)} \frac{dk}{2\pi} = N \; . \tag{5.6}$$

Thus a system of N_0 particles has a Fermi momentum

$$k_F = \frac{\pi N_0}{L} = \pi \rho_0 \; , \tag{5.7}$$

while a system of $N = N_0 + N_R + N_L$ particles has a Fermi momentum equal to $k_F + (\pi/L)(N_R + N_L)$. If the N particles occupy the momenta states symmetrically about zero momentum, the total momentum of that state is zero; in this state, both the right and left Fermi points have $(N_R+N_L)/2$ particles more than the original ground state. Now let us shift $(N_R - N_L)/2$ particles from the left Fermi point to the right Fermi point, so that the right Fermi point has N_R particles more and the left Fermi point has N_L particles more than the original system. We then see that the total momentum has changed from zero to $[k_F + (\pi/L)(N_R + N_L)](N_R - N_L)$; this is the last term in the expression for the momentum operator.

The form of the parameterization of the last two terms in the Hamiltonian in Eq. (5.4) can be understood as follows. (Note that these two terms vanish in the thermodynamic limit and do not contribute to the specific heat. However they are required for the completeness of the theory up to terms of order $1/L$, and for a comparison with conformal field theory). Specifically, we will prove that if the coefficients of $(\pi/2L)(N_R + N_L)^2$ and of $(\pi/2L)(N_R - N_L)^2$ in Eq. (5.4) are denoted by A and B respectively, then

$$AB = v^2 . \tag{5.8}$$

It will then follow that if A is equal to v/K, B must be equal to vK. Although the expressions in Eq. (5.4) are valid for lattice models also, let us for simplicity consider a continuum model which is invariant under Galilean transformations. First, let us set $N_R = N_L$, so that we have added $\Delta N = 2N_R$ particles to the system. The sound velocity v of a one-dimensional system is related to the density of particles $\rho = N/L$ (where $N = N_0 + N_R + N_L$), the particle mass m, and the pressure $\mathcal{P}$ as

$$m\rho v^2 = - L \left(\frac{\partial \mathcal{P}}{\partial L} \right)_N . \tag{5.9}$$

The pressure is related to the ground state energy by $\mathcal{P} = -(\partial E_0/\partial L)_N$. Hence

$$m\rho v^2 = - L \left(\frac{\partial^2 E_0}{\partial L^2} \right)_N = \frac{N^2}{L} \left(\frac{\partial^2 E_0}{\partial N^2} \right)_L , \tag{5.10}$$

where the second equality follows from the first because E_0 depends on N and L only through the combination N/L. Comparing Eqs. (5.4) and (5.10), we see that the coefficient of $(\pi/2L)(\Delta N)^2$ is given by

$$A = \frac{mv^2}{\pi \rho_0} . \tag{5.11}$$

(In certain expressions such as Eq. (5.11), we have ignored the difference between ρ and ρ_0 since $\Delta N \ll N_0$). Next, let us take $N_L = -N_R$; this corresponds to moving N_R particles from the left Fermi point $-k_F$ to the right Fermi point k_F keeping the total number of particles equal to N_0. The change in momentum is therefore given by $\Delta P = 2\pi\rho_0 N_R$. Since we can also view such an excitation as a center of mass excitation with momentum ΔP, the change in energy is given by $\Delta E = (\Delta P)^2/(2mN)$ since the total mass of the system is mN. It follows from this that the coefficient of $(\pi/2L)(N_R - N_L)^2$ satisfies

$$B = \frac{\pi \rho_0}{m} . \qquad (5.12)$$

We thus see that $AB = v^2$ independently of the nature of the interactions between the particles.

We now consider the other important property of a Luttinger liquid, namely, the absence of a discontinuity in $n(k)$ at the Fermi momenta or, equivalently, the absence of a pole in the one-particle propagator. Thus the effect of interactions is so drastic in one dimension that the self-energy Σ in Eq. (5.2) becomes non-analytic at the Fermi points. As a result, $n(k)$ becomes continuous at $k = \pm k_F$ with the form

$$n(k) = n(k_F) + \text{constant} \times \text{sign}(k - k_F) |k - k_F|^\beta , \qquad (5.13)$$

where $\text{sign}(z) \equiv 1$ if $z > 0$, -1 if $z < 0$ and 0 if $z = 0$. The exponent β is a positive number whose value depends on the strength of the interactions; for a non-interacting system, $\beta = 0$ and we recover the discontinuity in $n(k)$. Similarly, the density of states (DOS) is obtained by integrating Eq. (5.2) over all momenta; near zero energy it vanishes with a power-law form

$$\tilde{n}(\omega) \sim |\omega|^\beta , \qquad (5.14)$$

which signals the absence of one-particle states in the low-energy spectrum. We will see later how the exponent β can be calculated in an interacting system called the Tomonaga-Luttinger model.

5.2 Bosonization

The basic idea of bosonization is that there are certain objects which can be calculated either in a fermionic theory or in a bosonic theory, and the two calculations give the same answer [2, 1, 3, 5, 7, 8]. Further, a particular quantity may seem very difficult to compute in one theory and may be easily calculable in the other theory. Bosonization works best in two space-time dimensions although there have been some attempts to extend it to higher dimensions.

In two dimensions, bosonization can be studied in either real time (Minkowski space) or in imaginary time (Euclidean space). In both cases, there is a one-to-one correspondence between the correlation functions of some fermionic and bosonic operators. We will work in real time here because bosonization has an added advantage in that case, namely, that there is a direct relationship between the creation and annihilation operators for a boson in terms of the corresponding operators for a fermion [8]. To show this, we just need to consider a bosonic and a fermionic Fock space. A Hamiltonian is *not* needed at this stage; we need to introduce a Hamiltonian only when discussing interactions and time-dependent correlation functions.

5.2.1 Bosonization of a fermion with one chirality

Let us begin by considering just one component, say, right-moving, of a single species of fermions on a circle of length L with the following boundary condition on the one-particle wave functions $\tilde{\psi}(x)$,

$$\tilde{\psi}(L) = e^{-i\pi\sigma} \, \tilde{\psi}(0) \ . \tag{5.15}$$

Thus $\sigma = 0$ and 1 correspond to periodic and antiperiodic boundary conditions, but any value of σ lying in the range $0 \leq \sigma < 2$ is allowed in principle. (If we assume that the particles are charged, then $\pi\sigma$ can be identified with an Aharonov-Bohm phase and can be varied by changing the magnetic flux through the circle). The normalized one-particle wave functions are then given by

$$\begin{aligned}
\tilde{\psi}_{n_k} &= \frac{1}{\sqrt{L}} \, e^{ikx} \ , \\
k &= \frac{2\pi}{L} \left(n_k - \frac{\sigma}{2} \right) \ ,
\end{aligned} \tag{5.16}$$

5.2. Bosonization

where $n_k = 0, \pm 1, \pm 2, ...$ is an integer. We now introduce a second quantized Fermi field

$$\psi_R(x) = \frac{1}{\sqrt{L}} \sum_{k=-\infty}^{\infty} c_{R,k} \, e^{ikx} , \tag{5.17}$$

where the subscript R stands for right-moving, and

$$\{c_{R,k}, c_{R,k'}\} = 0 , \quad \text{and} \quad \{c_{R,k}, c_{R,k'}^\dagger\} = \delta_{kk'} . \tag{5.18}$$

Using the identity

$$\sum_{n=-\infty}^{\infty} e^{iny} = 2\pi \sum_{m=-\infty}^{\infty} \delta(y - 2\pi m) , \tag{5.19}$$

we obtain

$$\begin{aligned}\{\psi_R(x), \psi_R(x')\} &= 0, \quad \text{and} \\ \{\psi_R(x), \psi_R^\dagger(x')\} &= \delta(x - x') \quad \text{for} \quad 0 \leq x, x' \leq L.\end{aligned} \tag{5.20}$$

We define the vacuum or Fermi sea of the system to be the state $|0\rangle$ satisfying

$$\begin{aligned} c_{R,k} |0\rangle &= 0 \quad \text{for} \quad k > 0 , \\ c_{R,k}^\dagger |0\rangle &= 0 \quad \text{for} \quad k \leq 0 , \end{aligned} \tag{5.21}$$

as shown in Fig.5.3. (Following this definition of the vacuum state, some people prefer to write the particle annihilation operator $c_{R,k}$ as a hole creation operator $d_{R,-k}^\dagger$ for $k \leq 0$). Given any operator A which can be written as a product of a string of c's and $c^\dagger$'s, we denote its normal ordered form by the symbol $:A:$. This new operator is defined by moving all the c_k with $k > 0$ and $c_k^\dagger$ with $k \leq 0$ to the right of all the c_k with $k \leq 0$ and $c_k^\dagger$ with $k > 0$. This is achieved by transposing as many pairs of creation and annihilation operators as necessary, remembering to multiply by a factor of -1 for each transposition. (It is sometimes claimed that $:A: = A - \langle 0|A|0\rangle$. This is true if A is quadratic in the c's and $c^\dagger$'s, but it is not true in general).

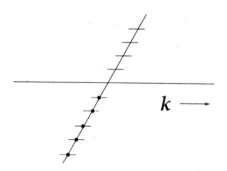

Figure 5.3: The one-particle states of a right-moving fermion showing the occupied states (filled circles) below zero energy and the unoccupied states above zero energy.

Next we define the fermion number operator

$$\hat{N}_R = \sum_{k=-\infty}^{\infty} : c^\dagger_{R,k} c_{R,k} : \; = \sum_{k>0} c^\dagger_{R,k} c_{R,k} - \sum_{k\leq 0} c_{R,k} c^\dagger_{R,k} . \quad (5.22)$$

Thus $\hat{N}_R |0\rangle = 0$. Now consider all possible states $|\Psi\rangle$ satisfying $\hat{N}_R |\Psi\rangle = 0$. Clearly, any such state can only differ from $|0\rangle$ by a certain number of particle-hole excitations, *i.e.*, it must be of the form

$$|\Psi\rangle = c^\dagger_{R,k_1} c_{R,k_2} c^\dagger_{R,k_3} c_{R,k_4} c^\dagger_{R,k_5} c_{R,k_6} \ldots |0\rangle , \quad (5.23)$$

where the k_i are all different from each other, $k_1, k_3, \ldots > 0$, and $k_2, k_4, \ldots \leq 0$. Two such excitations are shown in Fig.5.4. We will now see that all such states can be written in terms of certain bosonic creation operators acting on the vacuum. Let us define the operators

$$b^\dagger_{R,q} = \frac{1}{\sqrt{n_q}} \sum_{k=-\infty}^{\infty} c^\dagger_{R,k+q} c_{R,k} ,$$

$$b_{R,q} = \frac{1}{\sqrt{n_q}} \sum_{k=-\infty}^{\infty} c^\dagger_{R,k-q} c_{R,k} ,$$

$$q = \frac{2\pi}{L} n_q , \quad (5.24)$$

5.2. Bosonization

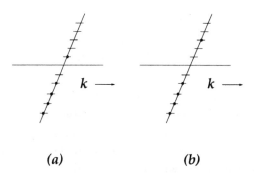

(a) (b)

Figure 5.4: Two possible particle-hole excitations of a right-moving fermionic system showing the occupied states.

where $n_q = 1, 2, 3, \ldots$. Note that we have defined the boson momentum label q to be positive. Also, the fermion boundary condition parameter σ does not appear in the definitions in Eq. (5.24). We can check that

$$\begin{aligned}
{[\hat{N}_R, b_{R,q}]} &= [\hat{N}_R, b_{R,q}^\dagger] = 0 \;, \\
[b_{R,q}, b_{R,q'}] &= 0 \;, \\
[b_{R,q}, b_{R,q'}^\dagger] &= \delta_{qq'} \;.
\end{aligned} \quad (5.25)$$

Checking the last identity for $q = q'$ is slightly tricky due to the presence of an infinite number of fermion momenta k. One way to derive the commutators is to multiply each c_k and $c_k^\dagger$ by a factor of $\exp[-\alpha|k|/2]$ in Eq. (5.24), and to let $\alpha \to 0$ at the end of the calculation. We should emphasize that the length scale α is not to be thought of as a short-distance cut-off like a lattice spacing; if we had introduced a lattice, the number of fermion modes would have been finite, and the bosonization formulas in Eq. (5.24) would not have given the correct commutation relations.

We see that the vacuum defined above satisfies $b_{R,q}|0\rangle = 0$ for all q. If we consider any operator A consisting of a string of b's and $b^\dagger$'s, we can define its bosonic normal ordered form $: A :$ by taking all the b_q's to the right of all the $b_q^\dagger$'s by suitable transpositions. Given an operator A which can be written in terms of either fermionic or bosonic operators, normal ordering it in the fermionic and bosonic ways do not always give the same result. However, it will always be clear from the context which normal ordering we mean.

We can now begin to understand why bosonization works. First of all, note that there is a one-to-one correspondence between the particle-hole excitations described in Eq. (5.23) and the bosonic excitations created by the $b^\dagger$'s [9]. For instance, consider a bosonic excitation in which states with the momenta labeled by the integers $n_1 \geq n_2 \geq ... \geq n_j > 0$ (following the convention in Eq. (5.24)) are excited. Some of these integers may be equal to each other; that would mean that particular momenta has an occupation number greater than 1. Now we can map this excitation to a fermionic excitation in which j fermions occupying the states labeled by the momenta integers $0, -1, -2, ..., -j+1$ (following the convention in Eq. (5.16)) are excited to momenta labeled by $n_1, n_2 - 1, n_3 - 2, ..., n_j - j + 1$ respectively. This is clearly a one-to-one map, and we can reverse it to uniquely obtain a bosonic excitation from a given fermionic excitation. This mapping allows us to show, once an appropriate Hamiltonian is defined, that thermodynamic quantities like the specific heat are identical in the fermionic and bosonic models.

The above mapping makes it plausible, although it requires more effort to prove, that *all* particle-hole excitations can be produced by combinations of $b^\dagger$'s acting on the vacuum. For instance, the state in Fig.5.4(a) is given by $b_{R,1}^\dagger|0\rangle$. However the state in Fig.5.4(b) has a more lengthy expression in terms of bosonic operators, although it is also a single particle-hole excitation just like (a); to be explicit, it is given by the linear combination $(1/6)[2b_{R,3}^\dagger + 3b_{R,2}^\dagger b_{R,1}^\dagger + (b_{R,1}^\dagger)^3]|0\rangle$.

Next, we define bosonic field operators and show that some bilinears in fermionic fields, such as the density $\rho_R(x)$, have simple expressions in terms of bosonic fields. Define the fields

$$\chi_R(x) = \frac{i}{2\sqrt{\pi}} \sum_{q>0} \frac{1}{\sqrt{n_q}} b_{R,q} \, e^{iqx - \alpha q/2} \, ,$$

5.2. Bosonization

$$\chi_R^\dagger(x) = -\frac{i}{2\sqrt{\pi}} \sum_{q>0} \frac{1}{\sqrt{n_q}} b_{R,q}^\dagger \, e^{-iqx-\alpha q/2},$$

$$\phi_R(x) = \chi_R(x) + \chi_R^\dagger(x) - \frac{\sqrt{\pi}x}{L}\hat{N}_R. \quad (5.26)$$

The last term in the definition of $\phi_R(x)$ has been put in for later convenience; it simplifies the expressions for the Hamiltonian and the fermion density in terms of ϕ_R. (Some authors prefer not to include that term in the definition of ϕ_R but add it separately in the Hamiltonian and density). Note that $\hat{N}_R$ commutes with both χ_R and $\chi_R^\dagger$. From the commutation relations in Eq. (5.25), we see that

$$[\chi_R(x), \chi_R(x')] = 0,$$
$$[\chi_R(x), \chi_R^\dagger(x')] = -\frac{1}{4\pi} \ln[1 - \exp(-\frac{2\pi}{L}(\alpha + i(x-x')))],$$
$$= -\frac{1}{4\pi} \ln[\frac{2\pi}{L}(\alpha + i(x-x'))]$$

in the limit $L \to \infty$. $\quad (5.27)$

Henceforth, the limit $L \to \infty$ will be assumed wherever convenient. We find that

$$[\phi_R(x), \phi_R(x')] = \frac{1}{4\pi} \ln\left[\frac{\alpha - i(x-x')}{\alpha + i(x-x')}\right],$$
$$= -\frac{i}{4} \,\text{sign}\,(x-x')$$

in the limit $\alpha \to 0$. $\quad (5.28)$

Thus the commutator of two ϕ's looks like a step function which is smeared over a region of length α.

Now we use the operator identity

$$\exp A \, \exp B = \exp\,(A + B + \frac{1}{2}[A,B]), \quad (5.29)$$

if $[A,B]$ commutes with both A and B. It follows that

$$\exp\,[i2\sqrt{\pi}\chi_R^\dagger(x)] \quad \exp\,[i2\sqrt{\pi}\chi_R(x)] \quad \exp\,[i\frac{2\pi x}{L}\hat{N}_R]$$
$$= \left(\frac{L}{2\pi\alpha}\right)^{1/2} \exp\,[i2\sqrt{\pi}\phi_R(x)]. \quad (5.30)$$

We observe that the left hand side of this equation is normal ordered while the right hand side is not; that is why the two sides are related through a divergent factor involving L/α.

We can show that the fermion density operator is linear in the bosonic field, namely,

$$\begin{aligned}\rho_R(x) &= :\psi_R^\dagger(x)\psi_R(x): \\ &= \frac{1}{L}\sum_{q>0}\sqrt{n_q}\,(\,b_{R,q}e^{iqx}+b_{R,q}^\dagger e^{-iqx}\,) \\ &+ \frac{1}{L}\sum_k :c_{R,k}^\dagger c_{R,k}: = -\frac{1}{\sqrt{\pi}}\frac{\partial\phi_R}{\partial x}.\end{aligned} \quad (5.31)$$

We now go in the opposite direction and construct fermionic field operators from bosonic ones. To do this, we first define the Klein factors η_R and $\eta_R^\dagger$ which are unitary operators satisfying

$$\begin{aligned}[\hat{N}_R,\eta_R^\dagger] &= \eta_R^\dagger,\quad [\hat{N}_R,\eta_R]=-\eta_R, \\ [\eta_R,b_{R,q}] &= [\eta_R,b_{R,q}^\dagger]=0.\end{aligned} \quad (5.32)$$

Pictorially, in terms of Figs.5.3 and 5.4, the action of $\eta_R^\dagger$ is to raise all the occupied fermion states by one unit of momentum, while the action of η_R is to lower all the fermion occupied states by one unit of momentum. Although these actions are easy to describe in words, the explicit expressions for η_R and $\eta_R^\dagger = \eta_R^{-1}$ in terms of the c's and $c^\dagger$'s are rather complicated[8]. The Klein factors will be needed to ensure the correct anticommutation relations between the fermionic operators constructed below.

We observe that

$$\begin{aligned}[b_{R,q},\psi_R(x)] &= -\frac{e^{-iqx}}{\sqrt{n_q}}\,\psi_R(x), \\ [b_{R,q}^\dagger,\psi_R(x)] &= -\frac{e^{iqx}}{\sqrt{n_q}}\,\psi_R(x).\end{aligned} \quad (5.33)$$

Since $b_{R,q}$ annihilates the vacuum, we have

$$b_{R,q}\,\psi_R(x)\,|0\rangle = -\frac{e^{-iqx}}{\sqrt{n_q}}\,\psi_R(x)\,|0\rangle. \quad (5.34)$$

5.2. Bosonization

Thus $\psi_R(x)|0\rangle$ is an eigenstate of $b_{R,q}$ for every value of q, namely, it is a coherent state. We therefore make the ansatz

$$\psi_R(x)|0\rangle = Q(x) \exp\left[-\sum_{q>0} \frac{e^{-iqx}}{\sqrt{n_q}} b_{R,q}^\dagger |0\rangle\right]|0\rangle,$$

$$= Q(x) \exp\left[-i2\sqrt{\pi}\,\chi_R^\dagger(x)\right]|0\rangle, \qquad (5.35)$$

where $Q(x)$ is some operator which commutes with all the b's and $b^\dagger$'s. Since ψ_R reduces the fermion number by one, Q must contain a factor of η_R. Let us try the form $Q(x) = F(x)\eta_R$, where $F(x)$ is a c-number function of x. The form of F is determined by computing

$$\begin{aligned} F(x) &= \langle 0|\,\eta_R^\dagger \eta_R F(x)\,|0\rangle \\ &= \langle 0|\,\eta_R^\dagger \psi_R(x)\,|0\rangle \\ &= \frac{e^{-i\pi\sigma x/L}}{\sqrt{L}}. \end{aligned} \qquad (5.36)$$

(The last line in Eq. (5.36) has been derived by using the actions of $\eta_R^\dagger$ above and of ψ_R in Eq. (5.17). To see this explicitly, note that $\langle 0|\eta_R^\dagger$ is the conjugate of the state in which the top most fermion has been removed from the vacuum. Hence, in $\psi_R|0\rangle$, we only have to consider the state in which the top most fermion has been removed; so we require the wave function of the state with $n_k = 0$ in Eq. (5.16)). We now obtain

$$\psi_R(x)|0\rangle = \frac{e^{-i\pi\sigma x/L}}{\sqrt{2\pi\alpha}}\,\eta_R\,e^{-i2\sqrt{\pi}\phi_R(x)}|0\rangle, \qquad (5.37)$$

where we have used Eq. (5.30) and the fact that $\chi_R(x)$ and $\hat{N}_R$ annihilate the vacuum. We are thus led to the plausible conjecture

$$\psi_R(x) = \frac{e^{-i\pi\sigma x/L}}{\sqrt{2\pi\alpha}}\,\eta_R\,e^{-i2\sqrt{\pi}\phi_R(x)}. \qquad (5.38)$$

To prove this, we need to show that the two sides of this equation have the same action on *all* states, not just the vacuum. Such a proof is given in Ref. [7]. Eq. (5.38) is one of the most important identities in bosonization.

We next introduce a non-interacting Hamiltonian by defining the energy of the fermion mode with momentum k to be

$$\epsilon_k = v_F k \qquad (5.39)$$

for all values of k. The Hamiltonian is

$$\begin{aligned} H_0 &= v_F \sum_{k=-\infty}^{\infty} k \; :c_{R,k}^\dagger c_{R,k}: \; + \frac{\pi v_F}{L} \hat{N}_R^2 \\ &= -v_F \int_0^L dx \; :\psi_R^\dagger i \partial_x \psi_R: \; + \frac{\pi v_F}{L} \hat{N}_R^2 \; . \end{aligned} \qquad (5.40)$$

This defines the chiral Luttinger model. (The term proportional to $\hat{N}_R^2$ has been introduced in Eq. (5.40) so as to reproduce similar terms in Eq. (5.4) after we introduce left-moving fields in the next section). We can check that $H_0|0\rangle = 0$ and

$$\begin{aligned} [H_0, b_{R,q}] &= -v_F q \, b_{R,q} \; , \\ [H_0, b_{R,q}^\dagger] &= v_F q \, b_{R,q}^\dagger \; , \end{aligned} \qquad (5.41)$$

To reproduce these relations in the bosonic language, we must have

$$\begin{aligned} H_0 &= v_F \sum_{q>0} q b_{R,q}^\dagger b_{R,q} + \frac{\pi v_F}{L} \hat{N}_R^2 \\ &= v_F \int_0^L dx \; :(\partial_x \phi)^2: \; . \end{aligned} \qquad (5.42)$$

We can introduce an interaction in this model which is quadratic in the fermion density. Let us consider the interaction

$$V = \frac{1}{2} \int_0^L g_4 \, \rho_R^2(x) = \frac{g_4}{2\pi} \sum_{q>0} q \, b_{R,q}^\dagger b_{R,q} \; . \qquad (5.43)$$

Physically, such a term could arise if there is a short-range (i.e., screened) Coulomb repulsion or a phonon mediated attraction between two fermions. We will therefore not make any assumptions

5.2. Bosonization

about the sign of the interaction parameter g_4. If we add Eq. (5.43) to Eq. (5.42), we see that the only effect of the interaction in this model is to renormalize the velocity from v_F to $v_F+(g_4/2\pi)$.

In the next section, we will consider a model containing fermions with opposite chiralities; we will then see that a density-density interaction can have more interesting effects than just renormalizing the velocity.

5.2.2 Bosonisation of a fermion with two chiralities

Let us consider a fermion with both right- and left-moving components as depicted in Figs. 5.3 and 5.5 respectively. For the left-moving fermions in Fig.5.5, we define the momentum label k as increasing towards the left; the advantage of this choice is that the vacuum has the negative k states occupied and the positive k states unoccupied for both chiralities. We introduce a chirality label ν, such that $\nu = R$ and L refer to right- and left-moving particles respectively. Sometimes we will use the numerical values $\nu = 1$ and -1 for R and L; this will be clear from the context. Let us choose periodic boundary conditions on the circle so that $\sigma = 0$. Then the Fermi fields are given by

$$\psi_\nu(x) = \frac{1}{\sqrt{L}} \sum_{k=-\infty}^{\infty} c_{\nu,k}\, e^{i\nu k x},$$

$$k = \frac{2\pi}{L} n_k, \qquad (5.44)$$

where $n_k = 0, \pm 1, \pm 2, ...,$ and

$$\{c_{\nu,k}, c_{\nu',k'}\} = 0,$$
$$\{c_{\nu,k}, c^\dagger_{\nu',k'}\} = \delta_{\nu\nu'}\, \delta_{kk'}. \qquad (5.45)$$

The vacuum is defined as the state satisfying

$$c_{\nu,k}\,|0\rangle = 0 \quad \text{for} \quad k > 0,$$
$$c^\dagger_{\nu,k}\,|0\rangle = 0 \quad \text{for} \quad k \leq 0. \qquad (5.46)$$

258   5. Introduction to Bosonization

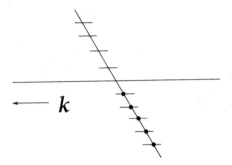

Figure 5.5: The one-particle states of a left-moving fermion showing the occupied states below zero energy and unoccupied states above zero energy. Note that the momentum label k increases towards the left.

We can then define normal ordered fermion number operators $\hat{N}_\nu$ in the usual way.

Next we define bosonic operators

$$\begin{aligned} b^\dagger_{\nu,q} &= \frac{1}{\sqrt{n_q}} \sum_{k=-\infty}^{\infty} c^\dagger_{\nu,k+q} c_{\nu,k} \,, \\ b_{\nu,q} &= \frac{1}{\sqrt{n_q}} \sum_{k=-\infty}^{\infty} c^\dagger_{\nu,k-q} c_{\nu,k} \,. \end{aligned} \quad (5.47)$$

Note that $b^\dagger_{R,q}$ and $b^\dagger_{L,q}$ create excitations with momenta q and $-q$ respectively, where the label q is always taken to be positive. We can show as before that

$$[b_{\nu,q}, b_{\nu',q'}] = 0 \,, \quad \text{and} \quad [b_{\nu,q}, b^\dagger_{\nu',q'}] = \delta_{\nu\nu'} \delta_{qq'} \,. \quad (5.48)$$

The unitary Klein operators η_ν ($\eta_\nu^\dagger$) are defined to be operators which raise (lower) the momentum label k of all the occupied states for fermions of type ν. We then have

$$\begin{aligned} \{\eta_R, \eta_L\} &= \{\eta_R, \eta_L^\dagger\} = 0 \,, \\ [\hat{N}_\nu, \eta^\dagger_{\nu'}] &= \delta_{\nu\nu'} \, \eta^\dagger_{\nu'} \,, \quad [\hat{N}_\nu, \eta_{\nu'}] = -\delta_{\nu\nu'} \, \eta_{\nu'} \,, \\ [\eta_\nu, b_{\nu',q}] &= [\eta_\nu, b^\dagger_{\nu',q}] = 0 \,. \end{aligned} \quad (5.49)$$

5.2. Bosonization

We now define the chiral creation and annihilation fields

$$\chi_\nu(x) = \frac{i\nu}{2\sqrt{\pi}} \sum_{q>0} \frac{1}{\sqrt{n_q}} b_{\nu,q}\, e^{i\nu qx - \alpha q/2},$$

$$\chi_\nu^\dagger(x) = -\frac{i\nu}{2\sqrt{\pi}} \sum_{q>0} \frac{1}{\sqrt{n_q}} b_{\nu,q}^\dagger\, e^{-i\nu qx - \alpha q/2}. \quad (5.50)$$

Then

$$[\chi_\nu(x), \chi_{\nu'}(x')] = 0,$$

$$[\chi_\nu(x), \chi_{\nu'}^\dagger(x')] = -\frac{1}{4\pi} \delta_{\nu\nu'} \ln\left[\frac{2\pi}{L}(\alpha - i\nu(x-x'))\right]$$

in the limit $L \to \infty$. (5.51)

The chiral fields

$$\phi_\nu(x) = \chi_\nu(x) + \chi_\nu^\dagger(x) - \frac{\sqrt{\pi}x}{L}\hat{N}_\nu, \quad (5.52)$$

satisfy

$$[\phi_\nu(x), \phi_{\nu'}(x)] = -\frac{i\nu}{4}\delta_{\nu\nu'}\, \text{sign}\,(x-x') \quad (5.53)$$

in the limit $\alpha \to 0$. Finally, we can define two fields dual to each other

$$\phi(x) = \phi_R(x) + \phi_L(x),$$
$$\theta(x) = -\phi_R(x) + \phi_L(x), \quad (5.54)$$

such that $[\phi(x), \phi(x')] = [\theta(x), \theta(x')] = 0$, while

$$[\phi(x), \theta(x')] = \frac{i}{2}\,\text{sign}\,(x-x'). \quad (5.55)$$

The fermion density operators $\rho_\nu(x) =: \psi_\nu^\dagger(x)\psi_\nu(x):$ satisfy $\rho_\nu = \partial_x \phi_\nu / \sqrt{\pi}$. Hence the total density and current operators are given by

$$\rho(x) = \rho_R + \rho_L = -\frac{1}{\sqrt{\pi}} \partial_x \phi,$$

$$j(x) = v_F(\rho_R - \rho_L) = \frac{v_F}{\sqrt{\pi}} \partial_x \theta, \quad (5.56)$$

where v_F is a velocity to be introduced shortly.

We can again show that the fermionic fields are given in terms of the bosonic ones as

$$\psi_R(x) = \frac{1}{\sqrt{2\pi\alpha}} \eta_R \, e^{-i2\sqrt{\pi}\phi_R} ,$$
$$\psi_L(x) = \frac{1}{\sqrt{2\pi\alpha}} \eta_L \, e^{i2\sqrt{\pi}\phi_L} . \qquad (5.57)$$

As before, we introduce a linear dispersion relation $\epsilon_{\nu,k} = v_F k$ for the fermions. The non-interacting Hamiltonian then takes the form

$$\begin{aligned} H_0 &= v_F \sum_{k=-\infty}^{\infty} k[\ :c_{R,k}^\dagger c_{R,k} + c_{L,k}^\dagger c_{L,k}: \ ] + \frac{\pi v_F}{L}(\hat{N}_R^2 + \hat{N}_L^2) \\ &= -v_F \int_0^L dx[\ :\psi_R^\dagger(x)i\partial_x\psi_R(x) - \psi_L^\dagger(x)i\partial_x\psi_L(x): \ ] \\ &\quad + \frac{\pi v_F}{L}(\hat{N}_R^2 + \hat{N}_L^2) \end{aligned} \qquad (5.58)$$

in the fermionic language, and

$$\begin{aligned} H_0 &= v_F \sum_{q>0} q(b_{R,q}^\dagger b_{R,q} + b_{L,q}^\dagger b_{L,q}) + \frac{\pi v_F}{L}(\hat{N}_R^2 + \hat{N}_L^2) \\ &= v_F \int_0^L dx \ [\ :(\partial_x\phi_R)^2 + (\partial_x\phi_L)^2: \ ] \\ &= \frac{v_F}{2} \int_0^L dx \ [\ :(\partial_x\phi)^2 + (\partial_x\theta)^2: \ ] \end{aligned} \qquad (5.59)$$

in the bosonic language. If we use this Hamiltonian to transform all the fields to time-dependent Heisenberg fields, we find that ψ_R, ϕ_R become functions of $x_R = x - v_F t$ while ψ_L, ϕ_L become functions of $x_L = x + v_F t$.

From Eq. (5.55), we see that the field canonically conjugate to ϕ is given by

$$\Pi = \partial_x \theta . \qquad (5.60)$$

Thus

$$[\phi(x), \Pi(x')] = i\delta(x - x') , \qquad (5.61)$$

5.2. Bosonization

and
$$H_0 = \frac{v_F}{2} \int_0^L dx \, [\, \Pi^2 + (\partial_x \phi)^2 \,] . \tag{5.62}$$

We now study the effects of four-fermi interactions. In the beginning it is simpler to work in the Schrödinger representation in which the fields are time-independent; we will transform to the Heisenberg representation later when we compute the correlation functions. Let us consider an interaction of the form

$$V = \frac{1}{2} \int_0^L dx \, [\, 2g_2 \, \rho_R(x)\rho_L(x) + g_4 \, (\, \rho_R^2(x) + \rho_L^2(x) \,) \,] . \tag{5.63}$$

Physically, we may expect an interaction such as $g :\rho^2(x): $, so that $g_2 = g_4 = g$. However, it is instructive to allow g_2 to differ from g_4 to see what happens. For reasons explained before, we will again not assume anything about the signs of g_2 and g_4. In the fermionic language, the interaction takes the form

$$\begin{aligned} V = \frac{1}{2L} \sum_{k_1, k_2, k_3 = -\infty}^{\infty} & [2g_2 c_{R,k_1+k_3}^\dagger c_{R,k_1} c_{L,k_2+k_3}^\dagger c_{L,k_2} \\ & + g_4 (c_{R,k_1+k_3}^\dagger c_{R,k_1} c_{R,k_2-k_3}^\dagger c_{R,k_2} \\ & + c_{L,k_1+k_3}^\dagger c_{L,k_1} c_{L,k_2-k_3}^\dagger c_{L,k_2})] . \end{aligned} \tag{5.64}$$

From this expression we see that g_2 corresponds to a two-particle scattering involving both chiralities; in this model, we can call it either forward scattering or backward scattering since there is no way to distinguish between the two processes in the absence of some other quantum number such as spin. The g_4 term corresponds to a scattering between two fermions with the same chirality, and therefore describes a forward scattering process.

The quartic interaction in Eq. (5.64) seems very difficult to analyze. However we will now see that it is easily solvable in the bosonic language; indeed this is one of the main motivations behind bosonization. The bosonic expression for the total Hamiltonian $H = H_0 + V$ is found to be

$$H = \sum_{q>0} q \, [v_F \, (b_{R,q}^\dagger b_{R,q} + b_{L,q}^\dagger b_{L,q}) + \frac{g_2}{2\pi} (b_{R,q}^\dagger b_{L,q}^\dagger + b_{R,q} b_{L,q})$$

$$+ \frac{g_4}{2\pi}(b^\dagger_{R,q}b_{R,q} + b^\dagger_{L,q}b_{L,q})] + \frac{\pi v_F}{L}(\hat{N}_R^2 + \hat{N}_L^2)$$
$$+ \frac{g_2}{L}\hat{N}_R\hat{N}_L + \frac{g_4}{2L}(\hat{N}_R^2 + \hat{N}_L^2). \tag{5.65}$$

The g_4 term again renormalizes the velocity. The g_2 term can then be rediagonalized by a Bogoliubov transformation. We first define two parameters

$$v = \left[\left(v_F + \frac{g_4}{2\pi} - \frac{g_2}{2\pi}\right)\left(v_F + \frac{g_4}{2\pi} + \frac{g_2}{2\pi}\right)\right]^{1/2},$$
$$K = \left[\left(v_F + \frac{g_4}{2\pi} - \frac{g_2}{2\pi}\right) / \left(v_F + \frac{g_4}{2\pi} + \frac{g_2}{2\pi}\right)\right]^{1/2}. \tag{5.66}$$

Note that $K < 1$ if g_2 is positive (repulsive interaction), and > 1 if g_2 is negative (attractive interaction). (If g_2 is so large that $v_F + g_4/(2\pi) - g_2/(2\pi) < 0$, then our analysis breaks down. The system does not remain a Luttinger liquid in that case, and is likely to go into a different phase such as a state with charge density order). The Bogoliubov transformation now takes the form

$$\tilde{b}_{R,q} = \frac{b_{R,q} + \gamma b^\dagger_{L,q}}{\sqrt{1-\gamma^2}},$$
$$\tilde{b}_{L,q} = \frac{b_{L,q} + \gamma b^\dagger_{R,q}}{\sqrt{1-\gamma^2}},$$
$$\gamma = \frac{1-K}{1+K}, \tag{5.67}$$

for each value of the momentum q. The Hamiltonian is then given by the quadratic expression

$$H = \sum_{q>0} vq\,[\tilde{b}^\dagger_{R,q}\tilde{b}_{R,q} + \tilde{b}^\dagger_{L,q}\tilde{b}_{L,q}]$$
$$+ \frac{\pi v}{2L}\left[\frac{1}{K}(\hat{N}_R + \hat{N}_L)^2 + K(\hat{N}_R - \hat{N}_L)^2\right]. \tag{5.68}$$

Equivalently,

$$H = \frac{1}{2}\int_0^L dx\,[vK\Pi^2 + \frac{v}{K}(\partial_x\phi)^2]. \tag{5.69}$$

5.2. Bosonization

The old and new fields are related as

$$\phi_R = \frac{(1+K)\tilde{\phi}_R - (1-K)\tilde{\phi}_L}{2\sqrt{K}},$$

$$\phi_L = \frac{(1+K)\tilde{\phi}_L - (1-K)\tilde{\phi}_R}{2\sqrt{K}},$$

$$\phi = \sqrt{K}\,\tilde{\phi} \quad \text{and} \quad \theta = \frac{1}{\sqrt{K}}\,\tilde{\theta}. \tag{5.70}$$

Note the important fact that the vacuum changes as a result of the interaction; the new vacuum $|\tilde{0}\rangle$ is the state annihilated by the operators $\tilde{b}_{\nu,q}$. Since the various correlation functions must be calculated in this new vacuum, they will depend on the interaction through the parameters v and K. In particular, we will see in the next section that the power-laws of the correlation functions are governed by K.

Given the various Hamiltonians, it is easy to guess the forms of the corresponding Lagrangians. For the non-interacting theory ($g_2 = g_4 = 0$), the Lagrangian density describes a massless Dirac fermion,

$$\mathcal{L} = i\psi_R^\dagger(\partial_t + v_F\partial_x)\psi_R + i\psi_L^\dagger(\partial_t - v_F\partial_x)\psi_L \tag{5.71}$$

in the fermionic language, and a massless real scalar field,

$$\mathcal{L} = \frac{1}{2v_F}(\partial_t\phi)^2 - \frac{v_F}{2}(\partial_x\phi)^2 \tag{5.72}$$

in the bosonic language. For the interacting theory in Eq. (5.69), we find from Eq. (5.70) that

$$\mathcal{L} = \frac{1}{2vK}(\partial_t\phi)^2 - \frac{v}{2K}(\partial_x\phi)^2 = \frac{1}{2v}(\partial_t\tilde{\phi})^2 - \frac{v}{2}(\partial_x\tilde{\phi})^2. \tag{5.73}$$

The momentum operator in Eq. (5.4) has the same expression in terms of the old and new fields, namely,

$$P = k_F(\hat{N}_R - \hat{N}_L) + \int_0^L dx\, \partial_x\phi\partial_x\theta. \tag{5.74}$$

We can check that $[P, \phi] = -i\partial_x\phi$ and $[P, \theta] = -i\partial_x\theta$.

Let us now write down the fields $\tilde{\phi}$ and $\tilde{\theta}$ in the Heisenberg representation. This is simple to do once we realize that the right- and left-moving fields must be functions of $x_R = x - vt$ and $x_L = x + vt$ respectively. We find that

$$\begin{aligned}\tilde{\phi}(x,t) &= \frac{i}{2\sqrt{\pi}} \sum_{q>0} \frac{1}{\sqrt{n_q}} [\tilde{b}_{R,q}\, e^{iq(x_R + i\alpha/2)} - \tilde{b}^\dagger_{R,q} e^{-iq(x_R - i\alpha/2)} \\ &\quad - \tilde{b}_{L,q} e^{-iq(x_L - i\alpha/2)} + \tilde{b}^\dagger_{L,q} e^{iq(x_L + i\alpha/2)}] \\ &\quad - \frac{\sqrt{\pi}}{L}\left[\frac{x}{\sqrt{K}}(\hat{N}_R + \hat{N}_L) - \sqrt{K} vt (\hat{N}_R - \hat{N}_L)\right], \\ \tilde{\theta}(x,t) &= \frac{i}{2\sqrt{\pi}} \sum_{q>0} \frac{1}{\sqrt{n_q}} [-\tilde{b}_{R,q} e^{iq(x_R + i\alpha/2)} + \tilde{b}^\dagger_{R,q} e^{-iq(x_R - i\alpha/2)} \\ &\quad - \tilde{b}_{L,q} e^{-iq(x_L - i\alpha/2)} + \tilde{b}^\dagger_{L,q} e^{iq(x_L + i\alpha/2)}] \\ &\quad + \frac{\sqrt{\pi}}{L}\left[\sqrt{K} x (\hat{N}_R - \hat{N}_L) - \frac{vt}{\sqrt{K}}(\hat{N}_R + \hat{N}_L)\right]. \quad (5.75)\end{aligned}$$

We observe that the coefficients of $\hat{N}_R$ and $\hat{N}_L$ have terms which are linear in t. This is necessary because we want the conjugate momentum field to satisfy

$$\tilde{\Pi} = \frac{1}{v}\partial_t\tilde{\phi} = \partial_x\tilde{\theta}. \quad (5.76)$$

We note that a dual equation holds, namely,

$$\frac{1}{v}\partial_t\tilde{\theta} = \partial_x\tilde{\phi}. \quad (5.77)$$

(One can check from Eq. (5.75) that $\tilde{\phi}_R$ and $\tilde{\phi}_L$ are functions of x_R and x_L alone). In terms of θ, the Lagrangian density is

$$\mathcal{L} = \frac{K}{2v}(\partial_t\theta)^2 - \frac{Kv}{2}(\partial_x\theta)^2 = \frac{1}{2v}(\partial_t\tilde{\theta})^2 - \frac{v}{2}(\partial_x\tilde{\theta})^2. \quad (5.78)$$

Although the Lagrangians in Eq. (5.73) and Eq. (5.78) have opposite signs, the Hamiltonians derived from the two are identical.

5.2. Bosonization

Before ending this section, let us comment on a global symmetry of all these models. It is known that fermionic systems with a conserved charge are invariant under a global phase rotation

$$\psi_R \to e^{i\lambda}\psi_R, \quad \text{and} \quad \psi_L \to e^{i\lambda}\psi_L, \qquad (5.79)$$

where λ is independent of (x,t). Eq. (5.57) then implies that the corresponding bosonic theories must remain invariant under

$$\phi \to \phi, \quad \text{and} \quad \theta \to \theta + \frac{\lambda}{\sqrt{\pi}}. \qquad (5.80)$$

This provides a constraint on the kinds of terms which can appear in the Lagrangians of such theories.

5.2.3 Field theory of modes near the Fermi momenta

In the last section, we discussed bosonization for a model of fermions which has the following properties.
(i) There are an infinite number of right- and left-moving modes with the momenta going from $-\infty$ to ∞, and
(ii) the relation between energy and momentum is linear for all values of the momentum.
Neither of these properties is true in condensed matter systems which typically are non-relativistic and have a finite (though possibly very large) number of states. The question is the following: can bosonization give useful results even if these two properties do not hold? We will see that the answer is yes, provided that we are only interested in the long-wavelength, low-frequency and low-temperature properties of such systems.

In an experimental system, the fermions may be able to move either on a discrete lattice of points such as in a crystal, or in a continuum such as the conduction electrons in a metal. For instance, non-interacting fermions moving in a continuum have a dispersion $\epsilon_k = k^2/2m$, while fermions hopping on a lattice have a dispersion such as $\epsilon_k = -t\cos(ka)$ if a is the lattice spacing and t is the nearest neighbor hopping amplitude. In either case, a non-interacting system in one-dimension will, at zero temperature, have a Fermi surface consisting of two points in momentum space given by $k = \pm k_F$ (see Fig.5.2). As stated before, we define the one-particle energy to be zero at the Fermi points. At low temperatures T or low frequencies ω, the only modes which can contribute are the ones lying close to those points, *i.e.*, with excitation energies of the order of or smaller than $k_B T$ or ω. Near the Fermi points, we can approximate the dispersion relation by a linear one, with the velocity being defined to be $v_F = (d\epsilon_k/dk)_{k=k_F}$. We thus restrict our attention to the right-moving modes with momenta lying between $k_F - \Lambda$ and $k_F + \Lambda$, and the left-moving modes with momenta lying between $-k_F - \Lambda$ and $-k_F + \Lambda$. Here Λ is taken to be much smaller than the full range of the momentum (which is $2\pi/a$ on a lattice if the lattice spacing is a), but $v_F \Lambda$ is much larger than the temperatures or frequencies of interest. If we include only these regions of momenta, then the second quantized Fermi field can be written in the approximate form

$$\psi(x,t) = \psi_R(x,t)\, e^{ik_F x} + \psi_L(x,t)\, e^{-ik_F x}, \qquad (5.81)$$

where ψ_R and ψ_L vary slowly over spatial regions which are large compared to the distance scale $1/\Lambda$. The momentum components of these slowly varying fields are related to those of ψ as

$$\begin{aligned}\psi_{R,k}(t) &= \psi_{k+k_F}(t), \\ \psi_{L,k}(t) &= \psi_{-k-k_F}(t),\end{aligned} \qquad (5.82)$$

where $-\Lambda \leq k \leq \Lambda$. These long-wavelength fields are the ones to which the technique of bosonization can be applied.

The definitions in Eqs. (5.81-5.82) tell us the forms of the various terms in a microscopic model and also tell us which of them survive in the long-wavelength limit. For instance, the density is given by

$$\begin{aligned}\rho &= :\psi^\dagger \psi := \psi_R^\dagger \psi_R + \psi_L^\dagger \psi_L + e^{-i2k_F x}\psi_R^\dagger \psi_L + e^{i2k_F x}\psi_L^\dagger \psi_R : \\ &= -\frac{1}{\sqrt{\pi}}\frac{\partial \phi}{\partial x} + \frac{1}{2\pi\alpha}[\eta_R^\dagger \eta_L e^{i(2\sqrt{\pi}\phi - 2k_F x)} + \eta_L^\dagger \eta_R e^{-i(2\sqrt{\pi}\phi - 2k_F x)}].\end{aligned}$$
$$(5.83)$$

5.2. Bosonization

The terms containing $\exp(\pm i 2k_F x)$ in Eq. (5.83) vary on a distance scale k_F^{-1} which is typically of the same order as the inverse particle density ρ^{-1}. These terms can therefore be ignored if we are only interested in the asymptotic behavior of correlation functions at distances much larger than k_F^{-1}. In a lattice model, we have to be more careful about this argument since the lattice momentum only needs to be conserved modulo $2\pi/a$ in any process. However, since $0 < k_F < \pi/a$ in general, and x/a is an integer, we see that the last two terms in Eq. (5.83) vary on the scale of the lattice unit a; we can therefore ignore those terms if we are only interested in phenomena at distance scales which are much larger than a.

On the other hand, there are situations when a density term like $\rho \cos(2k_F x)$ is generated in the model; for instance, this happens below a Peierls transition if the fermions are coupled to lattice phonons. We then find that the slowly varying terms in the continuum field theory are given by

$$\begin{aligned}\cos(2k_F x)\,\rho &= \frac{1}{2}[\,\psi_R^\dagger \psi_L + \psi_L^\dagger \psi_R\,] \quad \text{in general}\,, \\ &= \psi_R^\dagger \psi_L + \psi_L^\dagger \psi_R \quad \text{if} \quad e^{i4k_F x} = 1\,. \end{aligned} \quad (5.84)$$

The second possibility can arise in a lattice model if $4k_F a = 2\pi$, i.e., at half-filling; we then call it a dimerized system. We will call the term on the right hand sides of Eq. (5.84) the mass operator. We will see below that for any value of $K < 2$, this term produces a gap in the low-energy spectrum. This is called the dimerization gap if it occurs in a lattice system.

We should emphasize an important difference between models defined in the continuum and those defined on a lattice. In the continuum, $\psi_R^2(x) = \psi_L^2(x) = 0$ due to the anticommutation relations. Therefore a term like $\psi_R^{\dagger 2}(x)\psi_L^2(x)$ is equal to zero in the continuum. However such a term need not vanish on a lattice, if we take the two factors of $\psi_R^\dagger$ (or ψ_L) as coming from two neighboring sites separated by a distance a. In fact, this term is allowed by momentum conservation on a lattice if $4k_F a = 2\pi$, and it leads to umklapp scattering.

5.3 Correlation functions and dimensions of operators

We will now use bosonization to compute the correlation functions of some fermionic operators in the interacting theory discussed above. The power-law fall-offs of the correlation functions will tell us the dimensions of those operators.

The bosonic correlation function can be found from the commutation relations in Eq. (5.51), remembering that all normal-orderings have to be done with respect to the new vacuum $|\tilde{0}\rangle$. (Henceforth we will omit the tilde denoting the new vacuum, but we will continue to use the tilde for the new ϕ fields). For instance,

$$\langle 0|T\tilde{\phi}(x,t)\tilde{\phi}^\dagger(0,0)|0\rangle = -\frac{1}{4\pi}\ln\left[\left(\frac{2\pi}{L}\right)^2\left(x^2 - (vt - i\alpha\,\text{sign}(t))^2\right)\right]. \tag{5.85}$$

We can use the expressions in Eq. (5.57) and identities like Eq. (5.30) to obtain the correlation functions of various operators. For instance,

$$\langle 0|T e^{i2\sqrt{\pi}\beta\tilde{\phi}_R(x,t)} e^{-i2\sqrt{\pi}\beta\tilde{\phi}_R(0,0)}|0\rangle \sim \left(\frac{\alpha}{vt - x - i\alpha\,\text{sign}(t)}\right)^{\beta^2},$$

$$\langle 0|T e^{i2\sqrt{\pi}\beta\tilde{\phi}_L(x,t)} e^{-i2\sqrt{\pi}\beta\tilde{\phi}_L(0,0)}|0\rangle \sim \left(\frac{\alpha}{vt + x - i\alpha\,\text{sign}(t)}\right)^{\beta^2}, \tag{5.86}$$

in the limit $L \to \infty$; we will assume henceforth that this limit is taken in the calculation of all correlation functions. Consider now the positive-chirality fermion field; according to Eq. (5.57),

$$\psi_R = \frac{1}{\sqrt{2\pi\alpha}}\eta_R e^{-i2\sqrt{\pi}\phi_R}, \tag{5.87}$$

where ϕ_R is given in Eq. (5.70) in terms of $\tilde{\phi}_R$ and $\tilde{\phi}_L$. Hence its time-ordered correlation function takes the form

$$\langle 0|T\psi_R(x,t)\psi_R^\dagger(0,0)|0\rangle$$
$$\sim \frac{\alpha^{(1-K)^2/2K}}{2\pi(vt - x - i\alpha(s_t))^{(1+K)^2/4K}(vt + x - i\alpha(s_t))^{(1-K)^2/4K}}. \tag{5.88}$$

5.3. Correlation functions and dimensions of operators

where s_t stands for sign(t). We see that the correlation function falls off at large space-time distances (i.e., large compared to α) with the power $(1 + K^2)/(2K)$. This means that the scaling dimension of the operator ψ_R or $\psi_R^\dagger$ is $(1 + K^2)/4K$; this agrees with the familiar value of $1/2$ for non-interacting fermions.

If we set $x = 0$ in Eq. (5.88), and Fourier transform over time, we find that the one-particle density of states (DOS) has a power-law form near zero frequency,

$$\tilde{n}(\omega) \sim |\omega|^\beta , \qquad (5.89)$$

where

$$\beta = \frac{(1-K)^2}{2K} . \qquad (5.90)$$

The same result holds for the DOS of the $-$ chirality fermions. We therefore see that for any non-zero interaction, either repulsive or attractive, the one-particle DOS vanishes as a power. (This result is not to be confused with the *bosonic* DOS which, from Eq. (5.68), is a constant near zero energy since the energy is linearly related to the momentum which has a constant density. That leads to a specific heat which is linear in the temperature at low temperatures). Alternatively, we may set $t = 0$ in Eq. (5.88) and Fourier transform over space, with a factor of $\exp(ik_F x)$ since the momentum of the right-chirality fermions is measured with respect to the Fermi momentum k_F. We then see that the momentum distribution function is continuous at k_F with a power-law form,

$$n(k) = n(k_F) + \text{constant} \cdot \text{sign}(k - k_F)|k - k_F|^\beta , \qquad (5.91)$$

as we have sketched in Fig.5.1(b). These expressions for $\tilde{n}(\omega)$ and $n(k)$ are characteristic features of a Luttinger liquid.

Next let us compute the correlation function of an operator which is bilinear in the fermion fields, namely, the mass operator

$$M = \psi_R^\dagger \psi_L + \psi_L^\dagger \psi_R = \frac{1}{2\pi\alpha} \left[\eta_R^\dagger \eta_L e^{i2\sqrt{\pi}\phi} + \eta_L^\dagger \eta_R e^{-i2\sqrt{\pi}\phi} \right] . \qquad (5.92)$$

Using the same technique as before, we find that

$$\langle 0|TM(x,t)M(0,0)|0\rangle \sim \frac{\alpha^{2(K-1)}}{4\pi^2((vt - i\alpha(s_t))^2 - x^2)^K} . \qquad (5.93)$$

This shows that the scaling dimension of the mass operator is K. For the non-interacting case $K = 1$, we see that the addition of such a term to the Lagrangian density in Eq. (5.71) makes the Dirac fermion massive; this is why we have called it the mass operator. (For convenience, we will sometimes omit the Klein factors when writing fermionic operators in the bosonic language. We will of course need to restore those factors when calculating the correlation functions; clearly, correlation functions will vanish if the numbers of η_R and $\eta_R^\dagger$ (or η_L and $\eta_L^\dagger$) are not equal).

An important operator to consider is the density ρ. From Eqs. (5.83), (5.85) and (5.93), we see that the density-density equal-time correlation function is asymptotically given by

$$\langle 0| \rho(x,0)\rho(0,0) |0\rangle = -\frac{K}{2\pi^2 x^2} + \text{const} \cdot \frac{\cos(2k_F x)}{x^{2K}} . \quad (5.94)$$

We should emphasize that this is only the asymptotic expression; the complete expression generally contains oscillatory terms like $\cos(2nk_F x)/x^{2n^2 K}$ for all positive integers n. However the form of the denominator shows that these terms decay rapidly with x as n increases.

In general, we can consider an operator of the form

$$O_{m,n} = e^{i2\sqrt{\pi}(m\phi + n\theta)} . \quad (5.95)$$

(Such an operator can arise from a product of several ψ's and $\psi^\dagger$'s if we ignore the Klein factors; then Eq. (5.57) implies that $m \pm n$ must take integer values). We then find the following result for the two-point correlation function

$$\langle 0|TO_{m,n}(x,t)O^\dagger_{m',n'}(0,0)|0\rangle \sim \delta_{mm'}\delta_{nn'}$$
$$\frac{\alpha^{2(m^2 K + n^2/K)}}{(vt - x - i\alpha(s_t))^{(m\sqrt{K} - n/\sqrt{K})^2}(vt + x - i\alpha(s_t))^{(m\sqrt{K} + n/\sqrt{K})^2}} , \quad (5.96)$$

where we have taken the limit $L \to \infty$ as usual. (If L had been kept finite, the correlation function in Eq. (5.96) would have been non-zero even if $m \neq m'$ or $n \neq n'$. This may seem surprising

5.3. Correlation functions and dimensions of operators

since the global phase invariance in Eqs. (5.79 - 5.80) should lead to the Kronecker δ's in Eq. (5.96) even for finite values of L. The resolution of this puzzle is that we need to include the appropriate Klein factors in the definition in (5.95) to show that the correlation function of a product of fermionic operators is zero if it is not phase invariant). We conclude that the scaling dimension of $O_{m,n}$ is given by

$$d_O = m^2 K + \frac{n^2}{K}. \qquad (5.97)$$

The appearance of the cut-off α in the expressions for the various correlation functions may seem bothersome. This may be eliminated by redefining the operators $O_{m,n}$ in Eq. (5.95) by multiplying them with appropriate K-dependent powers of α; then the two-point correlation function has a well-defined limit as $\alpha \to 0$. The important point to note is that all the correlation functions fall off as power-laws asymptotically, and that the exponents give the scaling dimensions of those operators. The significance of the scaling dimension will be discussed in the next section.

For certain applications of bosonization, it is useful to know the forms of the correlation functions in imaginary time. From the various expressions above, it is clear that if x is held fixed at some non-zero value, then the poles in the complex t plane are either in the first or in the third quadrant. We may therefore rotate t by $\pi/2$ without crossing any poles. After doing this, we write $t = i\tau$ where τ is a real variable. Eq. (5.96) then takes the form

$$\langle O_{m,n}(x,t) O^\dagger_{m',n'}(0,0) \rangle \sim \delta_{mm'} \delta_{nn'} e^{i4mn\zeta} \left(\frac{\alpha^2}{x^2 + v^2 \tau^2} \right)^{m^2 K + n^2/K}, \qquad (5.98)$$

where $\zeta = \tan^{-1}(v\tau/x)$, and we have dropped the α-dependent terms in the denominator since there are no longer any poles for non-zero values of x.

5.4 Renormalization group analysis of perturbed models

We will now study the effects of some perturbations on the low-energy properties of Luttinger liquids. A standard way to do this is to use the renormalization group (RG) idea. Suppose that we are given an action at a microscopic length scale which may be a lattice spacing a; the action contains some small perturbations proportional to certain dimensionless parameters λ_i, such that for $\lambda_i = 0$, we have a gapless system with an infinite correlation length ξ, i.e., all correlations fall off as power laws. Then the RG procedure typically consists of the following steps.

(i) First, a small range of high momentum modes of the various fields are integrated out. Specifically, we will assume that the momenta lie in the range $[-\Lambda, \Lambda]$ while the frequencies go all the way from $-\infty$ to ∞. Then we will integrate out the modes with momenta lying in the two intervals $[-\Lambda, -\Lambda/s]$ and $[\Lambda/s, \Lambda]$ and with all frequencies from $-\infty$ to ∞. Here $s = e^{dl}$ where dl is a small positive number. The asymmetry between the momentum and frequency integrations is necessary to ensure that the action remains local in time at all stages. (Note that we are using sharp momentum cut-offs in this section, whereas we used a smooth momentum cut-off with the parameter α in the previous sections).

(ii) Secondly, the space-time coordinates, the fields and the various parameters are rescaled by appropriate powers of s so that the new action looks exactly like the old action. This new action is effectively at a larger length scale equal to ae^{dl}. Clearly, the changes in the parameters λ_i must be proportional to the small number dl. Since we are going to repeat the process of integrating out high momentum modes, we introduce the idea of an effective length scale $a(l) = ae^l$; we also define length scale dependent parameters $\lambda_i(l)$, where $\lambda_i(0)$ denote the values of λ_i in the original action. We then define the β-functions

$$\beta(\lambda_i) = \frac{d\lambda_i(l)}{dl} . \qquad (5.99)$$

These are functions of all the $\lambda_i(l)$'s so that we get a set of coupled

5.4. RG analysis of perturbed models

non-linear equations in general. In principle, the β-functions are given by infinite power series in the λ_i, but in practice, we can only obtain the first few terms depending on the number of loops of the various Feynman diagrams that we can compute. The RG analysis is therefore usually limited to small values of λ_i.

(iii) Finally, we integrate the RG equations, *i.e.*, the differential equations described by the β-functions, in order to obtain the functions $\lambda_i(l)$. For simplicity, let us consider the case of a single perturbation with a coefficient λ. Then one of three things can happen as l increases from 0. Either $\lambda(l)$ goes to zero in which case we recover the unperturbed theory at long distances; or λ does not change with l; or $\lambda(l)$ grows with l till its value becomes of order 1. In the last case, the RG equation cannot be trusted beyond that length scale since the β-functions are generally only known up to some low order in the λ's. All that we can say is that beyond the length scale ae^l where $\lambda(l)$ becomes of order 1, a completely new kind of action is likely to be required to describe the system. Large perturbations of a gapless system often (but not always) correspond to a gapped system whose correlation length ξ (which governs the exponential decay of various correlation functions) is of the same order as that length scale ae^l. Thus, although the blowing up of a parameter λ at some scale does not tell us what the new action must be beyond that scale, it can give us an idea of the correlation length of that new theory. This is the main use that is made of RG equations. To complete the picture and find the new theory beyond the scale ξ, one usually has to do some other kind of analysis.

Let us now examine in a little more detail the various kinds of RG equations which can arise at low orders. Suppose that to first order, the RG equation for a single perturbing term is given by

$$\frac{d\lambda}{dl} = b_1 \lambda , \quad (5.100)$$

where b_1 is some constant. If $b_1 < 0$, any non-zero value of λ at $l = 0$ flows to 0 as l increases. Such a perturbation is called irrelevant. If $b_1 > 0$, it is called a relevant perturbation. A small perturbation then grows exponentially with l and reaches a number of order 1 at a distance scale given by $e^{b_1 l} \sim 1/\lambda(0)$. In many models, this gives an estimate of the correlation length ξ and of the energy gap ΔE of the system, namely,

$$\xi = ae^l = \frac{a}{\lambda(0)^{1/b_1}},$$

$$\text{so} \quad \Delta E = \frac{v}{\xi} = \frac{v\lambda(0)^{1/b_1}}{a}. \quad (5.101)$$

Finally, if $b_1 = 0$, the perturbation is called marginal. One then has to go to second order in λ. If the RG equation takes the form

$$\frac{d\lambda}{dl} = b_2 \lambda^2, \quad (5.102)$$

then a small perturbation of one particular sign flows to zero and is called marginally irrelevant, while a small perturbation of the opposite sign grows and is called marginally relevant. For instance, suppose that $b_2 > 0$. Then the above equation gives

$$\lambda(l) = \frac{\lambda(0)}{1 - b_2 \lambda(0) l}. \quad (5.103)$$

If we start with a negative value of $\lambda(0)$, $\lambda(l)$ flows to 0. For large l, $\lambda(l)$ goes to zero logarithmically in the distance scale as $-1/(b_2 l)$ independently of the starting value. (It turns out that this produces logarithmic corrections to the power-law fall-offs of the correlation functions at large distances and the various excitation energies [10]). On the other hand, if we start with a small positive value of $\lambda(0)$, then $\lambda(l)$ grows and becomes of order 1 at a distance scale which we identify with a correlation length

$$\xi = ae^{1/(b_2 \lambda(0))}. \quad (5.104)$$

The corresponding energy gap $\Delta E = v/\xi$ is extremely small for small values of $\lambda(0)$; it may be very hard to distinguish this kind of a system from a gapless system by numerical studies. This is in sharp contrast to the situation with a relevant perturbation where the gap scales as a power of $\lambda(0)$.

5.4. RG analysis of perturbed models

There is a simple relation between the scaling dimension of an operator O (assumed to be hermitian for simplicity), and the first-order coefficient b_1 in its β-function. We recall that the scaling dimension d_O is defined as half of the exponent appearing in the two-point correlation function at large distances, namely,

$$\langle O(x,0) O(0,0) \rangle = |x|^{-2d_O} . \tag{5.105}$$

It is convenient to define the normalization of O in such a way that the right hand side of Eq. (5.105) has a prefactor equal to 1. Consequently, O has the engineering dimensions of a^{-d_O}. Let us now add a perturbation to the Hamiltonian (or to the Lagrangian with a negative sign) of the form

$$\delta H = \lambda a^{d_O - 2} v \int dx\, O , \tag{5.106}$$

where the factors of a and v (the velocity of the unperturbed Luttinger liquid) are put in to make λ dimensionless; note that v/a has the dimensions of energy. Then the first-order RG equation for λ must take the form given in Eq. (5.100) with

$$b_1 = 2 - d_O . \tag{5.107}$$

This important statement will be proved below for the class of operators $O_{m,n}$ introduced in Eq. (5.95). If $d_O = 2$, the perturbation is marginal and we have to proceed to Eq. (5.102). It turns out that b_2 can be obtained from a *three-point* correlation function, but we will not pursue that here [10].

It will not come as a surprise that the RG equations for interacting quantum systems in one dimension can often be derived in two different ways, namely, using the fermionic theory or the bosonic one. Although both the derivations are limited in practice to small values of the perturbations λ_i, we will see that the bosonic derivation is superior because it can handle the interactions in Eq. (5.63) exactly. In the fermionic derivation, we have to assume that not only the λ's but also the interaction parameters g_2 and g_4 are small. We will now discuss some simple examples of β-function calculations to first order in the two kinds of theories.

As a particularly simple exercise, consider a non-interacting massive Dirac theory, where the mass term is to be treated as a perturbation. We define the Fourier components of ψ_ν as

$$\psi_R(x,t) = \int_{-\Lambda}^{\Lambda} \frac{dk}{2\pi} \int_{\infty}^{\infty} \frac{d\omega}{2\pi} e^{i(kx-\omega t)} \psi_R(k,\omega) ,$$

$$\psi_L(x,t) = \int_{-\Lambda}^{\Lambda} \frac{dk}{2\pi} \int_{\infty}^{\infty} \frac{d\omega}{2\pi} e^{-i(kx+\omega t)} \psi_L(k,\omega) . \quad (5.108)$$

Then the action takes the form

$$S[\psi_\nu, \psi_\nu^\dagger] = \int_{-\Lambda}^{\Lambda} \frac{dk}{2\pi} \int_{\infty}^{\infty} \frac{d\omega}{2\pi}$$
$$\left[\psi_R^\dagger(k,\omega)(\omega - vk)\psi_R(k,\omega) + \psi_L^\dagger(k,\omega)(\omega - vk)\psi_L(k,\omega) \right.$$
$$\left. -\mu\left(\psi_R^\dagger(-k,\omega)\psi_L(k,\omega) + \psi_L^\dagger(-k,\omega)\psi_R(k,\omega)\right) \right] . \quad (5.109)$$

Since μ has the dimensions of energy, the dimensionless parameter must be taken to be

$$\lambda = \frac{a\mu}{v} . \quad (5.110)$$

(The value of a is completely arbitrary here and it will not appear in any physical quantity as we will see). We consider the partition function in the functional integral representation,

$$Z = \int \mathcal{D}\psi_\nu \mathcal{D}\psi_\nu^\dagger \, e^{iS} . \quad (5.111)$$

We integrate out the modes in the momentum and frequency ranges specified in step (i) of the RG procedure outlined above. Since Eq. (5.109) describes non-interacting fermions, the mode integration produces an action which looks exactly the same, except that the momentum integrations go from $-\Lambda/s$ to Λ/s. To restore this to the original range of $[-\Lambda, \Lambda]$, we define the new (primed) quantities

$$\begin{aligned} k' &= sk , \\ \omega' &= s\omega , \\ \psi_\nu'(k',\omega') &= s^{-3/2} \psi_\nu(k,\omega) , \\ \lambda' &= s\lambda . \end{aligned} \quad (5.112)$$

5.4. RG analysis of perturbed models

The resultant action in terms of the new variables and fields looks exactly the same as the original action in terms of the old variables. Note that we had to rescale the mass parameter also in order to achieve this. Since $s = e^{dl}$, we obtain the RG equation

$$\frac{d\lambda(l)}{dl} = \lambda(l) . \tag{5.113}$$

Clearly, this describes a relevant perturbation, and $\lambda(l)$ grows to 1 at a length scale

$$\xi = ae^l = \frac{v}{\mu} . \tag{5.114}$$

The energy gap is $\Delta E = v/\xi = \mu$ as expected.

Now let us add density-density interactions as in Eq. (5.63) to the above massive theory. The question is the following: do ξ and ΔE scale in the same way with μ as they do in the non-interacting theory? Clearly, it is not easy to answer this in the fermionic language since the interactions themselves are not easy to handle in that language, and the mass perturbation is an additional complication. But bosonization comes to our rescue here since the bosonic theory remains quadratic even after including the four-fermi interactions; hence the mass perturbation is the only thing that needs to be studied.

Let us consider a more general perturbing operator of the form

$$O = O_{m,0} + O^\dagger_{m,0} , \tag{5.115}$$

where $O_{m,n}$ is defined in Eq. (5.95); the reason for setting $n = 0$ will be explained later. From Eq. (5.97), the scaling dimension of O is given by $d_O = m^2 K$; note that this contains the effects of the four-fermion interaction in a non-trivial way through the parameter K. In the bosonic language, the perturbed action has the sine-Gordon form,

$$S[\tilde{\phi}] = \int dx dt \Big[\frac{1}{2v}(\partial_t \tilde{\phi})^2 - \frac{v}{2}(\partial_x \tilde{\phi})^2 - \frac{v\lambda}{a^2} \cos(2m\sqrt{\pi K}\tilde{\phi})\Big], \tag{5.116}$$

where we have changed variables from ϕ to $\tilde{\phi}$ using Eq. (5.70). We now have to apply the RG procedure to this action. We introduce the Fourier components of $\tilde{\phi}$ as

$$\tilde{\phi}(x,t) = \int_{-\Lambda}^{\Lambda} \frac{dk}{2\pi} \int_{\infty}^{\infty} \frac{d\omega}{2\pi} e^{i(kx-\omega t)} \tilde{\phi}(k,\omega) \ . \qquad (5.117)$$

(In principle, the momentum cut-offs for fermion and boson fields need not be equal, but we will use the same symbol Λ for convenience). Next we consider the partition function

$$Z = \int \mathcal{D}\tilde{\phi} \ e^{iS} \ , \qquad (5.118)$$

and expand e^{iS} in powers of λ to obtain an infinite series. Let us write the field $\tilde{\phi}$ as the sum

$$\tilde{\phi} = \tilde{\phi}_< + \tilde{\phi}_> \ , \qquad (5.119)$$

where both $\tilde{\phi}_<$ and $\tilde{\phi}_>$ contain all frequencies, but $\tilde{\phi}_<$ only contains momenta lying in the range $[-\Lambda/s, \Lambda/s]$, whereas $\tilde{\phi}_>$ only contains momenta lying in the ranges $[-\Lambda, -\Lambda/s]$ and $[\Lambda/s, \Lambda]$. Following step (i) of the RG procedure, we have to perform the functional integration over $\tilde{\phi}_>$, and then re-exponentiate the infinite series to obtain the new action in terms of $\tilde{\phi}_<$. We will do this calculation only to first order in λ. This is not difficult since $e^{\pm i 2m\sqrt{\pi K}\tilde{\phi}}$ can be written as the product of exponentials of $\tilde{\phi}_<$ and $\tilde{\phi}_>$, while the quadratic part of the action decouples as $S_0[\tilde{\phi}] = S_0[\tilde{\phi}_<] + S_0[\tilde{\phi}_>]$. Let us denote the expectation value of a functional $F[\tilde{\phi}_>]$ as

$$\langle \ F[\tilde{\phi}_>] \ \rangle = \int \mathcal{D}\tilde{\phi}_> \ e^{iS_0[\tilde{\phi}_>]} \ F[\tilde{\phi}_>] \ . \qquad (5.120)$$

Now we have to compute

$$\langle \ e^{\pm i 2m\sqrt{\pi K}\tilde{\phi}_>(x,t)} \ \rangle \ . \qquad (5.121)$$

By translation invariance, the value of this is independent of the coordinates (x,t), so we can evaluate it at the point $(0,0)$. We then use the fact that $\langle \tilde{\phi}_>^n(0,0)\rangle = 0$ if n is odd, while

$$\langle \tilde{\phi}_>^n(0,0)\rangle = (n-1)(n-3)\cdots 1 \langle \tilde{\phi}_>^2(0,0)\rangle^{n/2} \qquad (5.122)$$

5.4. RG analysis of perturbed models

if n is even. Thus the expectation value in (5.121) is given by

$$\begin{aligned}\langle e^{\pm i 2m\sqrt{\pi K}\tilde{\phi}_>(0,0)}\rangle &= \sum_{n=0}^{\infty}\frac{1}{n!}(\pm i 2m\sqrt{\pi K})^n \langle \tilde{\phi}_>^n(0,0)\rangle \\ &= e^{-2m^2\pi K \langle \tilde{\phi}_>^2(0,0)\rangle}.\end{aligned} \quad (5.123)$$

Now we use the fact that

$$\langle \tilde{\phi}_>^2(0,0)\rangle = 2\int_{\Lambda/s}^{\Lambda}\frac{dk}{2\pi}\int_{-\infty}^{\infty}\frac{d\omega}{2\pi}\frac{i}{\omega^2/v - vk^2 + i\epsilon} = \frac{\ln s}{2\pi} \quad (5.124)$$

to show that the left hand side of Eq. (5.123) is equal to $s^{-m^2 K}$. Putting everything together, we find the new action to be

$$S[\tilde{\phi}_<] = \int dx dt \Big[\frac{1}{2v}(\partial_t \tilde{\phi}_<)^2 - \frac{v}{2}(\partial_x \tilde{\phi}_<)^2 \\ - \frac{v\lambda s^{-m^2 K}}{a^2}\cos(2m\sqrt{\pi K}\tilde{\phi}_<) \Big], \quad (5.125)$$

where the momentum integrals only go from $-\Lambda/s$ to Λ/s. To restore the range of the momentum to $[-\Lambda, \Lambda]$ and to recover the form of the action in Eq. (5.116), we have to define

$$\begin{aligned} k' &= sk, \text{ and } x' = s^{-1}x, \\ \omega' &= s\omega, \text{ and } t' = s^{-1}t, \\ \tilde{\phi}'(k',\omega') &= \tilde{\phi}_<(k,\omega), \\ \lambda' &= s^{2-m^2 K}\lambda, \end{aligned} \quad (5.126)$$

and write the action in terms of primed variables. Since $s = e^{dl}$, we see that $d\lambda = \lambda' - \lambda$ satisfies the RG equation

$$\frac{d\lambda}{dl} = (2 - m^2 K)\lambda. \quad (5.127)$$

This proves the relation between the first-order β-function coefficient b_1 and the scaling dimension d_O. Note that the β-functions of the parameters v and K remain zero up to this order in the perturbation. However they do get a contribution to second order in λ as shown in Ref.[3].

The mass perturbation corresponds to the special case of Eq. (5.115) with $m = 1$. We now see that it is marginal for $K = 2$ and is relevant if $K < 2$. In the latter case, $\lambda(l)$ grows till we reach a length scale $\xi = a/\lambda(0)^{1/(2-K)}$ where the length scale of the coefficient of the cosine term in the Lagrangian becomes of the same order as a; that is the appropriate point to stop the RG flow of λ. The expression for ξ implies that the energy gap of the system is given by

$$\Delta E = \frac{v}{a} \lambda(0)^{1/(2-K)} . \tag{5.128}$$

Thus the effect of the renormalization is to produce a sine-Gordon theory with the Lagrangian density

$$\mathcal{L} = \frac{1}{2v}(\partial_t \tilde{\phi})^2 - \frac{v}{2}(\partial_x \tilde{\phi})^2 - \text{const} \cdot \frac{(\Delta E)^2}{v} \cos(2\sqrt{\pi K}\tilde{\phi}) , \tag{5.129}$$

where x and t in this expression denote the *original* coordinates, and it is understood that this Lagrangian is *not* to be renormalized any further. This theory is exactly solvable and its spectrum is known in detail [11]. It has both bosonic and fermionic (soliton) excitations, and both of them have energy gaps of the order of ΔE given in Eq. (5.128).

Finally, let us briefly consider some other relevant and marginal perturbations that can appear in a system which, at the microscopic model, involves fermions on a lattice. If the model has the global phase invariance discussed in Eqs. (5.79 - 5.80), then the operators $O_{m,n}$ appearing in the bosonized theory must necessarily have $n = 0$. The scaling dimension is then $d_O = m^2 K$. Since $m \geq 1$, there is only a finite number of relevant operators possible depending on the value of K. For $K > 2$, there are no relevant operators at all. For $1/2 < K < 2$, the mass operator is the only relevant term, and so on.

Turning to the possible marginal operators, we see that the umklapp operator $O_{2,0} = \psi_R^{\dagger 2} \psi_L^2$ is marginal for $K = 1/2$. This is a particularly important case to consider because a Luttinger liquid at $K = 1/2$ is known to have a global $SU(2)$ symmetry;

it therefore describes a large number of gapless systems involving spins. From conformal field theory, the value of b_2 in the RG equation Eq. (5.102) for the umklapp operator O is exactly known to be $4\pi/\sqrt{3}$ for the normalization given in Eq. (5.105). The coefficient of O in the Hamiltonian, namely λ, depends on the microscopic parameters of the model. In general, a system will have a non-zero value of λ. As discussed above, for one sign of λ, the system remains gapless but with logarithmic corrections to various physical quantities; for instance, a $1/\ln T$ term appears in the magnetic susceptibility of a spin system at low temperatures. For the other sign of λ, the system spontaneously dimerizes producing a finite correlation length and an energy gap; this leads to an exponentially vanishing susceptibility at low temperatures.

5.5 Applications of bosonization

We will now study various applications of the method of bosonization. The method, as you have learned, can only be applied in one dimension, so we restrict ourselves to one-dimensional models. As you have also seen, the main advantage of the method of bosonization is that many interacting fermion theories can often be recast (within some approximations) as non-interacting boson theories. This enables the explicit calculation of correlation functions. This is an advantage, even in Bethe ansatz solvable one-dimensional models, because it is often not possible to compute correlation functions using the Bethe ansatz.

We will concentrate on the applications of the bosonization technique in the following problems - (i) the quantum antiferromagnetic spin 1/2 chain, (ii) the Hubbard model in one dimension, (iii) transport in clean quantum wires and (iv) transport through isolated impurities. Since the physics of each of these applications is a huge subject by itself, here we will only concentrate on explaining the model and the quantities that we can obtain through the use of bosonization, rather than go into details of its phenomenology.

5.6 Quantum antiferromagnetic spin 1/2 chain

The model

The first problem that we shall study is the model of a spin 1/2 antiferromagnetic chain. We are picking this model, since you have already learned a lot about the model from the course on quantum spin chains and spin ladders. Here, we will restrict ourselves to just the study of the spin 1/2 anisotropic Heisenberg model with the Hamiltonian given by

$$H = \sum_{i=1}^{N} \left[\frac{J}{2} (S_i^+ S_{i+1}^- + S_i^- S_{i+1}^+) + J_z\, S_i^z S_{i+1}^z \right], \quad (5.130)$$

where the interactions are only between nearest neighbor spins, and $J > 0$. $S_i^+ = S_i^x + i S_i^y$ and $S_i^- = S_i^x - i S_i^y$ are the spin raising and lowering operators. Although this model can be exactly solved using the Bethe ansatz and one has the explicit result that the model is gapless for $-J \leq J_z \leq J$ and gapped for $J_z > J$, (there is a phase transition exactly at the isotropic point $J_z = J$), it is not easy to compute explicit correlation functions in that approach. Hence, it is more profitable to use field theory methods.

Symmetries of the model

Note that this spin model has a global $U(1)$ invariance, which is rotations about the S^z axis. Precisely when $J_z = J$, the $U(1)$ invariance is enhanced to an $SU(2)$ invariance, because at this point the model can simply be written as $H = J \sum_i \mathbf{S_i} \cdot \mathbf{S_{i+1}}$. The model also has discrete symmetries under $S^x \to -S^x$, $S^y \to S^y$, and under $S^z \to -S^z$. Note also that one can change the sign of the XY part of the Hamiltonian by making a rotation by π about the S^z axis on alternate sites, without affecting the J_z term, although this is not an extra symmetry of the model.

Aside on non-linear sigma models

Even using field theory methods, there are two distinct approaches to the problem. In the large-S limit, there exists a semi-classical field theory approach to this model, which leads to an

5.6. Quantum antiferromagnetic spin 1/2 chain

$O(3)$ non-linear sigma model ($NL\sigma M$), with integer and half-integer spins being distinguished by the absence or presence of a Hopf term in the action. In this approach, it is easy to see that integer spin models have a gap in the spectrum. However, it is less easy to study the effect of the Hopf term and show that 1/2-integer spin models are gapless. In fact, in this case, it was the spin model which gave information about field theories with the Hopf term!

Jordan-Wigner transformation

For spin 1/2 models, it is possible to fermionize and then bosonize the spin model and study its spectrum. That is the approach we will follow in the rest of this lecture. First, we will try to convince you that it is possible to rewrite the spin model in terms of spinless fermions. The spin 1/2 model has two states possible at every site - spin ↑ or spin ↓. Hence, it can be mapped to another two state model which we can construct in terms of fermions. We shall assume that an ↑ spin or ↓ can be denoted by the presence or absence of a fermion at that site. Since no more than a single spinless fermion can sit at a site, the degrees of freedom in both the models are the same. This mapping is implemented by introducing a fermion annihilation operator ψ_i at each site and writing the spin at the site as

$$
\begin{aligned}
S_i^z &= \psi_i^\dagger \psi_i - 1/2 = n_i - 1/2 \\
S_i^- &= (-1)^i \, \psi_i e^{i\pi \sum_j n_j} ,
\end{aligned} \quad (5.131)
$$

where the sum runs from one boundary of the chain up to the $(i-1)^{\text{th}}$ site and S_i^+ is the hermitian conjugate of S_i^-. So an ↑-spin is denoted by $n_i = 1$ and the ↓-spin by $n_i = 0$ at the site i. One might have naively guessed that the spin-lowering operator should be expressed by ψ_i which denotes annihilation of a fermion (with the spin raising operator being given by the hermitian conjugate). One can explicitly check that this gives the correct commutation relations of the spin operators at a site because $[S_i^+, S_i^-] = 2S_i^z$ just reproduces the correct anticommutation relations for the fermions $\{\psi_i, \psi_i^\dagger\} = 1$. The extra string factor has to be added in order to correct for different site statistics - the fermions at different sites anticommute, whereas the spin operators commute. In fact, it is instructive to check explicitly that the string operator changes the commutation relation on different sites.

Exercise 5.1 *Check the above explicitly.*

The Hamiltonian

Now, we rewrite the spin model in terms of the fermions. We find that

$$H = -\frac{J}{2}\sum_i [\psi_i^\dagger e^{i\pi n_i}\psi_{i+1} + h.c.] + J_z \sum_i [(n_i - 1/2)(n_{i+1} - 1/2)]$$

$$= -\frac{J}{2}\sum_i [\psi_i^\dagger \psi_{i+1} + h.c.] + J_z \sum_i [(n_i - 1/2)(n_{i+1} - 1/2)] \,. \tag{5.132}$$

The point to notice is that the string operator has cancelled out in the nearest neighbor interaction, except for a phase term, which also can be explicitly shown to be just 1 because $e^{i\pi n_i}$ precedes a creation operator $\psi_i^\dagger$ which can only act if $n_i = 0$. The spin-flip terms are like the hopping terms in the fermion Hamiltonian and give rise to motion of fermions whereas the S^z-S^z interaction term leads to a four fermion interaction between fermions on adjacent sites (the analog of the on-site Hubbard interaction for spinless fermions). So for non-zero J_z, the fermionic model is non-trivial. There exists a competition between the hopping term or kinetic energy term, which gains in energy when the electrons are free to hop from site to site, and the potential energy which costs J_z if there are electrons present on adjacent sites. So naively, for large J_z, one expects the potential energy to win and electrons to be localized on non-adjacent sites, and for small J_z, one expects the kinetic energy to win and to have delocalized fermions. Let us see whether this expectation is true and how it comes about.

5.6. Quantum antiferromagnetic spin 1/2 chain

Set $J_z = 0$

To make the problem simpler, we first consider the case where $J_z = 0$ or where there are no interaction terms. Then this is just the model of free spinless fermions. By Fourier transforming the fermions, - $\psi_j = \sum_k \psi_k e^{ikja}/\sqrt{N}$, (a is the lattice spacing) where the k sum is over momentum values in the first Brillouin zone, - we find that the Hamiltonian is given by

$$H = -J \sum_k \cos(ka)\, \psi_k^\dagger \psi_k \,. \tag{5.133}$$

Exercise 5.2 *Obtain the above Hamiltonian explicitly.*

The discrete symmetry of the model under $S_i^- \leftrightarrow -S_i^+$ and $S_i^z \to -S_i^z$ implies a particle-hole symmetry $\psi_i \leftrightarrow \psi_i^\dagger$ in the fermion language. Thus, the ground state has to have total spin $M \equiv \sum_i S_i^z = 0$ or equivalently in the fermionic language, the ground state is precisely half-filled. This symmetry can be broken by the addition of a magnetic field term that couples linearly to S^z. In the fermionic language, this is equivalent to adding a chemical potential term (which couples to n_i which is the S^z term) in which case, the ground state no longer has $M = 0$ and the fermion model is no longer half-filled. Thus, for $M = 0$, the band is precisely half-filled and the Fermi surface ($E = 0$) occurs exactly at $ka = \pm\pi/2 \equiv k_F a$ (because the density of states is symmetric about $E = 0$.) Low energy excitations are particle-hole excitations about the Fermi surface, which can occur either at a single Fermi point *i.e.*, $k \sim 0$ modes, or across Fermi points, which are the $k \sim 2k_F = \pi/a$ modes.

Effective field theory

The next step is to write down an effective field theory for the low energy modes. Now comes the approximation. Let us make the assumption that it is only the modes near the Fermi surface (or here, the Fermi points), which are relevant at low energies. Hence, we are only interested in ka values near $ka = \pm\pi/2$ and we may approximate the dispersion relation around the Fermi points to be linear - *i.e.*, $\cos(ka) = \cos(\pm k_F a + k'a) = \cos(\pm \pi/2 + k'a) = \mp \sin(k'a) = \mp(k'a)$. We introduce the labels left and right to denote fermion modes near $ka = -\pi/2$ and $ka = \pi/2$ respectively and henceforth drop the primes on the momenta and assume that they are always measured from the Fermi points; as before, we take k as increasing towards the right near the right Fermi point, and increasing towards the left near the left Fermi point. If we want to solve the problem without any approximations, we have to allow for excitations about the Fermi points with arbitrary k. The approximation that is made is that we only allow small values of k compared to k_F. This is why the excitations around the left and right Fermi points can be thought of as independent excitations. In this approximation, the Hamiltonian breaks up into

$$H = Ja \sum_k k \; (\psi^\dagger_{R,k} \psi_{R,k} + \psi^\dagger_{L,k} \psi_{L,k}) \; . \tag{5.134}$$

(Note that we have incorporated the change in sign mentioned below Eq. (5.133)). The fermions around the Fermi points are Dirac fermions since we have linearized the dispersion. These fields do not contain any high momentum modes. In real space, the original non-relativistic fermion field, which has high energy modes (rapidly oscillating factors), has been split up as exponential prefactors times smoothly varying fields -

$$\begin{aligned}\psi_j &\sim e^{-ik_F ja} \int_{-k_F a - \Lambda}^{-k_F a + \Lambda} \frac{d(ka)}{2\pi} e^{ikja} \psi_k \\ &\quad + e^{ik_F ja} \int_{k_F a - \Lambda}^{k_F a + \Lambda} \frac{d(ka)}{2\pi} e^{ikja} \psi_k \\ &\equiv e^{-ik_F ja} \psi_{Lj} + e^{ik_F ja} \psi_{Rj} \; . \end{aligned} \tag{5.135}$$

We assume that the $\Lambda \ll k_F a$ and that it is sufficient to keep just these modes, if we are interested in physics at length scales much greater than $1/\Lambda$, which is of course much greater than the lattice spacing. (The real physical cutoff is the lattice length or in momentum space, the Fermi momentum. As a low energy approximation, we are introducing the larger length cutoff $1/\Lambda$ or

5.6. Quantum antiferromagnetic spin 1/2 chain

the smaller momentum cutoff Λ). For both R and L fermions, states with $k > 0$ are empty and correspond to electron operators (c_k), while states with $k < 0$ are filled and correspond to hole operators ($d_k^\dagger$). (See Fig.5.2). In terms of these operators, the Hamiltonian can be rewritten as

$$H = Ja \sum_{k>0} k \, (c_{L,k}^\dagger c_{L,k} + d_{L,k}^\dagger d_{L,k} + c_{R,k}^\dagger c_{R,k} + d_{R,k}^\dagger d_{R,k}) \,. \quad (5.136)$$

We now introduce continuum fermion fields made up of particle (electron) and anti-particle (hole) operators at the left and right Fermi points as

$$\psi_R(x,t) = \frac{1}{\sqrt{Na}} \sum_{k>0} [c_{R,k} e^{-ik(vt-x)} + d_{R,k}^\dagger e^{ik(vt-x)}]$$

$$\psi_L(x,t) = \frac{1}{\sqrt{Na}} \sum_{k>0} [c_{L,k} e^{-ik(vt+x)} + d_{L,k}^\dagger e^{ik(vt+x)}] \,, \quad (5.137)$$

where $v = Ja$ is defined to be the velocity. Note that the factor of $1/\sqrt{a}$ is needed to relate continuum fermions to lattice fermions. (The factor of $\sqrt{a}$ is needed to get the dimensions right. The lattice fermions satisfy $\{\psi_i, \psi_j^\dagger\} = \delta_{ij}$, whereas continuum fermions satisfy $\{\psi(x), \psi^\dagger(y)\} = \delta(x-y)$ where the Dirac δ-function has the dimension of 1/length. Also $Na = L$ gives the conventional box normalization of the continuum fermions). Note also that the standard inclusion of the e^{-ikvt} for the particle fields and the e^{ikvt} for the anti-particle fields, show that the right-movers are a function only of $x_R = x - vt$ and the left-movers are a function only of $x_L = x + vt$. This observation will come in useful when we compute correlation functions. In general, we only need to compute equal time correlation functions. The time-dependent correlations are then obtained by replacing x by x_R for right-movers and by x_L for left-movers.

In terms of the continuum fields, the Hamiltonian is obtained as

$$H = iv \int dx \, [-\psi_R^\dagger \frac{d}{dx} \psi_R + \psi_L^\dagger \frac{d}{dx} \psi_L] \,. \quad (5.138)$$

Exercise 5.3 *Check that this Hamiltonian reduces to the one in Eq. (5.136) using Eq. (5.137).*

We see that the corresponding Lagrangian density is just the standard one for free fermions given by

$$L = i\psi_R^\dagger(\partial_t + v\partial_x)\psi_R + i\psi_L^\dagger(\partial_t - v\partial_x)\psi_L \ . \tag{5.139}$$

Using the standard rules of bosonization, this Lagrangian can also be rewritten as

$$L = \frac{1}{2v}(\partial_t\phi)^2 - \frac{v}{2}(\partial_x\phi)^2 = \frac{1}{2}\partial_\mu\phi\partial^\mu\phi \ , \tag{5.140}$$

where the last equality requires setting $v = 1$.

Note: It is also worth checking to see that the same Hamiltonian in Eq. (5.138) is obtained by directly starting with the real space lattice model given in Eq. (5.132), rewriting the lattice fermions in terms of the continuum fermions remembering the $\sqrt{a}$ conversion factor, using $\sum_i a = \int dx$ and using $\psi_{i+1} = \psi_i + a\partial_x\psi_i$.

Correlation functions

Thus, we have a Lorentz-invariant massless Dirac fermion field theory in the low energy approximation. All low energy properties can be obtained from the field theory, which in fact are trivially computed, since this is a free massless field theory. As far as fermionic correlation functions are concerned, one does not even require bosonization. However, for the spin correlations, it depends on how the spins can be expressed in terms of fermions. For instance, we can explicitly obtain the following spin-spin correlation function

$$G^{zz}(x,t) \equiv <S^z(x,t)S^z(0,0)> \tag{5.141}$$

simply using Wick theorem. We start by writing S_j^z in term of the fermions as $S_j^z = n_j - 1/2 = \psi_j^\dagger\psi_j - 1/2 =: \psi_j^\dagger\psi_j :$, since the expectation value of n_j is half. Since the lattice fermion can be written in terms of the continuum fermions as

$$\psi_j = \sqrt{a}\left[e^{ik_Fj}\psi_R(x=ja) + e^{-ik_Fj}\psi_L(x=ja)\right] \ , \tag{5.142}$$

5.6. Quantum antiferromagnetic spin 1/2 chain

and since $e^{i2k_F j} = e^{(i\pi)x/a} = (-1)^{x/a}$, we find that the spin operator can be written as

$$\begin{aligned} S_j^z/a &= S^z(x=ja,t) =: \psi_L^\dagger \psi_L : + : \psi_R^\dagger \psi_R : \\ &\quad + (-1)^{x/a}[\psi_R^\dagger \psi_L + \psi_L^\dagger \psi_R]. \end{aligned} \qquad (5.143)$$

Directly using the Wick theorem and the fermion correlators

$$\begin{aligned} <T\psi_L(x,t)\psi_L^\dagger(0,0)> &= \frac{-i}{2\pi(x_L - i\alpha\,(s_t))} \\ \text{and} \quad <T\psi_R(x,t)\psi_R^\dagger(0,0)> &= \frac{i}{2\pi(x_R + i\alpha\,(s_t))}, \end{aligned} \qquad (5.144)$$

we see that

$$G^{zz}(x,t) = -\frac{a^2}{4\pi^2}\left[\left(\frac{1}{x_R^2} + \frac{1}{x_L^2}\right) - (-1)^{x/a}\frac{1}{x_R x_L}\right], \qquad (5.145)$$

where $x_R = x - t$ and $x_L = x + t$ and as before, $s_t = \text{sign}(t)$.

Exercise 5.4 *Obtain this explicitly.*

This can also be computed using bosonization. Note that even without doing the calculation, one could have guessed that the four-point correlation of the fermions must go as $1/l^2$, where l is a distance, because in 1+1 dimensions, the fermion field has a mass dimension of $1/2$ or distance dimension of $-1/2$. So, in the absence of any other scale in the problem (the fermion field is massless and there are no interactions to cause divergences or introduce any anomalous mass scale), as long as the spin correlations can be expressed purely in terms of local fermion fields, no calculations are needed to see that correlations go as $1/l^2$. But we do need to calculate to get the explicit coefficients of $1/x_R^2$, etc., because they could be multiplied by dimensionless quantities like $f(x_R/x_L)$, etc.

However, to obtain the off-diagonal correlation function $<S^+(x,t)S^-(0,0)>$ in the fermionic language is more difficult because of the non-local string operator. Here, simple dimensional analysis is not sufficient to give the answer and one actually needs bosonization. The correlation function can be written as

$$\begin{aligned} G^{+-}(x,t) &= <S^+(x,t)S^-(0,0)> \\ &= (-1)^{x/a}\,[e^{-ik_Fx/a}\psi_R^\dagger(x,t) + e^{ik_Fx/a}\psi_L^\dagger(x,t)] \times \\ & \quad [e^{i\pi\int_0^x(:\psi^\dagger(x',t)\psi(x',t):+1/2a)dx'} + h.c.] \times \\ & \quad [\psi_R(0,0) + \psi_L(0,0)]\,, \end{aligned} \qquad (5.146)$$

where the string operator stretches between the two positions of the spin operator. (The other terms cancel out between S^- and S^+). Also, we have explicitly made the string operator hermitian, since it is hermitian in the lattice model. The reason bosonization comes in handy here is because the non-local operator when written in terms of bosons, turns out to be perfectly simple. We just use the bosonization identity

$$\begin{aligned} \int_0^x dx' \,:\psi^\dagger(x',t)\psi(x',t): &= -\frac{1}{\sqrt{\pi}}\int_0^x dx'\,\partial_{x'}\phi \\ &= -\frac{1}{\sqrt{\pi}}[\phi(x,t) - \phi(0,t)] \\ &= -\frac{1}{\sqrt{\pi}}[\phi_R(x,t) + \phi_L(x,t) - \phi_R(0,t) - \phi_L(0,t)]\,. \end{aligned} \qquad (5.147)$$

Substituting this in Eq. (5.146), and substituting for the other fermion operators in terms of bosons, we get

$$\begin{aligned} G^{+-}(x,t) &= (-1)^{x/a}\,\frac{a}{2\pi\alpha}[\eta_R^\dagger e^{-ik_Fx/a}e^{i2\sqrt{\pi}\phi_R(x,t)} \\ & \quad +\eta_L^\dagger e^{ik_Fx/a}e^{-i2\sqrt{\pi}\phi_L(x,t)}\,] \times \\ & \quad [e^{ik_Fx/a-i\sqrt{\pi}(\phi_R(x)+\phi_L(x)-\phi_R(0)-\phi_L(0))} \\ & \quad + e^{-ik_Fx/a+i\sqrt{\pi}(\phi_R(x)+\phi_L(x)-\phi_R(0)-\phi_L(0))}\,] \times \\ & \quad [\eta_R e^{-i2\sqrt{\pi}\phi_R(0,0)} + \eta_L e^{i2\sqrt{\pi}\phi_L(0,0)}\,] \end{aligned} \qquad (5.148)$$

fully in terms of bosons. Now we use the operator identity $e^{A+B} = e^A e^B e^{-[A,B]/2}$ to write each of the 8 terms that appear in the above equation in terms of products of exponential factors. Just

5.6. Quantum antiferromagnetic spin 1/2 chain

for illustration, we explicitly write the first term which appears by multiplying the first term in each of the square brackets in the above equation.

$$G^{+-}(x,t) = \frac{a}{2\pi\alpha} [\mathcal{A} e^{i\sqrt{\pi}\phi_R(x,t)} e^{-i\sqrt{\pi}\phi_L(x,t)} e^{-i\sqrt{\pi}\phi_R(0,0)} e^{i\sqrt{\pi}\phi_L(0,0)} \\ + 7 \text{ other terms}]. \quad (5.149)$$

with $\mathcal{A}$ denoting $\eta_R^\dagger \eta_R$. Now, we use the standard commutators $[\phi_{R/L}(x), \phi_{R/L}(y)] = (-/+) i \operatorname{sign}(x-y)/4$, and $[\phi_{R/L}(x), \phi_{L/R}(y)] = 0$ (since we are using Klein factors), and the standard algorithm for computing the correlation functions

$$< e^{i2\sqrt{\pi}m_1\phi_L(x)} e^{-i2\sqrt{\pi}m_2\phi_L(0)} > \sim \lim_{\alpha \to 0} \left(\frac{\alpha}{x_L - i\alpha(s_t)}\right)^{m_1 m_2}$$
$$< e^{i2\sqrt{\pi}m_1\phi_R(x)} e^{-i2\sqrt{\pi}m_2\phi_R(0)} > \sim \lim_{\alpha \to 0} \left(\frac{\alpha}{x_R + i\alpha(s_t)}\right)^{m_1 m_2}, \quad (5.150)$$

when m_1 and m_2 have the same sign and vanish when they have opposite signs [6]. This implies that of the 8 terms above, four of them give zero contribution. Adding up the contributions of the remaining four, we obtain

$$G^{+-}(x,t) \sim \frac{1}{(x_R x_L)^{1/4}} [(-1)^{x/a} + \text{const} \left(\frac{1}{x_R^2} + \frac{1}{x_L^2}\right)]. \quad (5.151)$$

Note that the Klein factors always come as $\eta_i^\dagger \eta_i = 1$ in this correlation function. Also note that one cannot fix the arbitrary constant that can appear between the uniform and the alternating parts of the correlation function because of the normal ordering ambiguities. It is only the exponents which can be found.

Exercise 5.5 *Obtain the above explicitly.*

Thus even for the non-interacting theory or purely the XY model, bosonization comes in handy to compute the correlation functions. As we have already said, the reason the correlation functions are not obtainable just by naive scaling arguments is because the expression for the 'off-diagonal' spin correlations in terms of the fermion operators is non-trivial, because of the presence of the string term. These are the only non-zero correlators in the theory. The other correlators such as G^{z+} or G^{++} are zero by symmetry - i.e., because of $U(1)$ invariance in the spin model or because of charge conservation in the fermion model.

Case when $J^z \neq 0$

We now consider the Hamiltonian in Eq. (5.132). In the fermionic language, the last term is given by

$$\delta H = J_z \sum_j : \psi_j^\dagger \psi_j :: \psi_{j+1}^\dagger \psi_{j+1} : . \tag{5.152}$$

At very large J_z, we would expect electrons to be localized on every alternate site so that adjacent sites are not occupied. However, this will not be true for small J_z, so the point of the exercise is to see when this happens and what the ground state looks like, for both small J_z and large J_z. In the low energy limit, we can rewrite this term in terms of the continuum Dirac fermions at the Fermi points (use Eq. (5.142)) as

$$\delta H = aJ_z \int dx [: \psi_R^\dagger(x)\psi_R(x) + \psi_L^\dagger(x)\psi_L(x) + (-1)^{x/a} M(x) :] \times$$
$$: \psi_R^\dagger(x+a)\psi_R(x+a) + \psi_L^\dagger(x+a)\psi_L(x+a)$$
$$+ (-1)^{x/a+1} M(x+a) :] ,$$
$$\text{where} \quad M(x) = \psi_R^\dagger(x)\psi_L(x) + \psi_L^\dagger(x)\psi_R(x) . \tag{5.153}$$

Using the notation $\rho_L(x) = \psi_L^\dagger(x)\psi_L(x)$ and $\rho_R(x) = \psi_R^\dagger(x)\psi_R(x)$, ($\rho_R + \rho_L$ is the charge density, and $\rho_R - \rho_L$ is the current density; ρ_L and ρ_R are also called the left and right moving currents respectively), we can rewrite Eq. (5.153) as

$$\delta H = aJ_z \int dx \ [\ \rho_R(x)\rho_R(x+a) + \rho_L(x)\rho_L(x+a)$$
$$+ \rho_R(x)\rho_L(x+a) + \rho_L(x)\rho_R(x+a)$$
$$- M(x)M(x+a)] . \tag{5.154}$$

5.6. Quantum antiferromagnetic spin 1/2 chain

Here we have used the fact that oscillatory factors integrate to zero. (More precisely, they give rise to higher dimension operators, which, however, are irrelevant and ignored in this analysis). In the current-current terms, we can use the expansions $\rho_L(x+a) = \rho_L(x) + a\partial_x\rho_L(x)$, $\psi_L(x+a) = \psi_L(x) + a\partial_x\psi_L(x)$, etc., and the fact that square of a Fermi field vanishes, e.g., $\psi_L^2(x) = 0$, to deduce that terms of the form $\rho_L(x)\rho_L(x+a)$ are higher dimension operators (they have four fermion operators and at least one derivative term) and renormalize to zero in the $a \to 0$ limit. So among the current-current terms, we are only left with ρ_L-ρ_R cross terms of the form $\rho_L(x)\rho_R(x)$ as the lowest dimension operators. For the four fermion terms in the second line also, we apply the same expansion. Dropping higher derivative terms, we see that the only term which survives in the product of the curly brackets is of the form $-\rho_R(x)\rho_L(x) - \rho_L(x)\rho_R(x)$. The extra negative sign is because we need to anticommute one of the fields. This adds to the $\rho_R\rho_L$ term coming from the first line and we are finally left with

$$\delta H = 4J_z a \int dx\, \rho_L(x)\rho_R(x)\,. \tag{5.155}$$

This is a four fermion term which in continuum quantum field theory is called the Thirring term. In the fermionic language, this is an interacting quantum field theory. However, it is easy to solve by bosonization.

By the standard rules of bosonization for non-interacting fermions, we can write

$$\rho_L = \frac{1}{2\sqrt{\pi}}\left(\frac{1}{v}\partial_t + \partial_x\right)\phi\,, \quad \text{and} \quad \rho_R = \frac{1}{2\sqrt{\pi}}\left(-\frac{1}{v}\partial_t + \partial_x\right)\phi\,. \tag{5.156}$$

In that case, using the units that $Ja = v = 1$, we get

$$\delta L = -\delta H = \frac{J_z}{J\pi}\partial_\mu\phi\partial^\mu\phi\,, \tag{5.157}$$

where δL denotes the change in the Lagrangian. This is precisely of the same form as the bosonization of the free fermion Hamiltonian. So the new Lagrangian is given by

$$L = \frac{1}{2K} \partial_\mu \phi \partial^\mu \phi , \qquad (5.158)$$

where

$$\frac{1}{K} = 1 + \frac{2J_z}{J\pi} . \qquad (5.159)$$

This can be made to look like the free term by redefining the field ϕ - i.e., we define a new field $\tilde{\phi} = \phi/\sqrt{K}$, so that in terms of $\tilde{\phi}$, the Lagrangian is just $\frac{1}{2}(\partial_\mu \tilde{\phi} \partial^\mu \tilde{\phi})$. However, the canonical momentum obtained from the rescaled Lagrangian is just $\tilde{\Pi} = \partial_0 \tilde{\phi}$ whereas the momentum obtained from the Lagrangian in Eq. (5.158) $\Pi = \partial_0 \phi/K$. Hence, the momentum gets rescaled compared to the original momentum as $\tilde{\Pi} = \sqrt{K}\Pi$. Clearly, the new coordinate and momenta satisfy the canonical commutation relations, since they are rescaled in opposite ways. (Remember that ϕ takes values on a compact circle, since the original spin operators are defined in terms of exponential of the boson fields and are invariant under periodic changes of ϕ). But since the right and left mover fields are defined by taking both the field and the canonical momentum, and they scale in different ways, one can no longer write the right and left moving fields in the tilde representation as just scaled versions of the right and left moving fields of the original theory - in fact, they mix up the left and right moving fields. Explicitly,

$$\begin{aligned}
\tilde{\phi}_R(t,x) &= \frac{1}{2}[\tilde{\phi}(t,x) - \int_{-\infty}^x dx'\, \tilde{\Pi}(t,x')] \\
&= \frac{1}{2}[\phi(t,x)/\sqrt{K} - \int_{-\infty}^x dx'\, \sqrt{K}\Pi(t,x')] \\
&= \frac{(\phi_R + \phi_L)}{2\sqrt{K}} - \frac{\sqrt{K}(\phi_L - \phi_R)}{2} \\
&= \frac{1}{2}(\frac{1}{\sqrt{K}} + \sqrt{K})\, \phi_R + \frac{1}{2}(\frac{1}{\sqrt{K}} - \sqrt{K})\, \phi_L \\
&= \cosh\beta\, \phi_R + \sinh\beta\, \phi_L , \qquad (5.160)
\end{aligned}$$

where $e^{-\beta} = \sqrt{K}$. Similarly

$$\tilde{\phi}_L(t,x) = \cosh\beta\, \phi_L + \sinh\beta\, \phi_R . \qquad (5.161)$$

5.6. Quantum antiferromagnetic spin 1/2 chain

One can now express the spin fields in terms of the $\tilde{\phi}$ fields. They are given by

$$S^z(x,t) \simeq \sqrt{\frac{K}{\pi}}\partial_x\tilde{\phi} + (-1)^{x/a}\,\text{const}\;e^{i2\sqrt{\pi K}\tilde{\phi}}$$
$$S^-(x,t) \simeq (-1)^{x/a}e^{i\sqrt{\pi/K}(\tilde{\phi}_R-\tilde{\phi}_L)} + \text{const}\times$$
$$[e^{i(2\sqrt{\pi K}(\tilde{\phi}_R+\tilde{\phi}_L)+\sqrt{\pi/K}(\tilde{\phi}_R-\tilde{\phi}_L))}$$
$$+e^{i(-2\sqrt{\pi K}(\tilde{\phi}_R+\tilde{\phi}_L)+\sqrt{\pi/K}(\tilde{\phi}_R-\tilde{\phi}_L))}]\;. \quad (5.162)$$

With these substitutions, it is trivial (albeit algebraically more tedious!) to recalculate the spin-spin correlators $G^{zz}(x,t)$ and $G^{+-}(x,t)$. Since the method is exactly the same as for the free case, we just quote the answers here.

$$G^{zz}(x,t) \simeq -\frac{K}{4\pi}(\frac{1}{x_L^2}+\frac{1}{x_R^2}) + (-1)^{x/a}\,\text{const}\,(x_R x_L)^{-K}$$
$$G^{+-}(x,t) \simeq (-1)^{x/a}(x_R x_L)^{-1/4K}$$
$$+\,\text{const}\,(x_R x_L)^{-(\frac{1}{2\sqrt{K}}-\sqrt{K})^2}(\frac{1}{x_L^2}+\frac{1}{x_R^2})\;. \quad (5.163)$$

Exercise 5.6 *Obtain the above expressions.*

Note that at $K = 1/2$, the two correlations above are the same.

Limitations of this calculation

So the end result is that we have now obtained spin-spin correlation functions even including J_z. But since we have made a low energy continuum approximation and included only a few low-lying modes around the Fermi point, this derivation of the correlation functions is not true for arbitrary J_z. For instance, we left out terms that were irrelevant by naive power counting, which only works in the non-interacting case. Once we have interactions, some of those operators could acquire anomalous dimensions and hence become relevant. In other words, we have seen that interactions change the dimensions of operators. However, we have only studied operators of the form $\rho_L(x)\rho_R(x)$, which were marginal to start with and seen how they evolved. But we did not keep all the irrelevant operators and see how they evolved. Sometimes, they will also become relevant with sufficiently strong interactions.

A more general effective action approach

However, one can try to understand what can possibly change if we include other corrections that we left out in our approximation. One way of doing this is to look at all possible relevant terms that can appear consistent with the symmetries of the problem. The idea is not to try and derive these terms but to write them down in the effective Lagrangian assuming that if they are not explicitly prohibited by a symmetry, then they will appear. This is the philosophy behind what are called effective field theories.

Aside on how to 'read off' dimensions of operators

We know that to see whether an operator is relevant or irrelevant, we have to compute its correlation function and find out its scaling dimension. Then, we have to check whether the scaling dimension is such that the coefficient of the operator grows or becomes smaller as the energy scale is reduced. So given any operator O_i in terms of bosons, we first compute the correlation function $< O_i(x,t) O_i(x',0) >$ which goes as $1/(x-x')^{2d_i}$. For a free fermion theory with no interactions, (equivalently a boson theory with the interaction parameter $K=1$) we know that $< O_i(x,t) O_i(x',0) > = 1/(x-x')^{2\tilde{d}_i}$ where $\tilde{d}_i$ is just the naive scaling dimension or the engineering dimension of the operator O_i. The difference between $\tilde{d}_i$ and the d_i that appears when we actually compute the correlation function is because of the interactions and is called the anomalous dimension of the operator. As was explained in the other courses in this school [13], the extra dimensional parameter comes from the cutoff scale. We shall use the term scaling dimension to mean d_i itself. For an operator of the form $O_i \sim e^{i2\sqrt{\pi}\beta(\phi_L+\phi_R)}$, the scaling dimension is given by $d_i = \beta^2$, for the standard (non-interacting) form of the Hamiltonian. Since the space-time dimension is two, it is clear that $d_i > 2$ implies that the coefficient λ_i of the operator has to have dimension $2 - d_i < 0$. So each time the cutoff is scaled down by

5.6. Quantum antiferromagnetic spin 1/2 chain

a factor Λ, $\lambda_i \to \lambda_i \Lambda^{2-d_i}$. Hence, after successive rescalings, this term in the action is irrelevant and scales to zero. On the other hand, $d_i < 2$ denotes relevant operators, whose coefficients grow under scaling downs of the cutoff. $d_i = 2$ is a marginal operator, whose coefficient remains unchanged under rescalings. (We will come back to this when we study impurity scattering and scaling dimensions of 'boundary operators').

Back to the effective action

The only possible Lorentz-invariant relevant terms that can be added to the Lagrangian is either $\cos 2\sqrt{\pi}\beta \, (\tilde{\phi}_L + \tilde{\phi}_R)$ or $\cos 2\sqrt{\pi}\beta \, (\tilde{\phi}_L - \tilde{\phi}_R)$; both of these have dimension β^2 and are thus relevant for $\beta < \sqrt{2}$. (The real problem on a lattice, of course, does not have Lorentz invariance. However, in the long distance or low energy limit, all such Lorentz non-invariant interactions will probably be irrelevant). Of these, the $U(1)$ symmetry under rotations about the z-axis in fermion language implies that ψ_L and ψ_R have to be multiplied by the same phase (because S_z which has terms of the form $\psi_L^\dagger \psi_R + h.c.$ should not change). This in turn means that $\tilde{\phi}_L \to \tilde{\phi}_L + c$ and $\tilde{\phi}_R \to \tilde{\phi}_R - c$ so that $\tilde{\phi}_L + \tilde{\phi}_R \to \tilde{\phi}_L + \tilde{\phi}_R$ and $\tilde{\phi}_L - \tilde{\phi}_R \to \tilde{\phi}_L - \tilde{\phi}_R$+constant. Thus, to be consistent with this symmetry, we can only allow $\cos 2\sqrt{\pi}\beta(\tilde{\phi}_L + \tilde{\phi}_R)$.

Furthermore, since the spin operators are all expressed in terms of exponentials of the boson fields, (see Eq. (5.162)) the boson fields need to be 'compactified on a circle'. This only means that the boson fields are periodic -

$$\tilde{\phi} \leftrightarrow \tilde{\phi} + \sqrt{\frac{\pi}{K}} \qquad (5.164)$$

since the spin fields cannot distinguish between $\tilde{\phi}$ and $\tilde{\phi} + \sqrt{\pi/K}$. This restricts β in $\cos 2\sqrt{\pi}\beta(\tilde{\phi}_L + \tilde{\phi}_R)$ to be of the form $n\sqrt{K}$ where n is an integer.

Finally, we use an unusual feature which occurs in the continuum field theories of many lattice spin models. The translational symmetry of the lattice spin model by one site (or more sites for more general models) maps to a discrete symmetry in the continuum model, which is distinct from translational symmetry. This can be seen from the continuum definition of the spin in terms of the Dirac fermions - Eq. (5.143). When we change j to $j+1$, the oscillatory factor $(-1)^j \to -(-1)^j$. This is a drastic change from site to site. So if in the continuum version, we want to define smooth fields without having this rapid oscillations, we need to define one field for every pair of sites. Thus invariance under translation by $2a$ on the lattice becomes translational invariance in the continuum model. But from Eq. (5.143), we see that translational symmetry through a single site corresponds to the discrete symmetry

$$\psi_L \to i\psi_L, \quad \psi_R \to -i\psi_R . \tag{5.165}$$

In the bosonic language, this corresponds to

$$\tilde{\phi}_L \to \tilde{\phi}_L + \frac{1}{2}\sqrt{\frac{\pi}{K}}, \quad \text{and} \quad \tilde{\phi}_R \to \tilde{\phi}_R + \frac{1}{2}\sqrt{\frac{\pi}{K}} . \tag{5.166}$$

This symmetry implies that the only terms that can be added to the Lagrangian are of the form $\cos 2\sqrt{\pi}2n\sqrt{K}(\tilde{\phi}_L + \tilde{\phi}_R)$, so that $\beta = 2n\sqrt{K}$. This is relevant when $K < 1/2$ when $n = 1$. So the system is in a massless phase till K reaches $1/2$ below which it develops a relevant interaction, and a mass gap.

However, we cannot use our low energy approximate result to estimate the point at which the spin model develops a relevant interaction. Besides adding the possible relevant term mentioned above, the most general thing the other terms that we have neglected can do is to change the relation between K and J_z in an unpredictable way. In fact, the low energy result relating K to the perturbation J_z is only true to lowest order in J_z/J. This particular spin-chain model is, in fact, solvable by Bethe ansatz and the exact answer is

$$\frac{1}{K} = 1 + \frac{2}{\pi}\sin^{-1}(\frac{J_z}{J}) \tag{5.167}$$

which, to lowest order in J_z/J, gives us the relation in Eq. (5.159). From this, we see that $1/K \to 2$ at $J_z = J$. This is precisely the

5.6. Quantum antiferromagnetic spin 1/2 chain

K value for which the cosine interaction term becomes relevant. To prove that a relevant interaction necessarily leads to a mass gap is non-trivial, but it is certainly plausible. Once, there is a relevant interaction, its coefficient grows under renormalization. It becomes divergent as we make the energy scale lower and lower, so we have to cut it off at some scale, which is the mass scale associated with the theory. However, it could also lead us to a new fixed point, which may not have a mass gap.

To get higher orders in J_z/J in this effective field theory approach is not easy, because one needs to go beyond the region of linear dispersion. Also, once one starts including modes with $k \simeq k_F$, we need to be careful to put in the restriction that $-k_F < k < k_F$. Also, since the Bethe ansatz already gives the exact answer to all orders, there may not be much point in trying to do this for this problem.

So what does all this formalism gain us? How does the ground state evolve as J_z changes? For small values of J_z all that happens is that the spin-spin correlations have a slightly different power law fall-off with anomalous non-integer exponents. Does this continue for all values of J_z? No. Once, J_z reaches $J_z = J$, the isotropic point, there exists a phase transition to a massive phase where spin-spin correlations fall off exponentially fast at large separations. In this particular problem, of course, one knew this answer from the Bethe ansatz, but the point is that the bosonization method can be used even for other models, which are not exactly solvable by the Bethe ansatz. But without the Bethe ansatz, one cannot analytically find the value of the parameter where the phase transition into a massive phase occurs. The other important gain that we have in this method is that it allows the computation of correlation functions, which is not possible using the Bethe ansatz. Finally, since it is a symmetry analysis, it tells us that for any Hamiltonian of this type, the model is likely to be massless only when the theory has $U(1)$ symmetry and the Z_2 symmetry of translation by a single site and even then, only for some restricted values of the parameter space.

The best reference for this application is Affleck's lectures [5] on field theories and critical phenomena, which we have followed fairly faithfully.

5.7 Hubbard model

The Hubbard model is one of the simplest realistic models that one can study which has a competition between the kinetic energy and the potential energy. The kinetic energy or the hopping term gains, or rather the energy gets lowered, if the fermions are delocalized - free to move throughout the sample. In this model, the potential energy represents screened Coulomb interactions between electrons and the model is constructed so that it costs energy to put two electrons at the same place. So the potential energy prefers each electron to sit at its own site. The model is given by

$$H = -\frac{t}{2}\sum_{j\alpha}(\psi_{j\alpha}^\dagger \psi_{j+1\alpha} + h.c.) + U\sum_j n_{j\uparrow}n_{j\downarrow} + \mu\sum_{j,\alpha}\psi_{j\alpha}^\dagger\psi_{j\alpha},$$
(5.168)

where t is the hopping parameter, U is a positive constant denoting the repulsion between two electrons at a site, μ is the chemical potential and α is the spin index which can be $\uparrow$ or $\downarrow$. This model is very similar to the fermion model we studied for the spin chain except that the electrons have spin and the chemical potential term allows for arbitrary fillings. The U term or Hubbard term is analogous to the nearest neighbor J_z interaction term for spinless electrons.

At half-filling (one electron/site, since a filled band implies two electrons/site), for large U, the model is expected to describe an insulator. One can easily understand this, because at infinite U, the ground state will have one electron at every site. Any excitation will cost an energy of U. So there is a gap to excitations and the model behaves as an insulator. It is called a Mott-Hubbard insulator (as opposed to other band insulators) because here the insulating gap is created by interactions.

The question that one would like to ask is, at what value of U does the Mott-Hubbard gap open, because naively one may think

5.7. Hubbard model

that at very small values of U, the model allows free propagation of electrons and describes a metal. Using bosonization, we will show that in one dimension, this expectation is wrong. The Mott-Hubbard gap opens up for any finite U if the filling is half and not otherwise. For any other filling, the model at low energies is an example of a Luttinger liquid with separate spin and charge excitations. The spin modes are always gapless whereas the charge modes are gapless at any filling other than half-filling; precisely at half-filling a charge gap opens up. The model for arbitrary filling (other than half-filling) and positive U is said to be in the Luttinger liquid phase. Spin and charge correlations fall-off as power laws and we expect power law transport. At half-filling, the model is in a charge-gapped phase called the charge-density wave phase.

Aside: For negative U, it is found that the spin excitations are always gapped. Here, the model is said to be in the Luther-Emery phase or spin gapped phase.

Bosonization of the model without interactions

How do we go about seeing all that we have described above? In higher dimensions, we would do a mean field theory, but in one space dimension, we know that a mean field analysis is not very useful because of the infrared divergences of the low energy fluctuations. (In other words, if we write down a mean field theory and then try to do systematic corrections about the mean field theory, then order by order in perturbation theory, we find that the integrals which appear in the corrections are divergent). So it seems like a good idea to try and use bosonization. In fact, the way this model is analyzed is very similar to the way we analyzed the spinless fermion model in the previous section. We first switch off the interactions and start with the Fourier decomposition

$$\psi_{j\alpha} = \frac{1}{\sqrt{N}} \sum_k \psi_{k\alpha} e^{ikja} . \qquad (5.169)$$

We rewrite the Hamiltonian as

$$H_0 = \sum_{k\alpha} (\mu - t\cos ka)\, \psi_{k\alpha}^\dagger \psi_{k\alpha} , \qquad (5.170)$$

where the k values go from $-\pi/a$ to π/a. In the ground state, all states with $|k| < k_F$ are filled, where k_F is determined by the chemical potential from the equation $\mu = \cos k_F a$. Just as in the spinless case, we will look only at the low energy modes near the Fermi surface, so that each fermion is written as

$$\psi_{j\alpha} \sim e^{-ik_F ja} \int_{-k_F a - \Lambda}^{-k_F a + \Lambda} \frac{d(ka)}{2\pi} e^{ikja} \psi_{k\alpha}$$
$$+ e^{ik_F ja} \int_{k_F a - \Lambda}^{k_F a + \Lambda} \frac{d(ka)}{2\pi} e^{ikja} \psi_{k\alpha}$$
$$\equiv e^{-ik_F ja} \psi_{Lj\alpha} + e^{ik_F ja} \psi_{Rj\alpha}, \quad (5.171)$$

so that the $\psi_{Lj\alpha}$ and $\psi_{Rj\alpha}$ do not contain high energy modes. Substituting this expression in Eq. (5.168), we get

$$H_0 = -\frac{t}{2} \sum_{j\alpha} [(e^{-ik_F a} \psi^\dagger_{Lj\alpha} \psi_{Lj+1\alpha} + e^{ik_F a} \psi^\dagger_{Rj\alpha} \psi_{Rj+1\alpha} +$$
$$e^{-ik_F a(2j+1)} \psi^\dagger_{Lj\alpha} \psi_{Rj+1\alpha} + e^{ik_F a(2j+1)} \psi^\dagger_{Rj\alpha} \psi_{Lj+1\alpha}) + h.c.]$$
$$+ \mu \sum_{j\alpha} [\psi^\dagger_{Lj\alpha} \psi_{Lj\alpha} + \psi^\dagger_{Rj\alpha} \psi_{Rj\alpha} +$$
$$e^{-i2k_F ja} \psi^\dagger_{Lj\alpha} \psi_{Rj\alpha} + e^{i2k_F ja} \psi^\dagger_{Rj\alpha} \psi_{Lj\alpha}]. \quad (5.172)$$

The oscillatory terms do not contribute because they have the form

$$\sum_j e^{i2k_F ja} \psi^\dagger_{Rj\alpha} \psi_{Lj\alpha} \sim \sum_j e^{i2k_F ja} \sum_{k,k'} e^{-ikja} e^{ik'ja} \psi^\dagger_{k\alpha} \psi_{k'\alpha}$$
$$\sim \sum_{k,k'} \psi^\dagger_{k\alpha} \psi_{k'\alpha} \sum_j e^{i2k_F ja - ikja + ik'ja}, (5.173)$$

and the sum over j in the last expression produces $\delta_{2k_F, k-k'}$ which cannot be satisfied for small values of k, k'. (Note that even for the half-filled case where $k_F a = \pi/2$, this cannot be satisfied). Hence, we may drop these terms and we are left with only

$$H_0 = -\frac{t}{2} \sum_{j\alpha} [(e^{-ik_F a} \psi^\dagger_{Lj\alpha} \psi_{Lj+1\alpha} + e^{ik_F a} \psi^\dagger_{Rj\alpha} \psi_{Rj+1\alpha}) + h.c.]$$
$$+ \mu \sum_{j\alpha} (\psi^\dagger_{Lj\alpha} \psi_{Lj\alpha} + \psi^\dagger_{Rj\alpha} \psi_{Rj\alpha}). \quad (5.174)$$

5.7. Hubbard model

Now, we expand $\psi_{j+1\alpha} \equiv \psi_\alpha(j+1) = \psi_\alpha(j) + a\partial_x \psi_\alpha(j) +$ higher order irrelevant terms and use the fact that $\cos k_F a = \mu$ (which means that part of the hopping term cancels with the μ term) to get

$$H = -\frac{at}{2} \sum_{j\alpha} (e^{-ik_F a} \psi^\dagger_{Lj\alpha} \partial_x \psi_{Lj\alpha} + e^{ik_F a} \psi^\dagger_{Rj\alpha} \partial_x \psi_{Rj\alpha} + h.c.)$$

$$= iat \sin(k_F a) \sum_{j\alpha} (\psi^\dagger_{Lj\alpha} \partial_x \psi_{Lj\alpha} - \psi^\dagger_{Rj\alpha} \partial_x \psi_{Rj\alpha}), \quad (5.175)$$

where we have also integrated the hermitian conjugate terms by part to get it in the form of the second equation above. Finally, we can rewrite this Hamiltonian as a continuum Hamiltonian in terms of continuum fields (defined with the usual factor of $\sqrt{a}$ as $\psi_\alpha(j)/\sqrt{a} = \psi_\alpha(x)$) and using $\sum_j a = \int dx$

$$H_0 = it \sin(k_F a) \sum_\alpha \int dx \, [\psi^\dagger_{L\alpha}(x) \partial_x \psi_{L\alpha}(x) - \psi^\dagger_{R\alpha}(x) \partial_x \psi_{R\alpha}(x)],$$
$$(5.176)$$

where we call $t \sin(k_F a) = v_F a$, the Fermi velocity times a. The derivation here is very similar to the one for spinless fermions, except that here we have carried it out in real space instead of momentum space. This Hamiltonian can be bosonized using the usual rules of bosonization and we get

$$H_0 = \frac{v_F a}{2} \sum_\alpha \int dx \, [\Pi_\alpha^2 + (\partial_x \phi_\alpha)^2]. \quad (5.177)$$

Exercise 5.7 *Derive the Hamiltonian in Eq. (5.176) through a momentum space derivation.*

Bosonization of the interaction term

The next step is to figure out the low energy part of the on-site Hubbard interaction. Here, again, the principle is the same. We rewrite the four-fermion term written in terms of the original fermions in terms of the low energy Dirac fermion modes. Just as in the spin model, the $S^z - S^z$ term or the four fermion term corresponded to a product of normal ordered bilinears, here also the four fermion term in Eq. (5.168) can be written in terms of the product of normal ordered bilinears if we subtract the average charge densities of the $\uparrow$ and $\downarrow$ fields. So we may write

$$H_{\text{int}} = U \sum_j n_{j\uparrow} n_{j\downarrow} = U \sum_i : n_{j\uparrow} :: n_{j\downarrow} : . \qquad (5.178)$$

In terms of the Dirac fields, this becomes

$$\begin{aligned} H_{\text{int}} &= U \sum_j [(: \psi^\dagger_{jL\uparrow} \psi_{jL\uparrow} : + : \psi^\dagger_{jR\uparrow} \psi_{jR\uparrow} : \\ &+ \psi^\dagger_{jR\uparrow} \psi_{jL\uparrow} e^{-i2k_F ja} + \psi^\dagger_{jL\uparrow} \psi_{jR\uparrow} e^{i2k_F ja}) \times (\uparrow \to \downarrow)]. \end{aligned} \qquad (5.179)$$

We now expand the products and keep only the terms with no oscillatory factor, to get

$$\begin{aligned} H_{\text{int}} &= U \sum_j (J_{jR\uparrow} + J_{jL\uparrow})(J_{jR\downarrow} + J_{jL\downarrow}) \\ &+ U \sum_j (\psi^\dagger_{jR\uparrow} \psi_{jL\uparrow} \psi^\dagger_{jL\downarrow} \psi_{jR\downarrow} + h.c.) . \end{aligned} \qquad (5.180)$$

The remaining terms have the oscillatory factors of either $e^{i2k_F ja}$ or $e^{i4k_F ja}$ and can be set to zero for arbitrary filling. Notice however, that $e^{i4k_F ja} = 1$ and is not oscillatory at half-filling since $k_F a = \pi/2$. We will come back to this point later. Now we first express these fields in terms of the continuum fields and just use the standard bosonization formulae to get

$$\begin{aligned} H_{\text{int}} &= Ua \int dx \, [\frac{1}{\pi} \partial_x \phi_\uparrow \partial_x \phi_\downarrow + \eta^\dagger_{R\uparrow} \eta_{L\downarrow} \eta^\dagger_{L\downarrow} \eta_{R\uparrow} \times \\ &\quad \frac{Ua}{2\pi^2 \epsilon^2} \cos \sqrt{4\pi}(\phi_{R\uparrow} + \phi_{L\uparrow} - \phi_{R\downarrow} - \phi_{L\downarrow})] . \end{aligned} \qquad (5.181)$$

Exercise 5.8 *Derive the above.*

The interesting point to note here is that the cosine term only depends on $\phi_\uparrow - \phi_\downarrow$. (We use the earlier defined notation that

5.7. Hubbard model

$\phi = \phi_L + \phi_R$ and $\theta = -\phi_R + \phi_L$). So if we define the charge and spin fields

$$\phi_c = \frac{\phi_\uparrow + \phi_\downarrow}{\sqrt{2}}, \quad \text{and} \quad \phi_s = \frac{\phi_\uparrow - \phi_\downarrow}{\sqrt{2}}, \quad (5.182)$$

the Hamiltonian is completely separable in terms of these two fields and we may write $H = H_0 + H_{\text{int}} = H_c + H_s$ with

$$\begin{aligned} H_c &= \frac{v_F}{2} \int dx \, [\Pi_c^2 + (1 + \frac{U}{\pi v_F})(\partial_x \phi_c)^2] \\ H_s &= \frac{v_F}{2} \int dx \, [\Pi_s^2 + (1 - \frac{U}{\pi v_F})(\partial_x \phi_s)^2 + \frac{Ua}{2\pi\epsilon^2} \cos\sqrt{8\pi}\phi_s] , \end{aligned}$$

(5.183)

where the bosonized form of the kinetic energy term given by H_0 in Eq. (5.177) along with the first term in Eq. (5.181) ($U\partial_x\phi_\uparrow \partial_x \phi_\downarrow/\pi$) can also be written in terms of the charge and spin fields as above. The charge sector is massless, but for the spin sector, one has a cosine term in the Hamiltonian. From our earlier experience of spinless models, we know that a cosine term can lead to a mass gap, when it becomes relevant. So we need to compute the dimension of the operator and see when it becomes relevant. Note that we have chosen the product of the Klein factors to be unity[1]. But we only know how to compute correlation functions when the quadratic Hamiltonian is in the standard form. To get that, we need to rescale the ϕ fields and their conjugate momenta (in the opposite way so that the commutation relations are preserved) as

$$\bar{\phi}_c = (1 + \frac{U}{\pi v_F})^{1/4} \phi_c, \quad \text{and} \quad \bar{\Pi}_c = (1 + \frac{U}{\pi v_F})^{-1/4} \Pi_c, \quad (5.184)$$

and similarly for the spin fields to get the Hamiltonian in the standard form, from which we can directly read out the dimensions

---

[1] For single chain problems, the Klein factors usually cause no problems and can be set to be unity, in most cases. The only care that we need to take is to remember the negative sign that one gets when two of them are exchanged. But for multi-chain models, when more than four explicit Klein factors exist, one needs to be more careful.

of the operators. In terms of the bar fields, we see that

$$\begin{aligned}
H_c &= (1 + \frac{U}{\pi v_F})^{1/2} \frac{v_F a}{2} \int dx [\bar{\Pi}_c^2 + (\partial_x \bar{\phi}_c)^2] \\
H_s &= (1 - \frac{U}{\pi v_F})^{1/2} \frac{v_F a}{2} \int dx [\bar{\Pi}_s^2 + (\partial_x \bar{\phi}_s)^2 \\
&\quad + \frac{Ua}{2\pi\epsilon^2} \frac{1}{(1 - \frac{U}{\pi v_F})^{1/2}} \cos \sqrt{\frac{8\pi}{(1 - \frac{U}{\pi v_F})^{1/4}} \bar{\phi}_s}].
\end{aligned}$$
(5.185)

The charge sector is purely quadratic (both before and after rescaling!) and remains massless, whereas for the spin sector, the rescaling was necessary to 'read off' the dimension of the cosine operator. Since its scaling dimension is given by $d = 2/(1 - \frac{U}{\pi v_F})^{1/4}$, it is irrelevant ($d < 2$) for any weak positive U and the spin sector is also massless. On the other hand, for any negative U, this term has dimension $d > 2$ and is relevant. As we explained in the spinless case, this means that the spin sector acquires a mass gap for all negative U.

Also note that the velocities of the charge and the spin modes have got renormalized in different ways. $v_c = (1 + \frac{U}{\pi v_F})^{1/2} v_F$ is the velocity of the charge mode and $v_s = (1 - \frac{U}{\pi v_F})^{1/2} v_F$ is the velocity of the spin mode. Thus, spin and charge move independently. This is one of the hallmarks of Luttinger liquid behavior in one-dimensional fermion models. It is only for $U = 0$, that the spin and the charge modes move together.

How does one look for such spin-charge separation in one-dimensional models? Experimentally, one has to look at different susceptibilities and measure the Wilson ratio, which is the ratio of the spin susceptibility to the specific heat coefficient. The specific heat coefficient depends both on spin and charge modes and is given by

$$\frac{\gamma}{\gamma_0} = \frac{1}{2} (\frac{v_F}{v_c} + \frac{v_F}{v_s}),$$
(5.186)

where γ_0 is the specific heat of non-interacting electrons with velocity v_F. However, spin susceptibility only depends on the spin

5.7. Hubbard model

mode and is given by

$$\frac{\chi}{\chi_0} = \frac{v_F}{v_s}. \tag{5.187}$$

Thus, the Wilson ratio is given by

$$R_W = \frac{\chi/\chi_0}{\gamma/\gamma_0} = \frac{2v_c}{v_c + v_s}. \tag{5.188}$$

Clearly, when there is no spin-charge separation, this is given by one. So deviations of the Wilson ratio from unity are a sign of spin-charge separation in real systems.

Finally, let us consider the case exactly at half-filling, $k_F a = \pi/2$. In this case, the $e^{i4k_F j a}$ term we neglected in Eq. (5.179) as oscillatory, is no longer oscillatory, since $e^{i4k_F a} = 1$. In this case, there exists a term in the Hamiltonian of the form

$$H_{\text{umklapp}} = U \sum_j (\psi_{jR\uparrow}^\dagger \psi_{jL\uparrow} \psi_{jR\downarrow}^\dagger \psi_{jL\downarrow} + h.c.) . \tag{5.189}$$

Note that this term destroys two right movers and creates two left movers or vice-versa. So there is an overall change in momentum by $4k_F a = 2\pi$, which has to be absorbed by the lattice. It is an umklapp process unlike the earlier interaction term for arbitrary filling which created and destroyed a particle at the left Fermi point and also created and destroyed a particle at the right Fermi point and did not change any momentum. It is easy to see that this term also gives rise to a cosine term by bosonizing, which, after rescaling gives

$$\frac{Ua}{2\pi\epsilon^2\sqrt{(1+U/\pi v_F)}} \cos\sqrt{\frac{8\pi}{(1+U/\pi v_F)^{1/4}}} \bar\phi_c$$

neglecting Klein factors. Thus it appears in the Hamiltonian of the charge sector. This term is irrelevant for any negative U, but relevant for any positive U. Thus, precisely at half-filling, the charge sector has a gap. This is similar to the case for spinless fermions where the spin model actually corresponded to a half-filled spinless fermion model. But unlike the case for the spinless fermions where the gap only opens up at $J = J_z$, here the gap opens up for any positive U, however small.

Is there any way one can understand these results in a physical way? For negative U, we found that the spin sector has a gap. This can be understood by saying that since there is an attractive interaction between the spin ↑ and the spin ↓ densities of fermions, they will like to form singlets and sit on a single site. So to make a spin excitation, one needs to break a pair and this costs energy. But charge excitations can move around as bound spin singlet pairs with no cost in energy. On the other hand, for positive U, there exists a repulsion between two electrons at a site. So each electron will tend to sit on a different site. At half-filling, hence, there is no way for an electron to move, without trying to sit at a site, at which an electron is already present. And this costs a repulsive energy U. Hence, there is a gap to charge excitations. But one can flip spins at a site and hence have spin excitations with no cost in energy.

So what results has bosonization given us here? We started with electrons with spin and charge moving together via a hopping term, but with a strong on-site Coulomb repulsion. This term could not be treated perturbatively. However, when we rewrote the theory in terms of bosons, with one boson for the ↑ spin and one for the ↓ spin, we found that the theory decoupled in terms of new spin and charge bosons. For a generic filling, the charge boson was just a massless free boson excitation, whereas the spin boson Hamiltonian had a cosine term, which was relevant when U was negative, but irrelevant for positive U. But at half-filling, the charge excitations develop a gap, for any positive U. The most important thing to note here is that the charge and spin degrees of freedom have completely decoupled. Since the two fields are scaled differently, they move with different velocities in the system. This is a result that one could never have obtained perturbatively. Thus, at any filling other than half-filling, the low energy limit of the Hubbard model is a Luttinger liquid with massless spin and charge excitations moving with different velocities. A good reference for this part is Shankar's article [6] which also explains in great detail how to compute correlation functions.

5.8 Transport in a Luttinger liquid - clean wire

The last two applications involved the study of correlation functions, with the aim of finding out the different phases possible in a one-dimensional system of interacting fermions. In this part of the course, we will study another application of bosonization, which is to study transport, in particular the DC (or zero frequency) conductivity in one-dimensional wires of interacting fermions.

Firstly, are one-dimensional wires experimentally feasible? The general idea to make narrow wires is to 'gate' 2D electron gases. In recent times, technology has developed enough to make these wires so narrow, that they contain only one transverse channel. So these are good enough approximations to one-dimensional wires. Another good approximation to coupled chains of one-dimensional models are carbon nanotubes, though those are not the kind of models we will study here.

The next point to note is that even at a qualitative level, transport in low dimensional systems is extremely different from transport in higher dimensions. To understand this point, we will first make qualitative statements about transport and conductivity before we explicitly start computing it using bosonization. The usual aim is to compute the conductance as a function of the voltage, temperature, presence of impurities or disorder and so on. Normally, when currents are measured in wires, one does not worry about quantum effects, because wires are still macroscopic objects, but that is clearly not the case here, since we are interested in one-dimensional wires. In fact, whenever the physical dimensions of the conductor becomes small, (it need not be really one-dimensional), the usual Ohmic picture of conductance where the conductance is given by

$$G = \sigma \frac{W}{L} = \sigma \frac{\text{width of conductor}}{\text{length of conductor}}, \quad (5.190)$$

where σ is a material dependent quantity called conductivity, breaks down. A whole new field called 'mesoscopic physics' has now been created to deal with electronic transport in such systems. The term 'mesoscopic' in between microscopic and macroscopic is used for systems, where the sizes of the devices are such that it is comparable with a) the de Broglie wavelength (or kinetic energy) of the electron, b) the mean free path of the electron and c) the phase relaxation length (the length over which the particle loses memory of its phase) of the electron. Ohmic behavior is guaranteed only when all these length scales are small compared to the size which happens for any macroscopic object. These lengths actually vary greatly depending on the material and also on the temperature. Typically, at low temperatures, they vary between a nanometer for metals to a micrometer for quantum Hall systems.

For mesoscopic wires, in general, quantum effects need to be taken into account. One way of computing these conductances is by using the quantum mechanical formulation of transmission and reflection through impurities and barriers. This formulation is called the Landauer-Buttiker formulation and works for Fermi liquids. However, it does not include interactions. But for one dimensional wires, interactions change the picture dramatically, since the quasi-particles are no longer fermion-like. Hence the Landauer-Buttiker formalism cannot be directly applied and one needs to compute conductances in Luttinger wires taking interactions into account right from the beginning. One way of doing this is by using bosonization and this is the method that we will follow here.

The aim is to compute the conductance of a one-dimensional wire. First, we will compute the conductance through a clean wire (no impurities or barriers) and argue why the conductance is not renormalized by the interaction. Then we will study the conductance again after introducing a single impurity. Here, we will see that the interactions change the picture dramatically. For a non-interacting one-dimensional wire, from just solving usual one-dimensional quantum mechanics problems, we know that we can get both transmission and reflection depending on the strength of the scattering potential. But for an interacting wire, we shall find that for any scattering potential, however small, for repulsive interactions between the electrons, there is zero transmission and full reflection (implies conductance is zero, or that the wire is 'cut') and for attractive interactions between electrons (which is of course possible only for some renormalized 'effective' electrons), there is full transmission and zero reflection (implying perfect conductance or 'healing' of the wire).

Ballistic conductor

Let us first define the conductance of a mesoscopic ballistic conductor (*i.e.*, a conductor with no scattering) without taking interactions into account. We said earlier that the usual definition of conductance as $G = \sigma \frac{W}{L}$ breaks down for mesoscopic systems. For instance, it is seen that instead of the conductance smoothly going down as a function of the area or width of the wire W, it starts going down discretely in steps, each of height $2e^2/h$. Also as L decreases, instead of increasing indefinitely, G saturates at some limiting value G_c. The general understanding now, is that as the wire becomes thinner and thinner, the current is carried in a very few channels, each of them carrying a current of $2e^2/h$ (two for spin degeneracy) until we reach the lowest value which is just a single channel (which we interpret as the lowest eigenstate of the transverse Hamiltonian) carrying this current. Moreover, as the length decreases, the resistance does not decrease indefinitely but instead reaches a limiting value. One way of understanding this is to simply consider this to be a contact resistance, independent of the length of the wire, which arises simply because the conductor and the contacts are different. One cannot make the contacts the same as the conductor, because then our assumption that the voltage drop is across the conductor alone does not make sense. That makes sense only if we assume that the contacts are infinitely more conducting than the conductor. So we are finally left with a non-zero resistance and the wire does not become infinitely conducting. In fact, in this limit, the conductance or resistance of the wire is purely a 'boundary' property and the 'conductivity' of the wire has no real significance. In fact, whether we get a finite conductivity or infinite conductivity depends on how one defines it.

However, for a single channel wire, clearly, the wire is one-dimensional and we know that interactions can change the picture drastically. The question that we want to answer here is precisely that. What is the conductance of a clean one-dimensional interacting wire or Luttinger wire?

Computing conductance of a clean one-dimensional (mesoscopic) wire

(a) Without leads

First, we shall perform a calculation to compute the conductance of a Luttinger liquid without any consideration of contacts or leads. (We shall restrict ourselves to spinless fermions since spin only increases the degrees of freedom and gives an overall multiplicative factor of two in the conductance). The conductance of a wire is calculated by applying an electric field to a finite region L of an infinitely long wire and the current I is related to the field as

$$I(x) = \int_0^L dx' \int \frac{d\omega}{2\pi} \, e^{-i\omega t} \sigma_\omega(x, x') E_\omega(x') \,, \quad (5.191)$$

where $E_\omega(x')$ is the frequency ω component of the time Fourier transform of the electric field. The conductivity $\sigma_\omega(x, x')$, in turn, is related to the (imaginary time) current-current correlation function by the usual Kubo formula as

$$\sigma_\omega(x, x') = -\frac{e^2}{\bar{\omega}} \int_0^\beta d\tau \, <T_\tau j(x,\tau) j(x',0)> e^{-i\bar{\omega}\tau} \,, \quad (5.192)$$

where $\tau = it$, $\omega = i\bar{\omega} + \epsilon$, T_τ is the (imaginary) time ordering operator and $j(x, \tau)$ is the current operator. Both these formulae are standard in many books [14] on many body techniques, so here we will confine ourselves to just describing what they mean. The first equation describes the current as a response to an electric field (externally applied plus induced) of frequency ω. The proportionality function is the conductivity. To get the usual Ohmic formula, all we need to do is replace $\sigma = \sigma_0 \delta(x - x')$ or remember that the $\sigma(x, x')$ is generally a function which is centered around

5.8. Transport in a Luttinger liquid - clean wire

$x \simeq x'$ and which falls off sufficiently fast elsewhere. The point for mesoscopic systems is that the length of the wire is roughly comparable with the range of $\sigma(x, x')$. Hence, the current gets contributions from the electric field all over the wire, which is different from what happens in the usual case, where the current at a point gets contributions only from the electric field very near that point. The second equation tells us that the conductivity is related to the current-current correlation function. This is derived by computing the current $I(x)$ in a Hamiltonian formulation to first order in the perturbation which is the applied electric field. The Euclidean formulation is used so that the generalization to finite temperature calculations is straightforward, but we shall only work at zero temperature and hence take the $\beta \to \infty$ limit.

Our aim here will be to compute the current-current correlation function and hence the conductance for a Luttinger wire using bosonization. We shall denote the Euclidean time action of a generic Luttinger liquid as

$$S_E = \frac{1}{2K} \int d\tau \int dx \, [\frac{1}{v}(\partial_\tau \phi)^2 + v(\partial_x \phi)^2] \,. \tag{5.193}$$

(Note that in the spin model and Hubbard model, τ was replaced by it). The current can directly be expressed in terms of the boson operators as

$$j(x,\tau) \equiv v(\rho_R - \rho_L) = -\frac{i}{\sqrt{\pi}} \partial_\tau \phi \,. \tag{5.194}$$

(The extra factor of i is because we are now using imaginary time τ). Our first step is to obtain the correlation function $< j(x,\tau) j(x', 0) >$ which is similar to the correlation functions for spinless fermions that we computed earlier when we were studying spin models, except that we are now interested in the Euclidean correlation functions. Since we can pull out the ∂_τ outside the correlation function[2], all we have to do is compute the propagator

---

[2] See R. Shankar in [15] for subtleties in pulling out the derivative outside the time ordering operator. One gets an extra term which cancels another term that we have ignored here, a singular c-number term.

given by
$$G(\tau, x, x') = <T_\tau \phi(x,\tau)\phi(x',0)> , \qquad (5.195)$$
or equivalently, its Fourier transform
$$G_{\bar\omega}(x, x') = \int_0^\beta d\tau <T_\tau \phi(x,\tau)\phi(x',0)> e^{-i\bar\omega\tau} . \qquad (5.196)$$
The conductivity is then given by
$$\begin{aligned}\sigma_{\bar\omega}(x,x') &= \frac{e^2}{\bar\omega\pi}\int_0^\beta d\tau <T_\tau \partial_\tau\phi(x,\tau)\partial_\tau\phi(x',0)> e^{-i\bar\omega\tau} \\ &= \frac{e^2\bar\omega}{\pi}G_{\bar\omega}(x,x') .\end{aligned} \qquad (5.197)$$

So now, to compute the conductance, all we have to do is compute the propagator for the boson with a free Euclidean action. The propagator satisfies the equation
$$\frac{1}{K}\left(-v\partial_x^2 + \frac{\bar\omega^2}{v}\right)G_{\bar\omega}(x,x') = \delta(x-x') , \qquad (5.198)$$
from which upon integrating once, we get
$$\frac{v}{K}\partial_x G(x,x')|_{x=x'-0}^{x=x'+0} = -1 . \qquad (5.199)$$
The solution to the differential equation is given by
$$\begin{aligned}G(x,x') &= A e^{|\bar\omega|(x-x')/v}, \quad x < x' \\ &= A e^{-|\bar\omega|(x-x')/v}, \quad x > x' .\end{aligned} \qquad (5.200)$$
Using this in Eq. (5.199), we see that
$$\left(\frac{2\bar\omega}{K}\right)A = 1, \qquad (5.201)$$
$$\begin{aligned}\text{so that } G_{\bar\omega}(x,x') &= \frac{K}{2\bar\omega}e^{|\bar\omega|(x-x')/v}, \quad x < x' \\ &= \frac{K}{2\bar\omega}e^{-|\bar\omega|(x-x')/v}, \quad x > x'\end{aligned} \qquad (5.202)$$
$$\begin{aligned}\text{leading to } \sigma_{\bar\omega}(x,x') &= \frac{Ke^2}{2\pi}e^{|\bar\omega|(x-x')/v}, \quad x < x' \\ &= \frac{Ke^2}{2\pi}e^{-|\bar\omega|(x-x')/v}, \quad x > x' . \quad (5.203)\end{aligned}$$

5.8. Transport in a Luttinger liquid - clean wire

The point to note is that in the $\bar{\omega} \to 0$ limit or static limit, the conductivity is finite and does not drop down to zero even for large $|x - x'|$. This is the main difference from macroscopic conductivities which always decay to zero as $|x - x'| \to \infty$. Furthermore, for $x = x'$, even for arbitrary $\bar{\omega}$, the Green's function has a finite value, which is responsible for the saturation value of the conductance. This only happens for a one-dimensional Green's function. In any other dimension, the Green's function and hence conductivity will be divergent at $x = x'$. Using this in the equation for the current, Eq. (5.191) for a static electric field $\bar{E}_\omega(x) = 2\pi\delta(\omega)E(x)$, we finally get

$$I(x) = \frac{Ke^2}{2\pi} \int_0^L dx' \, E(x') = \frac{Ke^2}{2\pi}(V_L - V_0) \qquad (5.204)$$

which gives the final result for the conductance as

$$g = \frac{Ke^2}{2\pi}. \qquad (5.205)$$

There are several subtle points to note in this calculation. One is that we have taken the $\omega \to 0$ before $|x - x'| \to \infty$, which is opposite to the usual order of limits in the Kubo formula. The physical justification for the usual order of limits in the Kubo formula comes from the fact that if we first take ω to zero, then we have a static electric field, which is periodic in space. This means that the charge will seek an equilibrium distribution after which there will be no flow of current. Setting $|x - x'| \to \infty$, on the other hand, means taking the thermodynamic limit or infinite length limit first, which allows for an unlimited supply of electrons and is probably equivalent to having reservoirs even if we have do not really have infinite length wires. For the mesoscopic systems, however, it is not correct to take the thermodynamic limit first. The physical situation here, is that one applies a static electric field to a finite length of the wire L, which in fact, is comparable to the range of the conductivity. If we take the $|x - x'| \to \infty$ limit first, then it is as if we are looking at a long length of the wire beyond the range of the conductivity. This is the usual limit and we will get the usual Ohm's law, which however, is wrong in this context. In fact, it is instructive to try out the calculation with the other order of limits -*i.e.* by computing $\sigma_{\tilde{\omega}}(q)$ and taking the $q \to 0$ limit first.

Exercise 5.9 *Try the above.*

The second point is something we have mentioned earlier - *i.e.*, we have not taken contacts or leads into account. This was the initial computation by Kane and Fisher [16] and they obtained the answer in Eq. (5.205) that the conductance of the clean wire depends on the interaction parameter K.

(b) Including leads

When any experiment is done, however, one does have explicit contacts or leads. In fact, when a measurement was actually done under conditions where one expected to measure the Luttinger parameter K, it was found to $\simeq 1$, instead of 0.7, which was expected from other measurements of the K value of the wire. (We will see how else K can be measured after considering impurity scattering). So we need to understand what happens when we actually try to measure the conductance of a Luttinger wire.

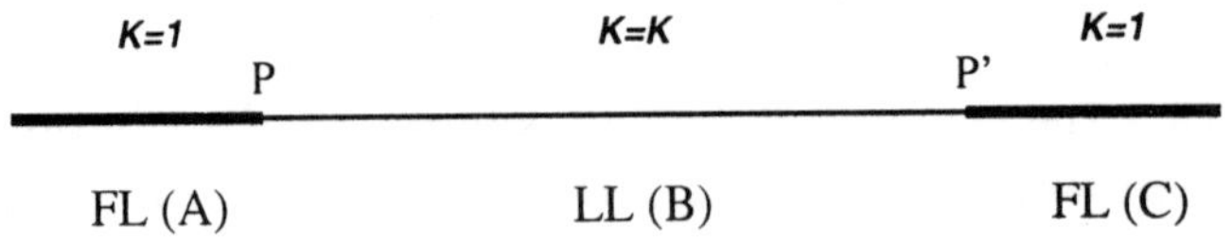

```
    K=1              K=K              K=1
         P                      P'
  ━━━━━━━━━━━━━━━━━━━━━━━━━━━━━━━━━━━━━━━

   FL (A)           LL (B)           FL (C)
```

Figure 5.6: The single channel quantum wire with Fermi liquid leads on the left and the right.

How do we model the leads? The simplest model to consider is that the Luttinger wire is connected to Fermi liquid leads on either side. (See Fig.5.6). So the regions A and C can be modelled by the same bosonic model with $K_L = 1$ and the wire in region B can be modelled as before as a Luttinger wire with $K = K$. But now,

5.8. Transport in a Luttinger liquid - clean wire

we have to put appropriate boundary conditions at the points P and P' between A and B and between B and C respectively. Note that we are making the assumption that one has the same ϕ field or same quasiparticle in all the three regions and it is only the LL parameters which are changing. Although, it is interesting to compute the conductance in this case, it is still not clear that this brings the calculation any closer to real experiments, because real experiments will have three dimensional reservoirs.

We start with the action in all the three regions in Euclidean space as

$$S_E = \frac{1}{2}\int_0^\beta d\tau \int_0^L dx \, [\frac{(\partial_\tau \phi)^2}{K(x)v(x)} + \frac{v(x)}{K(x)}(\partial_x \phi)^2], \quad (5.206)$$

with $K(x) = K_L$, $v(x) = v_L$ in regions A and C and $K(x) = K$, $v(x) = v$ in region B. This is just the free action of a scalar field in all the three regions. Fourier transforming the imaginary time variable with respect to $\bar{\omega}$, we obtain

$$S_E = \frac{1}{2}\int_0^\beta d\tau \int_0^L dx \, [\frac{\bar{\omega}^2 \phi^2}{K(x)v(x)} + \frac{v(x)}{K(x)}(\partial_x \phi)^2], \quad (5.207)$$

from which we see that the propagator satisfies the equation

$$\{-\partial_x(\frac{v(x)}{K(x)}\partial_x) + \frac{\bar{\omega}^2}{K(x)v(x)}\} G_{\bar{\omega}}(x,x') = \delta(x-x') . \quad (5.208)$$

Now let us consider the four regions. We assume that the interaction parameter changes abruptly at P and P', but that the Green's function is continuous and the derivative of the Green's function has the correct discontinuity at all the boundaries. So now, we need to solve the Green's function equation subject to these boundary conditions. Let us choose x' to lie between 0 and L. It is then easy to see that the solution is of the form

$$\begin{aligned} G_{\bar{\omega}}(x,x') &= A e^{|\bar{\omega}|x/v} \quad \text{for } x \leq 0 \\ &= B e^{|\bar{\omega}|x/v} + C e^{-|\bar{\omega}|x/v} \quad \text{for } 0 < x \leq x' \\ &= D e^{|\bar{\omega}|x/v} + E e^{-|\bar{\omega}|x/v} \quad \text{for } x' < x \leq L \\ &= F e^{-|\bar{\omega}|x/v} \quad \text{for } x > L \end{aligned} \quad (5.209)$$

for semi-infinite leads, because we have assumed that the lengths of the leads are sufficiently long compared to L so that we do not need to put any further boundary conditions on them. Note that here the Green's functions will no longer be functions of $x - x'$ since we have explicitly broken translational invariance. The constants $A, B, ..., F$ are found by matching the boundary conditions. Since we are interested in the DC conductance, we only need the solutions for $\bar{\omega} \to 0$ which are easy to obtain and are given by

$$A = F = \frac{K_L}{2\bar{\omega}}, \quad B = E = \frac{K_L + K}{4\bar{\omega}}, \quad C = D = \frac{K_L - K}{4\bar{\omega}}. \quad (5.210)$$

From this, we see that $\sigma_{\bar{\omega}}(x, x')$ is x and x' independent in the $\bar{\omega} \to 0$ limit and is equal to $K_L e^2 / 2\pi$ in all regions from which we find the conductance (using Eq. (5.191)) given by

$$g \equiv \frac{I}{V} = \frac{K_L e^2}{2\pi}. \quad (5.211)$$

is the same in all the regions. Thus, the conductance is determined by the K_L of the leads, which is just $K_L = 1$ for Fermi liquid leads and does not depend on interactions in the wire. This is a highly counter-intuitive answer! It is telling us that whether we measure the conductance in the leads or in the quantum wire, we get the same answer, so long as we take into account the fact that we are attaching leads, which allow for the fermions to enter and leave the quantum wire. At a very naive level, one may understand this by saying that since the wire itself has no impurities, the only source of resistance is the contact effect between the leads and the wire, which has nothing to to do with the interactions in the wire. However, remember that we have taken semi-infinite leads and abrupt contacts and we are only looking for DC conductance. If any of these assumptions are relaxed, certainly, there are differences in the three regions and one could get more interesting answers.

In fact, using a Landauer-Buttiker scattering approach [18], there has been some attempt to understand these results more intuitively.

Physically, the difference between this computation and the earlier one is that any real measurement requires Fermi liquid leads. So the end result is that the measurable conductance of an interacting one-dimensional wire is simply given by $g = e^2/h$ for spinless fermions and $g = 2e^2/h$ for fermions with spin [17].

5.9 Transport in the presence of isolated impurities

Computing conductance with a single impurity

Now let us consider the case when there is a single impurity at the origin. At first, we will model the impurity as a weak barrier and add a term to the action of the form

$$S_{\text{int}} = \int dx d\tau \, V(x) \psi^\dagger(x) \psi(x) \, . \qquad (5.212)$$

We assume that $V(x)$ is weak and is centred around the origin. For instance, we can choose $V(x) = \lambda \delta(x)$, where λ is much less than the Fermi energy.

First, let us think of what happens when we introduce such a perturbation in a non-interacting wire. In that case, all one has is a one-dimensional quantum mechanics problem with a δ-function potential at the origin. We can find the reflection and transmission probabilities for a single particle with momentum k as

$$R = \frac{\lambda^2}{\lambda^2 + k^2} \quad \text{and} \quad T = \frac{k^2}{\lambda^2 + k^2} \, . \qquad (5.213)$$

So for any λ, one gets both reflection and transmission. To get the total current, we just have to sum up the contributions of all the electrons close to the Fermi surface. But it is clear that there will be non-zero conductance for any potential, with the amount of current being transmitted depending on the strength of the potential. However, for the Luttinger wire, since there exists interactions between electrons in the wire, and no convenient quasiparticle picture, one cannot solve the problem this way. We have to use the bosonized field theory and include the impurity potential as a perturbation.

Let us first rewrite the impurity potential in terms of the left- and right- moving low energy Dirac modes. We find that

$$\begin{aligned}
\psi^\dagger(x)\psi(x) &= (\psi_R^\dagger e^{-ik_F x} + \psi_L^\dagger e^{ik_F x})(\psi_R e^{ik_F x} + \psi_L e^{-ik_F x}) \\
&= \psi_R^\dagger \psi_R + \psi_L^\dagger \psi_L + e^{-i2k_F x}\psi_R^\dagger \psi_L + e^{i2k_F x}\psi_L^\dagger \psi_R \\
&= -\frac{1}{\sqrt{\pi}}\partial_x \phi + \frac{1}{2\pi\alpha}(\eta_R^\dagger \eta_L e^{i2\sqrt{\pi}(\phi_R + \phi_L + 2k_F x)} \\
&\quad + \eta_L^\dagger \eta_R e^{-i2\sqrt{\pi}(\phi_R + \phi_L + 2k_F x)}),
\end{aligned}$$
(5.214)

where the last line is obtained using standard bosonization. So the full action is given by

$$\begin{aligned}
S &= S_E + S_{\text{int}} \\
&= S_E - \frac{\lambda}{\sqrt{\pi}}\partial_x \phi(0) + \frac{\lambda}{2\pi\alpha}\int d\tau \cos 2\sqrt{\pi}[\phi_R(0) + \phi_L(0)],
\end{aligned}$$
(5.215)

where S_E is given in Eq. (5.193) and we have incorporated the fact that the potential only acts at the origin. Moreover, we have simply set both $\eta_R^\dagger \eta_L$ and $\eta_L^\dagger \eta_R$ to be one, with the knowledge that in correlation functions, we will compute $O(\tau)O^\dagger(0)$ so that the Klein factors disappear using $\eta_{R/L}\eta_{R/L}^\dagger = \eta_{R/L}^\dagger \eta_{R/L} = 1$. The first term due to the interaction can be taken care of by a simple redefinition of $\partial_x \phi \rightarrow \partial_x \phi' = \partial_x \phi + \lambda/2\sqrt{\pi}$, which makes no difference to the conductance. This could have been seen even from the fermion terms from which it came. The $\psi_R^\dagger \psi_R + \psi_L^\dagger \psi_L$ term only causes scattering at the same Fermi point with momentum transfers $q \ll 2k_F$. This does not change the direction of propagation of the particles and hence does not affect conductance in any appreciable way. But the cosine term, on the other hand, occurs because of backscattering of fermions from the origin. These represent scattering with $q \sim |2k_F|$ - i.e., from the left branch to the right branch and vice-versa and change the direction of propagation of the particles. These scatterings will definitely affect the conductance. The action with this perturbation is no longer

5.9. Transport in the presence of isolated impurities

quadratic and cannot be exactly solved. However, since λ is a weak perturbation, one can try to use perturbation theory and the renormalization group approach to see the relevance of this perturbation at low energies.

What is the question that we want to answer? We want to compute the conductance through this barrier at low energies. One way to do that is to see whether this barrier coupling strength grows or becomes smaller as we go to lower energy scales. To check that, we need to perform the usual steps of a renormalization group analysis.

Here since the perturbation term is fixed in space, it is more convenient to first integrate out the variables away from the origin and write down the action purely in terms of the $\phi(x=0,\tau)$ variables. Since integrating out quadratic degrees of freedom is equivalent to using equations of motion for those degrees of freedom, we write down the equations of motion for the action S_0 as

$$\partial_x^2 \phi - \frac{\bar{\omega}^2}{v^2}\phi = 0 \Rightarrow \partial_x^2 \phi - k^2 \phi = 0 \ . \tag{5.216}$$

The solution to the above equations are given by

$$\begin{aligned} \phi &= A e^{|k|x}, \quad x > 0 \\ &= A e^{-|k|x}, \quad x < 0 \ , \end{aligned} \tag{5.217}$$

where $A \equiv \phi(x=0,\tau)$. Using this solution in the action, we get the effective action in terms of $\phi(\bar{\omega}) = \int \phi(x=0,\tau) e^{i\bar{\omega}\tau} d\tau$ as

$$\begin{aligned} S_{\text{eff}} &= \frac{1}{2K} \int \frac{d\bar{\omega}}{2\pi} \int_{-\infty}^{0} dx \, [v\phi^2 k^2 e^{2kx} + \frac{\bar{\omega}^2}{v} e^{2kx}] \\ &+ \frac{1}{2K} \int \frac{d\bar{\omega}}{2\pi} \int_{0}^{\infty} dx \, [v\phi^2 k^2 e^{-2kx} + \frac{\bar{\omega}^2}{v} e^{-2kx}] \\ &= \frac{1}{2K} \int \frac{d\bar{\omega}}{2\pi} \frac{2\phi^2 \bar{\omega}^2}{v} \, [\frac{e^{2kx}}{2k}|_{-\infty}^{0} + \frac{e^{-2kx}}{-2k}|_{0}^{\infty}] \\ &= \frac{1}{K} \int \frac{d\bar{\omega}}{2\pi} |\bar{\omega}|\phi^2 \end{aligned} \tag{5.218}$$

using $k = |\bar{\omega}|/v$. Notice the singular dependence on the Matsubara frequency $|\bar{\omega}|$. The reason for its appearance is the following. In real space, even for a quadratic action, all degrees of freedom (dof) are coupled. (It is only in Fourier space that every mode is decoupled). So when we integrate out all dof except the one at the origin, the dispersion of this degree of freedom can change and has changed. This is why we get the modulus factor in the effective action. When we Fourier transform back to imaginary time, we get

$$\sum_{\bar{\omega}} |\bar{\omega}| \phi_n^2 \rightarrow i \int \frac{d\omega}{2\pi} \int d\tau \int d\tau' \, e^{i\omega(\tau-\tau')} |\omega_n| \phi^*(\tau) \phi(\tau')$$
$$= -\int d\tau d\tau' \frac{2}{(\tau-\tau')^2} \phi^*(\tau) \phi(\tau') \,, \quad (5.219)$$

i.e., an explicitly non-local interaction in imaginary time.

So now, we have an action solely in terms of the variables at the origin with the action given by

$$S = \frac{1}{2K} \int \frac{d\bar{\omega}}{2\pi} |\bar{\omega}| \phi(\bar{\omega})^2 + \lambda \int \frac{d\bar{\omega}}{2\pi} \cos[2\sqrt{\pi}\phi(\bar{\omega})] \,. \quad (5.220)$$

The RG analysis now involves finding out how the coefficient λ behaves as we go to lower and lower energies. Before we perform the RG analysis, we may ask why would we want to go to lower energy scales? The general idea is that in spite of the fact that in different physical problems or models, the parameter λ may be slightly different, qualitatively many such models may have the same behavior. This is because they are all governed by the same fixed point Hamiltonian with the fixed point Hamiltonian being defined as the Hamiltonian one gets when the RG flow stops. So the aim is to keep reducing the energy scale till the RG flow stops so that we can find out the appropriate fixed point Hamiltonian for this model.

In this problem, we want to find out whether the fixed point Hamiltonian has a large barrier or a small barrier. To find that out, let us perform the three steps of the renormalization group transformation. We choose a high frequency cutoff Λ, which is the real physical cutoff of the theory. Then we rescale $\Lambda \rightarrow \Lambda/s$ with $s > 1$ and then divide $\phi(\bar{\omega})$ into $\phi_<(\bar{\omega})$ (slow modes) and $\phi_>(\bar{\omega})$

5.9. Transport in the presence of isolated impurities

(fast modes) for the modes with frequencies less than or greater than the cutoff Λ/s respectively. Finally, we integrate out the fast modes, which are the modes between Λ/s and Λ and rescale $\bar{\omega} \to \bar{\omega}' = s\bar{\omega}$ or $\tau \to \tau' = \tau/s$ to get back to the original range of integrations. To lowest order, (tree level contribution) we find that

$$\lambda \int d\tau \cos 2\sqrt{\pi}\phi_<(x=0,\tau) \to \lambda s^{1-d} \int d\tau \cos 2\sqrt{\pi}\phi_<(x=0,\tau) ,$$
(5.221)

where d is the dimension of the cosine operator. This was explicitly computed earlier and we found that $d = K$. The RG equation is now easily obtained by taking $s = 1 + dl$, for infinitesimal dl. We find that the new λ' after rescaling is given by

$$\lambda' = \lambda(1+dl)^{1-K}$$
$$\Rightarrow \quad \lambda' - \lambda = (1-K)\lambda dl$$
$$\Rightarrow \quad \frac{d\lambda}{dl} = (1-K)\lambda .$$
(5.222)

Normally, one would have had coupled RG equations for λ and K. But here since $1/K$ is the coefficient of a singular operator, it does not get renormalized to any order.

Notice that in Eq. (5.221), the coefficient of the operator gets rescaled by a factor s^{1-d} rather than s^{2-d_i} as we had mentioned earlier when we computed the dimension of the cosine operator in the spin model. The difference is that the operator in the spin model, in the action, required integration over both space and time. So we rescaled both the space and time (or equivalently both the momentum and the energy). However, in this case, the operator exists only at a fixed space point. So we only need to integrate over the time coordinate. Hence, the naive scaling dimension or engineering dimension of the operator is 1 and not 2. Such operators are called boundary operators. You will learn more about them in the course on boundary conformal field theory.

The RG equation is now trivial to analyze. For any $K > 1$, (which corresponds to attractive interactions between the electrons), the λ renormalizes to zero and for any $K < 1$, (corresponding to repulsive interactions), it grows stronger and stronger. In other words, for $K > 1$, the fixed point Hamiltonian is just the free boson Hamiltonian with no barrier and for $K < 1$, the fixed point Hamiltonian has two disconnected wires to the left and right of the origin. For $K = 1$, which is the limit of no interactions in the fermionic model, the coupling is marginal. (This was expected, since we know that for free fermions, both transmission and reflection occurs depending on the strength of the barrier potential). Thus, for attractive interactions, the barrier renormalizes to zero and the wire is 'healed', whereas for repulsive interactions, the barrier renormalizes to infinity and the wire is 'cut'. Note that both these answers are completely independent of the strength of the barrier potential [16].

Strong barrier limit

Since, we are doing perturbation theory, we cannot assume that this result holds for arbitrary λ. It is strictly valid only for $\lambda \simeq 0$. Once $\lambda \sim 1$, the perturbative analysis in λ breaks down. So it is worthwhile to try and see what happens in the other limit. Supposing we start with two decoupled wires and then allow a small hopping between the two wires. Will this hopping grow at low energies and heal the wire or will it renormalize to zero?

Here, we start with two semi-infinite Luttinger liquid wires and analyze the effect of adding a small hopping term coupling the two wires at $x = 0$. The models for $x < 0$ and $x > 0$ are given by the action

$$S_E = \frac{1}{2} \int_0^\beta d\tau \int_0^L dx \, [\frac{(\partial_\tau \phi_i)^2}{K(x)v(x)} + \frac{v(x)}{K(x)} (\partial_x \phi_i)^2] \qquad (5.223)$$

for $i = <$ and $i = >$ respectively. We can also write it in terms of the dual variables as

$$S_E = \frac{1}{2} \int_0^\beta d\tau \int_0^L dx \, [\frac{K(x)}{v(x)} (\partial_\tau \theta_i)^2 + K(x)v(x)(\partial_x \theta_i)^2] \, . \qquad (5.224)$$

Note that in terms of the dual variables, the action has $1/K$ in position of K. This is because the roles of the fields and the canonically conjugate momenta have interchanged. The fact that

5.9. Transport in the presence of isolated impurities

the wire is cut implies that at the point $x = 0$, there is zero density of either '<' or '>' particles - $\psi_<^\dagger \psi_<(x = 0) = 0$ and $\psi_>^\dagger \psi_>(x = 0) = 0$. In the bosonic language, this is imposed as $2\sqrt{\pi}\phi_<(x = 0) = 2\sqrt{\pi}\phi_>(x = 0) = \pi/2$ (and also $\partial_x \phi(x = 0) = 0$ as can be seen from Eq. (5.214)). Now a term which hops an electron from one wire to another in the Hamiltonian is just

$$\begin{aligned} H &= -t\, [\psi_<^\dagger \psi_> + h.c.] \\ &= -t\, [\psi_{R<}^\dagger \psi_{R>} + \psi_{L<}^\dagger \psi_{L>} + \psi_{R>}^\dagger \psi_{R<} + \psi_{L>}^\dagger \psi_{L<} \\ &\quad + \psi_{L<}^\dagger \psi_{R>} + \psi_{L>}^\dagger \psi_{R<} + \psi_{R<}^\dagger \psi_{L>} + \psi_{R>}^\dagger \psi_{L<}] \,, \end{aligned} \quad (5.225)$$

where the second equation involves the left and right moving fields and we have already set $x = 0$. Here, again, the terms that involve fields at one Fermi point are low energy forward scattering terms which do not affect the conductance. In terms of the bosonic fields too, they can be taken care of by trivial redefinitions. But the intra-Fermi point scatterings which will affect the conductance can be bosonized and written in the action as

$$\begin{aligned} \delta S &= -t \int d\tau\, [\eta_{L<}^\dagger \eta_{R>} e^{-i(\phi_{L<}+\phi_{R>})} + \eta_{L>}^\dagger \eta_{R<} e^{-i(\phi_{L>}+\phi_{R<})} \\ &\quad + \eta_{R<}^\dagger \eta_{L>} e^{-i(\phi_{R<}+\phi_{L>})} + \eta_{R>}^\dagger \eta_{L<} e^{-i(\phi_{R>}+\phi_{L<})}]\,. \end{aligned} \quad (5.226)$$

Now, we impose the boundary condition on the bosonic fields that we mentioned above, which constrains $\phi(0) = \phi_R(0) + \phi_L(0)$ to be equal to $\pi/2$. Using this, we can express the above equation solely in terms of the $\phi_L(0) - \phi_R(0) = \theta(0)$ fields and get

$$\delta S = 4t \int d\tau\, \cos(\theta_> - \theta_<)\,, \quad (5.227)$$

where once again, we have been able to drop the Klein factors after checking that they do not lead to any extra minus signs in the correlation functions. (Physically, the reason why we only get the θ_i term at the origin is because the constraint has set $\phi_i(x = 0) = \pi/2$). Computing the dimension of this operator, we see that to leading order, the RG equations are given by

$$\frac{dt}{dl} = (1 - \frac{1}{K})t\,. \quad (5.228)$$

(K has been replaced by $1/K$ because we now have to compute the dimensions in the dual action). Thus, for repulsive interactions ($K < 1$), the hopping term is irrelevant and flows to zero. This confirms the weak barrier calculation that the wire is insulating. On the other hand, for attractive interactions, the hopping strength grows, ultimately healing the wire. This again is in accordance with the weak coupling analysis.

Intermediate fixed points?

We have started from a wire with a weak barrier and shown that under repulsive interactions, the barrier strength grows. We have also started from two decoupled wires and shown that for repulsive interactions, any small hopping term renormalizes to zero. Hence, it seems plausible to conclude that for repulsive interactions in the wire, any barrier will cut the wire and the conductance goes to zero. However, one should keep in mind that our analysis is strictly true only for $\lambda, t \simeq 0$. Hence, it could happen that for intermediate values of the barrier strength, one could have a pair of non-trivial fixed points (see Fig.5.7).

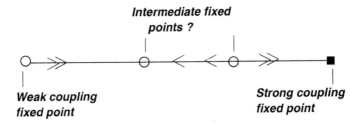

Figure 5.7: Renormalization group flow diagram for a quantum wire with repulsive interactions in the presence of an impurity or barrier. In the absence of any non-trivial fixed points, the stable fixed point is the strong coupling fixed point. But perturbative analyses at the strong and weak coupling fixed points cannot rule out a pair of non-trivial fixed points at intermediate strengths of the barrier potential

Conductance at finite voltage and temperature

The earlier analysis only tells us how the barrier strength or the tunneling amplitude grows or falls as we go to low energies. But instead of allowing the energy scale to become arbitrarily low, we can cut off the energy scale of renormalization at some finite energy scale, which could be the temperature T or the voltage V. Note that the energy scale at which we want to cutoff the integral is related to the initial high energy scale at which we start the RG as $E = E_0 e^{-l}$. So for attractive interactions for which weak barriers are irrelevant and for which one would expect perfect transmission at very low energies will have power law corrections when we put the lower energy cutoff as E. In that case, we have

$$\int_{\Lambda_0}^{\Lambda} \frac{d\lambda}{\lambda} = \int_0^{ln(E_0/E)} dl(1-K) \,, \qquad (5.229)$$

which means that the effective barrier strength Λ is proportional to $\Lambda_0 (E/E_0)^{K-1}$. So by choosing $E = T, V$, we see that one can get power law corrections to the naive conductance at $T \to 0, V \to 0$. In other words, if we measure the conductance at a finite temperature T, rather than at $T = 0$, instead of zero conductance for $K < 1$, we will get conductances which go as T^{1-K} (roughly the inverse of the barrier strength). Similarly, if instead of measuring conductances as $V \to 0$, we measure them at finite voltages, we find that the conductances go as V^{1-K}. On the other hand, for repulsive interactions, we need to start at the strong coupling limit with two decoupled Luttinger wires and allow for a small hopping, which is irrelevant in the RG sense. Here, again, if we cutoff the lower energy scale at E, we expect instead of zero transmission, power law corrections of the form $I \sim V^{1-1/K}$ and $I \sim T^{1-1/K}$. The only difference in the analysis at the strong coupling fixed point and the weak coupling fixed point is that K gets replaced by $1/K$ as we saw in the RG equations. This, in fact, is one way in which K can be measured in experiments. They could explicitly make a constriction in the quantum wire and measure conductances through it and extract K.

5.10 Concluding remarks

Almost any interacting quantum system in one dimension which is gapless and has a linear dispersion for the low-energy excitations can be described as a Luttinger liquid at low energies and long wavelengths. As we have seen, the properties of a Luttinger liquid are determined by the two parameters v and K. These in turn depend on the various parameters which appear in the microscopic Hamiltonian of the system. Some examples of systems where Luttinger liquid theory and bosonization can be applied are quantum spin chains (including some spin ladders), quasi-one-dimensional organic conductors and quantum wires (with or without impurities), edge states in a fractional quantum Hall system, and the Kondo problem. Some of these examples have been discussed above.

Antiferromagnetic spin-1/2 chains have a long history going back to their exact solution by the Bethe ansatz. In recent years, many experimental systems have been studied which are well-described by quasi-one-dimensional half-odd-integer spin models with isotropic (Heisenberg) interactions. Such systems behave at low energies as a $K = 1/2$ Luttinger liquid with an $SU(2)$ symmetry. It seems to be difficult to vary K experimentally in spin systems. In contrast, a single-channel quantum wire (which is basically a system of interacting electrons which are constrained to move along one particular direction) typically has two low-energy sectors, both of which are Luttinger liquids (except at special densities like half-filling). One of these is the spin sector which again has $K = 1/2$. The other one is the charge sector whose K value depends on a smooth way on the different interactions present in the system. Finally, the edge states in a fractional quantum Hall system behave as a chiral Luttinger liquid with K taking certain discrete rational values; the value of K can be changed by altering the electron density and the magnetic field in the bulk of the system. For all these systems, many properties have been measured such as the response to external electric and magnetic fields (conductivity or susceptibility) and to disorder, scattering of neutrons

5.10. Concluding remarks

or photons from these systems, and specific heat; so the two Luttinger parameters can be extracted from the experimental data. The measurements clearly indicate the Luttinger liquid-like behavior of these systems with various critical exponents depending in a non-universal way on the interactions in the system.

On the theoretical side, a large number of exactly solvable models in one dimension have been shown to behave as Luttinger liquids at low energies [2, 12]. These include

(i) models with short range interactions which are solvable by the Bethe ansatz, such as the XXZ spin-1/2 chain (where K can take a range of values from 1/2 to ∞; this includes the XY model with $K = 1$ and the isotropic antiferromagnet with $K = 1/2$ as special cases), and the repulsive δ-function Bose gas (where K can go from 1 in the limit of infinite repulsion to ∞ in the limit of zero repulsion), and

(ii) models with inverse-square interactions such as the Calogero-Sutherland model (where K can go from 0 to ∞) and the Haldane-Shastry spin-1/2 model (where $K = 1/2$).

The models of type (ii) are *ideal* Luttinger liquids in the sense that they are scale invariant; the coefficients of all the marginal operators vanish, and therefore their correlation functions and excitation energies contain no logarithmic corrections. This property makes it particularly easy to study these systems numerically since the asymptotic behaviors are reached even for fairly small system sizes.

What has been left out?

Finally, let us mention the various important things in this field which has been left out. We have only worked with spinless fermions in the transport analysis. When we include spin and do not destroy the $SU(2)$ spin symmetry of the system, the results are very similar to the spinless fermion case. For repulsive interactions, the barrier becomes infinite and for attractive interaction, the barrier is healed. However, when the $SU(2)$ symmetry is destroyed, there exists possibilities of intermediate (non-trivial) fixed points where either spin or charge can be transmitted and the other reflected. The other thing that has been left out is the phenomenon of resonant tunneling with two impurities. This is an interesting result, because it says that for repulsive interactions, a single impurity cuts the wire, but with two impurities, one can have particular energies, where there can be transmission. The reason, of course, is quantum mechanical tunneling. Here, the energy levels, are the energy levels of the quantum dot that is formed by the two impurities and one can have resonant tunneling at these energy levels. If we include interactions between the electrons on the island, (which is naturally included in the bosonized formalism), we can obtain the physics of the Coulomb blockade. The other important thing that we have left out, from a physical point of view, is what happens if there is a finite density of random impurities. In general, one would expect Anderson localization and no transport. But there are regimes of delocalization as well in the phase diagram. Finally, a very important application where the physics of the Luttinger liquids has actually been experimentally seen is in the edge states of the fractional Quantum Hall fluid. Since here, the edge states are chiral, a lot of the complications of backscattering due to impurities are avoided and it is possible to explicitly construct constrictions and allow tunneling through them. Here, both at the theoretical and experimental level, there are a lot of beautiful results that are worth understanding.

Another important topic that has not been covered here is non-abelian bosonization [3]. This is a powerful technique for studying one-dimensional quantum systems with a continuous global symmetry such as $SU(2)$. For instance, isotropic Heisenberg antiferromagnets and Kondo systems are invariant under spin rotations, and they can be studied more efficiently using non-abelian bosonization.

To conclude, let us just say that low dimensional systems and mesoscopic systems have gained in importance in the last few years. Although currently, much of the theoretical work in mesoscopic systems has only involved conventional Fermi liquid theories, it is clear that there are regimes where strong interactions are very important. We expect that bosonization will be one of the

5.10. Concluding remarks

important non-perturbative tools to analyze such problems for a few more years to come.

Acknowledgments

DS thanks J. Srivatsava for making Figures 1 to 5. Both of us thank the participants of the school for making us work for the lectures!

Bibliography

[1] H. J. Schulz, G. Cuniberti and P. Pieri, to appear in *Field Theories for Low Dimensional Condensed Matter Systems*, edited by G. Morandi, A. Tagliacozzo and P. Sodano, Springer Lecture Notes in Physics (2000), cond-mat/9807366; H. J. Schulz, in Proceedings of Les Houches Summer School LXI, edited by E. Akkermans, G. Montambaux, J. Pichard and J. Zinn-Justin (Elsevier, Amsterdam, 1995), cond-mat/9503150.

[2] F. D. M. Haldane, Phys. Rev. Lett. **45**, 1358 (1980); Phys. Rev. Lett. **47**, 1840 (1981); J. Phys. C **14**, 2585 (1981).

[3] A. O. Gogolin, A. A. Nersesyan and A. M. Tsvelik, *Bosonization and Strongly Correlated Systems* (Cambridge University Press, Cambridge, 1998).

[4] V. Ya. Krivnov and A. A. Ovchinnikov, Sov. Phys. JETP **55**, 162 (1982); A. V. Zabrodin and A. A. Ovchinnikov, Sov. Phys. JETP **63**, 1326 (1986).

[5] I. Affleck, in *Fields, Strings and Critical Phenomena*, edited by E. Brezin and J. Zinn-Justin (North-Holland, Amsterdam, 1989).

[6] R. Shankar, Lectures given at the BCSPIN School, Kathmandu, 1991, in *Condensed Matter and Particle Physics*, edited by Y. Lu, J. Pati and Q. Shafi (World Scientific, Singapore, 1993).

[7] S. T. Tomonaga, Prog. Theor. Phys. **5**, 544 (1950); J. M. Luttinger, J. Math. Phys. **4**, 1154 (1963); D. C. Mattis and E. H. Lieb, J. Math. Phys. **6**, 304 (1965).

[8] J. von Delft and H. Schoeller, Ann. der Physik **7**, 225 (1998), cond-mat/9805275.

[9] K. Schönhammer and V. Meden, Am. J. Phys. **64**, 1168 (1996).

[10] I. Affleck, D. Gepner, H. J. Schulz and T. Ziman, J. Phys. A **22**, 511 (1989).

[11] R. Rajaraman, *Solitons and Instantons* (North-Holland, Amsterdam, 1982).

[12] N. Kawakami and S.-K. Yang, Phys. Rev. Lett. **67**, 2493 (1991).

[13] See Chapter 3 by S. M. Bhattacharjee and Chapter 4 by D. Kumar.

[14] G. D. Mahan, *Many Particle Physics* (Plenum Press, New York).

[15] R. Shankar, Int. J. of Mod. Phys. B **4**, 2371 (1990).

[16] C. L. Kane and M. P. A. Fisher, Phys. Rev. B **46**, 15233 (1992).

[17] D. L. Maslov and M. Stone, Phys. Rev. B **52**, R5539 (1995).

[18] I. Safi and H. J. Schulz, Phys. Rev. B **52**, R17040 (1995); I. Safi, Phys. Rev. B **55**, R7331 (1997); I. Safi and H. J. Schulz, Phys. Rev. B **59**, 3040 (1999).

Chapter 6

Quantum Hall Effect and Composite Particles

R. Rajaraman

School of Physical Sciences
Jawaharlal Nehru University
New Delhi – 110 067

(Manuscript authored by Arun Paramekanti (TIFR) and Sumithra Sankararaman (IMSc), based on lecture notes)

In this set of lectures, we give a brief introduction to the quantum Hall effect (QHE). We recall the classical Hall effect and then study the quantum Hall effect by solving the Landau level problem in quantum mechanics. We introduce the Laughlin wave-function as a many-body wave-function which leads, through the plasma analogy, to fractional fillings and thus a solution of the fractional QHE problem. We briefly touch upon the idea of localisation leading to plateau formation. We discuss a (modified) Chern-Simons theory of the QHE and finally, we introduce Jain's composite fermions.

6.1 Classical Hall effect

Before discussing the quantum Hall effect, let us briefly recall the elementary classical problem of electrons moving in a plane in the presence of a magnetic field $\mathbf{B} = -B\hat{\mathbf{z}}$ and an electric field $\mathbf{E} = E_x\hat{\mathbf{x}}$, and being confined in the $\hat{\mathbf{y}}$ direction with a boundary. The Lorentz force due to the magnetic field pushes the electrons towards the boundary and leads to accumulation of charge along the boundary leading to a transverse electric field $E_y\hat{\mathbf{y}}$ which eventually balances the Lorentz force. Thus the electrons end up moving along the $\hat{\mathbf{x}}$ direction under the total electric field $\mathbf{E} = E_x\hat{\mathbf{x}} + E_y\hat{\mathbf{y}}$. We then arrive at $\mathbf{j} = \underline{\sigma}\cdot\mathbf{E}$ where $\underline{\sigma}$ is the conductivity matrix, with the diagonal elements $\sigma_{xx} = \sigma_{yy} = ne^2\tau/\mu$ (μ = electron mass). The off-diagonal elements can also be easily obtained since $\underline{\sigma}$ is defined as the matrix inverse of the resistivity matrix $\underline{\rho}$, with off-diagonal elements $\rho_{xy} = -\rho_{yx} = B/nec$. The current of particles along the $\hat{\mathbf{y}}$ direction is zero since the Hall field E_y balances the Lorentz force.

Exercise 6.1 *Consider the equation of motion for the electrons given by*

$$\dot{\mathbf{v}} = -e\frac{(\mathbf{E} + \mathbf{v}\times\mathbf{B})}{\mu} - \frac{\mathbf{v}}{\tau}$$

The first term on the right is the acceleration caused by the applied electric and magnetic fields (Lorentz force), while the second term is like a "viscous drag" arising from random scattering off impurities. Considering the "steady-state" solution where $\dot{\mathbf{v}} = 0$ and confining the electrons to move along the $\hat{\mathbf{x}}$ direction so that $v_y = 0$ in this "steady state", calculate the current $\mathbf{j} = -ne\mathbf{v}$ (n=electron density) to arrive at the above σ_{xx}. Note that in 2D, conductance and conductivity have the same dimensions.

6.2 Quantized Hall effect

We first define the filling fraction ν as the ratio of the electron density to the fluxon density where the latter is the amount of flux measured (in units of $\phi_0 = hc/e$) per unit area of the sample. Since the flux density is just B, $\nu = \frac{n}{(B/\phi_0)}$. In terms of ν, the classical Hall result can also be written as

$$\rho_{xx} = (\frac{ne^2\tau}{\mu})^{-1} \quad ; \quad \rho_{xy} = \frac{h}{e^2}\nu \; . \tag{6.1}$$

When Hall effect experiments were done on a 2D electron gas at very low temperatures and high magnetic fields, it was found that σ_{xy} as a function of ν contained extraordinarily flat plateaux. These plateaux were first found to occur at integer filling fractions with conductance values quantized to be the integers in units of e^2/h. Furthermore, at those filling fractions where σ_{xy} had plateaux, the diagonal resistivity ρ_{xx} was found to be zero. This was called the Integer quantum Hall effect (IQHE). Subsequently, in experiments involving higher magnetic fields and higher mobility samples, the same phenomenon of pleateaux in σ_{xy} and of vanishing of ρ_{xx} was also found at fractional values of the filling factor. These fractions (with one exception, still being understood) corresponded to odd integers in their denominator. This is often called the fractional QHE (FQHE). IQHE is relatively easy to understand and we will begin our discussion with this phenomenon. FQHE will be considered later.

6.3 Landau problem

To understand IQHE let us begin by considering the Landau Level problem in quantum mechanics of a charged particle in 2D moving in a perpendicular magnetic field.

Consider a Hamiltonian given by

$$H = \frac{1}{2\mu}\sum_i \left[(\mathbf{p}_i + \frac{e\mathbf{A}(\mathbf{r}_i)}{c})^2 + V_{\text{impurity}}(\mathbf{r}_i)\right]$$
$$+ \sum_{ij} V_{\text{coulomb}}(\mathbf{r}_{ij}) - g\sum_i \sigma_i \cdot \mathbf{B} . \quad (6.2)$$

The eigenfunctions of this full Hamiltonian cannot be obtained analytically. Let us therefore begin with the case of the clean (no impurity), non-interacting, spinless electron gas. That leaves us with the sum of one-electron Hamiltonians of the form

$$h^{(1)} = \frac{1}{2\mu}(\mathbf{p} + \frac{e\mathbf{A}}{c})^2 . \quad (6.3)$$

Consider the magnetic field $\mathbf{B} = -B\hat{\mathbf{z}}$ and choose the symmetric gauge $\mathbf{A} = \frac{By}{2}\hat{\mathbf{x}} - \frac{Bx}{2}\hat{\mathbf{y}}$. We transform to dimensionless variables

$$z = \frac{x+iy}{2l} \; ; \; \bar{z} = \frac{x-iy}{2l} , \quad (6.4)$$

where l is the magnetic length defined by $2\pi l^2 B = \phi_0 = hc/e$. Defining $\partial = l(\partial_x - i\partial_y)$ and $\bar{\partial} = l(\partial_x + i\partial_y)$ and $\omega = eB/\mu c$, the Hamiltonian $h^{(1)}$ reduces to

$$h^{(1)} = \frac{\hbar\omega}{2}\left[-\bar{\partial}\partial + \bar{z}z - (z\partial - \bar{z}\bar{\partial})\right]. \quad (6.5)$$

Exercise 6.2 *Derive the above $h^{(1)}$.*

Defining raising and lowering operators

$$a = \frac{\bar{\partial}+z}{\sqrt{2}} \; ; \; a^\dagger = \frac{-\partial+\bar{z}}{\sqrt{2}} \quad (6.6)$$

6.3. Landau problem

obeying the commutation relations $[a, a^\dagger] = 1$ and $[a, a] = 0 = [a^\dagger, a^\dagger]$, we get $h^{(1)} = \hbar\omega(a^\dagger a + 1/2)$. This is thus a 1D simple harmonic oscillator.

However, the electron moving in a plane must have two degrees of freedom, and hence we need 2 more operators to completely describe the system. This is achieved by defining

$$b = \frac{\partial + \bar{z}}{\sqrt{2}} \;;\; b^\dagger = \frac{-\bar{\partial} + z}{\sqrt{2}} \qquad (6.7)$$

satisfying the same algebra as the a-operators with $[a, b] = 0 = [a, b^\dagger]$. Note that $[h^{(1)}, b] = 0$. The eigenstates of $h^{(1)}$ are labelled by the quantum numbers of the a, b harmonic oscillators. There are two independent sets of oscillators, those with frequency ω corresponding to the operator a and those with frequency zero corresponding to the operator b. The eigenvalues of $h^{(1)}$ are independent of the b-harmonic oscillator.

What do the opertors $a, a^\dagger$ and $b, b^\dagger$ physically correspond to? Classically the electron trajectory in a magnetic field is a circle. Since $a, a^\dagger$ decrease or increase the energy, different eigenfunctions of this oscillator correspond to orbits of different radii. It can also be shown that the states of the $b, b^\dagger$ oscillator correspond to the guiding centers of these circles. This is given as an exercise below.

Exercise 6.3 *Solve the Lorentz force equation and get the velocity, radius, etc. Show that $z = i\dot{z}/\omega + C$. In the quantum problem, writing $i\dot{z} = [z, h]$, show that we get $C = z - i\dot{z}/\omega = b^\dagger/\sqrt{2}$. We thus interpret $b, b^\dagger$ as shifting the location of the circular orbit.*

The eigenstates of $h^{(1)}$:

Each eigenstate is labelled by two quantum numbers n (index of the a-harmonic oscillator) and m (index of the b-harmonic oscillator). We denote them by φ_{nm}. Consider the state φ_{00} such that

$$a\varphi_{00} = 0 = b\varphi_{00}. \qquad (6.8)$$

This has $n = 0$ and is therefore a ground state. Solving for this, we get $\varphi_{00} = D_{00} \exp(-|z|^2)$.

Exercise 6.4 *Solve the above equations to get the given solution for φ_{00} and normalize to get D_{00}.*

The states in the Lowest Landau Level (LLL) correspond to $n = 0$. These states φ_{0m} are created by $(b^\dagger)^m \varphi_{00}$ and are degenerate since $[h^{(1)}, b^\dagger] = 0$.

Exercise 6.5 *Show that acting $(b^\dagger)^m$ on φ_{00} gives $\varphi_{0m} = D_{0m} z^m \exp(-|z|^2)$.*

Exercise 6.6 *Plot $|\varphi_{0m}|^2$ in the plane for a few values of m. See that these wavefunctions correspond to radial patterns with increasing radius for larger m.*

Higher Landau Level states are created by $\varphi_{nm} = (a^\dagger)^n (b^\dagger)^m \varphi_{00}$ and their energies are given by $E_{nm} = (n + 1/2)\hbar\omega$.

Exercise 6.7 *Obtain the normalized form for φ_{nm}.*

6.4 Degeneracy counting

Formally on an infinite plane, the degeneracy of each Landau level is infinite. But in a real sample this degeneracy is not infinite but is limited by the area of the sample. Consider the LLL with normalized wavefunctions $\varphi_{0m} = D_{0m} z^m \exp(-|z|^2)$. In this state $<r^2>_{0m} = <\varphi_{0m}|r^2|\varphi_{0m}> = 2l^2(m+1)$. Thus, the circular wavefunctions have larger average radius with increasing m. For higher LL, $<r^2>_{nm} = 2l^2(n+m+1)$. For fixed n and large m, $<r^2> \simeq 2l^2 m$. Equating the maximum radius corresponding to m_{max} with the sample radius R, we get $2l^2 m_{max} = R^2 = \text{Area}/\pi$. This leads to $m_{max} = M = \phi/\phi_0$ (ϕ is the total flux through the sample) which is a macroscopic quantity. Here, M is just the Landau level degeneracy.

Exercise 6.8 *Using the earlier definition of the filling fraction ν, show that $\nu = $ (number of electrons)$/M$.*

6.5 Laughlin wavefunction

Recall that the wavefunctions of one electron in the symmetric gauge is

$$\phi_{nm} = (a^\dagger)^n (b^\dagger)^m \phi_{00} \qquad (6.9)$$

and the energy is given by

$$E_{nm} = (n + 1/2)\hbar\omega . \qquad (6.10)$$

What happens for a many electron system? Consider $\nu = 1$ which corresponds to a filled LLL. For n-electrons, ignoring spin and interactions, the wavefunction is a Slater determinant made of single particle LLL states, namely $\phi_{0m} = D_{0m} z^m e^{-|z|^2}$. This is just the standard van der Monde determinant whose value is also given by the product

$$\Psi_{\nu=1} = \Pi_{i<j}(z_i - z_j) e^{-\sum_i |z_i|^2} . \qquad (6.11)$$

Laughlin suggested that this would be a very good wavefunction even when electron interactions are included and this is the first of the family of the famous Laughlin wavefunctions. The following arguments support his claim:

(1) The wavefunction is antisymmetric under exchange of particle indices and hence describes a system of fermions.

(2) It is an exact eigenstate of the total angular momentum.

(3) The wavefunction has zeros whenever two particles coincide. This serves to keep the electrons apart and significantly reduces the repulsive Coulomb energy.

Exercise 6.9 *Prove statement (2) above by looking at the effect of rotations about the z-axis and showing that the wavefunction is unchanged except for an extra overall phase.*

Laughlin also proposed that for $\nu = 1/m$ (m odd, called primary fractions)

$$\Psi_{\nu=1/m} = e^{-\sum_i |z_i|^2} \Pi_{i<j}(z_i - z_j)^m \qquad (6.12)$$

is a good n-electron wavefunction.

This wavefunction shares all the virtues of the $\nu = 1$ wavefunction listed above. That it corresponds to a filling of $1/m$ can be seen as follows. Compute $<\Psi_{1/m}|r_i^2|\Psi_{1/m}>$. This matrix element will have dominant contribution from the maximum power of z_i in $\Psi_{1/m}$, which is $(N-1)m$. Thus, following earlier calculations of $<r^2>$, we find $<r_i^2>_{1/m} = 2l^2(N-1)m \sim 2l^2 Nm$. The maximum distance of the ith particle in the wavefunction $\Psi_{1/m}$ is thus $l\sqrt{2Nm}$. Setting this equal the radius of the sample leads to $m = 1/\nu$.

Exercise 6.10 *Prove the above assertion.*

6.6 Plasma analogy

The plasma analogy offers a powerful technique for obtaining the density distribution associated with various Laughlin wavefunctions. Consider the probability density of the Laughlin wavefunction in the many particle space -

$$|\psi(z_1 \ldots z_n)|^2 = |D|^2 \Pi_i |z_i - z_j|^{2m} e^{-2\sum_i |z_i|^2}$$
$$= |D|^2 \exp\left[2m \sum_{i<j} \log(\frac{r_{ij}}{2l}) - \frac{1}{2l^2} \sum_i r_i^2\right] \equiv e^{-\beta F} .$$
(6.13)

Notice that this expression has the same form as the free energy of a one component plasma in 2D in the presence of a neutralising background. Upto constants, we have

$$F = -\frac{1}{\beta}\left[2m \sum_{i<j} \log(\frac{r_{ij}}{2l}) - \frac{1}{2l^2} \sum_i r_i^2\right] . \quad (6.14)$$

In this analogy, there is a one-to-one mapping between the original electrons and a set of *fictitious* classical plasma particles. The probability density of an electron configuration (given by $|\psi|^2$) is mapped to the statistical (Gibbs) weight $\exp(-\beta F)$ of the plasma

6.6. Plasma analogy

particles. Configurations with large probability density in the electron problem correspond to the plasma particles being in configurations with the minimum "free energy" F.

The first term of F resembles the 2D Coulomb interaction between two charged particles. The second term will be shown to have the same form as the interaction of these particles with a neutralising background. Note that the kinetic energy in classical statistical mechanics only results in an overall constant which can be ignored for our purposes.

Exercise 6.11 *Using Gauss theorem in 2D, show that the interaction energy between 2 charged particles with charges q_1, q_2 separated by distance r_{12} is*
$V_{12} = -(q_1 q_2/2\pi) \log(r_{12})$.

This term can be identified with the first term of F. With $q_1 = q_2 = q$, this leads to $q^2/2\pi = 2m/\beta$. Therefore the charge of the analog plasma particle is $q = \sqrt{4\pi m/\beta}$.

Exercise 6.12 *Solving Laplace's equation in 2D, show that the potential energy of a particle of charge q in a background charge density of ρ_B is $V = q\rho_B r^2/4$. We assume the particles and the background charge to be confined to a circular disc.*

This term can be identified with the second term of F and leads to
$$\frac{q\rho_B}{4} = \frac{1}{2\beta l^2} \cdot \quad (6.15)$$

Using the above result for q, we now get the charge density of the neutralizing background as $\rho_B = (2/l^2)(1/\sqrt{4\pi m\beta})$. For the plasma particles in that background, their most favorable configuration must have charge density $\rho_P = \rho_B$. The corresponding number density of the plasma particles is

$$n \equiv \frac{\text{charge density}}{\text{charge per particle}} = \left(\frac{B}{hc/e}\right)\left(\frac{1}{m}\right). \quad (6.16)$$

We are trying to identify the set of $\{z_i\}$, for which $\psi(z_1, z_2, \ldots, z_n)$ is peaked. This will happen for the lowest free energy configurations of the classical plasma. However, there is a one-to-one mapping between the QH electrons and the plasma charges. Thus, the number density of plasma particles n is the same as the electron number density n_e. Thus, the above expression gives $\nu = 1/m$.

6.7 Quasi-holes and their Laughlin wavefunction

Recall that Laughlin's wavefunction for the ground state at $\nu = 1/m$ is

$$\Psi_{\nu=1/m} = \Pi_{i<j}(z_i - z_j)^m e^{-\sum_i |z_i|^2} . \quad (6.17)$$

The quasihole is the lowest, charged excitation over this ground state. Laughlin suggested the following wavefunction for a quasihole located at w. Notice that this wavefunction is also made up of LLL states.

$$\Psi_{quasihole} = \Pi_i(z_i - w)\Pi_{i<j}(z_i - z_j)^m e^{-\sum_i |z_i|^2} . \quad (6.18)$$

Let us now repeat the plasma analogy for $|\Psi_{quasihole}|^2$.

Exercise 6.13 *Construct the $|\Psi_{qh}|^2$ and equate it to $e^{-\beta F}$ and show that F can be written as*

$$F = -\frac{2m}{\beta}\log(r_{ij}) + \frac{1}{2l^2\beta}\sum_i r_i^2 - \sum_i \frac{2}{\beta}\log(|\mathbf{r}_i - \mathbf{R}|) .$$

We note that the above expression is identical to the one we derived when we did the plasma analogy for the electron case except for the extra term at the end. In the plasma analogy this corresponds to an impurity being present at $\mathbf{R}$. We have to find out the charge on this impurity. This is done as follows: The interaction between the plasma charge at site i, $q = \sqrt{(4\pi m/\beta)}$, and the impurity charge q_{imp} at $\mathbf{R}$ is given by $(1/2\pi)q\, q_{imp} \log(|\mathbf{r}_i - \mathbf{R}|)$. Now comparing with the free energy expression derived above we find that $(1/2\pi)q\, q_{imp} = 2/\beta$, which means that $q_{imp} = \sqrt{(4\pi/m\beta)}$

The impurity particle has the same charge as the plasma particles and will push out a compensating number of plasma particles around it. This means that in the real quantum hall system there is a depletion of electrons about some point which correponds to a hole.

How much is this depletion?

$$\delta n = -\frac{q_{imp}}{q} = -\frac{1}{m} = -\nu \ . \tag{6.19}$$

Notice the important result that this depletion in electron number is by a fraction $1/m$. In other words, the quasihole carries fractional charge. To find the statistics of these hole excitations, consider a hole at w_1 going around a circle of radius R. The phase change of the wavefunction is $2\pi N_R$ where N_R is the number of electrons enclosed in the circle. If there is another hole at w_2 around which the loop goes, the phase change is $2\pi N_R - 2\pi \frac{1}{m}$, since there is a depletion of electrons by $1/m$ around the hole w_2. The phase change when one particle goes around another identical particle will be twice the phase change when they are exchanged. The statistical angle is therefore π/m. The quasiholes are thus anyons for $m=3,5,\ldots$

6.8 Localization physics and the QH plateaux

The problem of electrons moving in a magnetic field in 2D in the absence of disorder leads to the picture of highly degenerate Landau levels which are seperated in energy by gaps of cyclotron energy $\hbar\omega$. In the presence of weak disorder, this picture changes considerably. Each Landau level is broadened into a band. Most of the states in the band are localised except in the middle of the band. We offer below an explanation of why this happens. Numerical studies also support this result.

Consider the Landau level problem in the presence of an electric field (E) along the $\hat{x}$ direction in the Landau gauge. It can be shown that the energy eigenfunctions are

$$\psi_{nk}(\mathbf{r}) = \frac{1}{\sqrt{L_y}} e^{(-iky)} H_n(x + kl^2 - \frac{eE}{\mu\omega^2}) e^{(-\frac{(x+kl^2 - eE/\mu\omega^2)^2}{2l^2})} \quad (6.20)$$

Exercise 6.14 *Work out (in the one-particle problem) the wavefunctions, eigenvalues and degeneracy in the Landau gauge $\mathbf{A} = -Bx\hat{\mathbf{y}}$ with no electric field. The wavefunctions in this case appear as a product of harmonic oscillators along the $\hat{\mathbf{x}}$ direction (labelled by a h.o. quantum number) and plane waves (labelled by the momentum label k) along the $\hat{\mathbf{y}}$ direction.*

Exercise 6.15 *Still working in the Landau gauge, with an electric field along the $\hat{\mathbf{x}}$ direction, show that the $\hat{\mathbf{x}}$-direction harmonic oscillator is displaced by $-eE/\mu\omega^2$ as given above.*

Looking at the solution of this problem, we find that equipotential curves in this problem lie along the $\hat{\mathbf{y}}$ direction and the potential increases linearly along the $\hat{\mathbf{x}}$ direction. Notice that the wavefunctions are harmonic oscillator solutions along the $\hat{\mathbf{x}}$ direction, centred at $-kl^2 - eE/\mu\omega^2$. This is multiplied by a plane wave along the $\hat{\mathbf{y}}$ direction. Thus, the new wavefunctions have a spread in the $\hat{\mathbf{x}}$ direction of the order of the magnetic length (l) and run along the equipotential curves of the electric field along the $\hat{\mathbf{y}}$ direction.

Having learnt that the QH wavefunctions tend to stay on equipotential surfaces in the presence of crossed electric and magnetic fields, and have a small spread of order magnetic length (l) at large B, we turn to the problem of disorder in the Landau level problem. The disorder potential is assumed to be weak and to vary slowly over distances of order l. In this case, a semiclassical approximation implies that in a small region, the electric field due to the disorder potential can be assumed to be constant and the above solution in constant electric field then leads to wavefunctions which lie on equipotential contours of the disorder potential.

6.8. Localization physics and the QH plateaux

We now consider the disorder potential as made up of hills and valleys. In a given Landau band, the low energy wavefunctions are then equipotential contours. Let us imagine filling up the band. These lowest energy wavefunctions will lie in deep valleys and thus are localized in small regions and cannot carry current across the sample. As we consider wavefunctions with increasing energy, such wavefunctions keep rising in the valleys and the spacing between these contours also decreases. It is not hard to imagine that at some critical value, the wavefunctions can overlap and span the entire sample, leading to the delocalized states in the middle of a Landau band. As the energy is further increased, the contours now begin to circle the hills in the disorder potential. Again, such states are localized. Thus, our arguments suggest that states below and above the unperturbed Landau level get localized and form the Landau band, but the states in the middle of this band continue to be extended and would contribute to transport.

Thus, the flatness of the Hall conductance over a region around the integer filling and jump between the plateaux is clear. As the filling fraction is increased from zero, we first encounter the localized states and no transport is possible. When we encounter the first set of extended states at the middle of the first Landau band, the Hall conductance jumps by an amount arising from these extended states. With increasing filling, we again encounter the localized states in the upper part of the Landau band and there is no increase in the Hall conductance until the next set of extended states in the next Landau band is reached. This leads to the plateau structure seen in experiments.

Note that, for the plateau to plateau transition to be sharp, the number of extended states has to actually be very small (microscopic).

Exercise 6.16 *Show that* $<j_y>(E=0)=0$. *(Note: $v_y = (p_y + eA_y/c)/\mu$ and $j_y = -ev_y$.) For $E \neq 0$, show that $<j_y>(E) = -ecE_x/B$; then $\sigma_{xy}(B) = -\sigma_{yx}(B) = e^2\nu/h$.*

6.9 Chern-Simons theory

Ignoring the electron spin, the many-electron spatial wavefunction has to be antisymmetric under the exchange of electron coordinates. Beginning with the wavefunction $\Psi(z_1,\ldots,z_N)$, construct the wavefunction

$$\Phi(r_1,\ldots,r_N) = e^{-im\sum_{i<j}\theta_{ij}}\Psi(r_1,\ldots,r_N)\,, \qquad (6.21)$$

where θ_{ij} is the angle of the vector pointing from the i-th to the j-th electron. Under exchange of two particles, say r and s, $\theta_{r,s} \to \theta_{r,s} + \pi$, while other θ's remain unchanged. Further, Ψ picks up a negative sign from its fermionic character. Thus, $\Phi(r_1,\ldots,r_N) \to e^{-i(m+1)\pi}\Phi$ under exchange of any two electron coordinates. Thus, for $m =$ odd integer, Φ represents a bosonic wavefunction, while it is a fermionic wavefunction for even integer m. For general real m, the wavefunction picks up a phase which is neither fermionic nor bosonic and this corresponds to anyonic statistics.

Now,

$$\begin{aligned}\nabla_i\Psi &= \nabla_i\left(e^{im\sum_{i<j}\theta_{ij}}\Psi\right) \\ &= e^{im\sum_{i<j}\theta_{ij}}\left(im\sum_{j\neq i}\nabla_i\theta_{ij} + \nabla_i\right)\Phi\,. \qquad (6.22)\end{aligned}$$

Defining $\mathbf{a}_i = (m\hbar c/e)\sum_{j\neq i}\nabla_i\theta_{ij}$, we see that $\mathbf{a}_i$ appears in the Hamiltonian for Φ exactly in the form of a vector potential! This fictitious vector potential is referred to as the Chern-Simons gauge field. We now try and compute the magnetic field $\mathbf{b} = \nabla \times \mathbf{a}$, corresponding to this gauge potential. Note that away from any singular points, $\nabla_i \times \mathbf{a}_i = \sum_{i<j}\nabla_i \times \nabla_i\theta_{ij} = 0$, since it is the curl of a gradient. However, there could be singular points where $\nabla_i\theta_{ij}$ diverges. In order to evaluate the contribution from such singular points, we have to consider the line integral $\oint_L \mathbf{a}_i \cdot d\mathbf{l}$ over any contour in the plane. This is just $mhc/e = m\phi_0$ whenever the i-th particle goes around any other particle in the system since $\oint_L \mathbf{a}_i \cdot d\mathbf{l} = \oint_S \nabla_i \times \mathbf{a}_i \cdot dS$ by Stokes theorem, where $\oint_S$ refers to

6.9. Chern-Simons theory

the surface integral bounded by the contour L of the line integral. Thus the effective magnetic field $\mathbf{b}$ can be represented by $\mathbf{b} = \nabla_i \times \mathbf{a}_i = m\phi_0 \sum_{j\neq i} \delta^{(2)}(\mathbf{r}_i - \mathbf{r}_j)$ and hence the i-th particle appears to experience a magnetic field due to a flux $m\phi_0$ localized at the position of each of the other particles. We have thus achieved what is sometimes referred to in the literature as "flux attachment". In the presence of an external vector potential, the wavefunction Φ satisfies

$$\left[\sum_i \left(-i\hbar\nabla_i + \frac{e}{c}\mathbf{A}(\mathbf{r}_i)\right)\right]\Psi = e^{im\sum_{i<j}\theta_{ij}} \times$$
$$\left[\sum_i \left(-i\hbar\nabla_i + \frac{e}{c}(\mathbf{A}(\mathbf{r}_i) + \mathbf{a}(\mathbf{r}_i))\right)\right]\Phi \ . \qquad (6.23)$$

For odd m, Φ is bosonic and we now examine the consequence of assuming complete Bose condensation of the bosons represented by Φ. In that case, we just have $\Phi = 1$ and this leads to

$$\Psi = e^{im\sum_{i<j}\theta_{ij}} = \Pi_{i<j}\left(e^{i\theta_{ij}}\right)^m \qquad (6.24)$$

which is reminescent of the angular factors in the Laughlin wavefunction. Notice that in order to get a wavefunction which is symmetric under particle exchange, any wavefunction of the form $\tilde{\Phi} = f(r_1,\ldots,r_N)\Phi$ could also have been used where f is a function symmetric under the interchange of any two particle coordinates, so that $\Psi = f^{-1}(r_1,\ldots,r_N)\exp\left(-im\sum_{i<j}\theta_{ij}\right)\tilde{\Phi}$. The Hamiltonian of the original fermionic system can be written as

$$H_{fermi} = \int d^2 r \Psi^\dagger(\mathbf{r})\left[-\frac{\hbar^2}{2\mu}(\nabla - \frac{ie}{\hbar c}\mathbf{A}(\mathbf{r}))^2 - eA_0(\mathbf{r})\right]\Psi(\mathbf{r})$$
$$+ \frac{1}{2}\int d^2 r\, d^2 r'\, \delta\rho(\mathbf{r})\, V(\mathbf{r}-\mathbf{r}')\, \delta\rho(\mathbf{r}') \qquad (6.25)$$

with $\rho(\mathbf{r}) = \Psi^\dagger(\mathbf{r})\Psi(\mathbf{r})$ being the density operator and $\delta\rho(\mathbf{r})$ denoting the deviation of the fermion density from its mean value $\bar{\rho}$.

We now try a more general transformation to get the full Laughlin wavefunction, and which also allows us to derive the

Jain phenomenology of composite fermions to obtain the full hierarchy of FQHE states. With this is mind, we define $\Phi(\mathbf{r}) = \exp(-J(\mathbf{r}))\Psi(\mathbf{r})$ and try to write $J(\mathbf{r})$ from our intuition based on the first quantized wavefunctions. We set

$$J(\mathbf{r}) = m \int d^2 r' \, \rho(\mathbf{r}') \, \log(z - z') - |z|^2. \qquad (6.26)$$

With a purely imaginary J, we can recover the usual Chern-Simons theory, but this transformation is more general.

Exercise 6.17 *With this definition of $J(\mathbf{r})$, and using $[\rho(x), \Psi(x')] = -\Psi(x)\delta^{(2)}(x - x')$, show the following identities:*

$$(a) \quad e^{-J(\mathbf{r})}\Psi(\mathbf{r}') = (z - z')^m \, \Psi(\mathbf{r}')e^{-J(\mathbf{r})}$$
$$(b) \quad e^{-J(\mathbf{r})}\Psi^\dagger(\mathbf{r}') = (z - z')^{-m}\Psi^\dagger(\mathbf{r}')e^{-J(\mathbf{r})}$$

Following this, it is also easy to derive the commutation rules obeyed by the Φ-fields.

Exercise 6.18 *Show that $\Phi(x)\Phi(x') = (-1)^{m+1}\Phi(x')\Phi(x)$.*

Thus, the Φ's satisfy bosonic commutation relations for odd m and fermionic anticommutation relations for even m. It is important to note that the transformation we have made is not unitary since $J(\mathbf{r})$ is not a purely imaginary quantity unlike the pure Chern-Simons case. Hence $\Phi^\dagger$ and Φ are no longer canonically conjugate variables. We have to define a new field $\Pi(\mathbf{r}) = \Phi^\dagger(\mathbf{r}) \exp(J(\mathbf{r}) + J^\dagger(\mathbf{r}))$ which is canonically conjugate to Φ.

Exercise 6.19 *Show that*

$$(a) \quad \Pi(x)\Pi(x') = (-1)^{m+1}\Pi(x')\Pi(x)$$
$$(b) \quad \Phi(x)\Pi(x') = (-1)^{m+1}\Pi(x')\Phi(x) + \delta^{(2)}(x - x').$$

6.9. Chern-Simons theory

It is easy to check that $\Pi(x)\Phi(x) = \Psi^\dagger(x)\Psi(x)$ is the density operator in terms of the new variables. Further, $[\hat{N}_{fermi}, \Pi(x)] = \Pi(x)$ and thus Π, which acts as a creation operator for the Φ field, also increases the number of electrons. Thus, the number of electrons in the original theory is the same as the number of bosons in the transformed theory. We note that defining the fermion vacuum by $\Psi|0>= 0$, we see that $\Phi|0>= 0$ as well. Now, following our earlier discussion for the pure Chern-Simons gauge field, we find that

$$\left[\sum_i \left(-i\hbar\nabla_i + \frac{e}{c}\mathbf{A}(\mathbf{r}_i)\right)\right]\Psi(\{\mathbf{r}\}) = e^{J(\{\mathbf{r}\})} \times$$
$$\left[\sum_i \left(-i\hbar\nabla_i + \frac{e}{c}\mathbf{A}_{eff}(\mathbf{r}_i)\right)\right]\Phi \quad (6.27)$$

where $\mathbf{A}_{eff}(\mathbf{r}_i) = \mathbf{A}(\mathbf{r}_i) - i(\hbar c/e)\nabla_i J(\{\mathbf{r}\})$. Writing $\mathbf{D} = (\nabla + i(e/\hbar c)\mathbf{A}_{eff})$ then leads to the following Hamiltonian

$$H_{boson} = \int_\mathbf{r} \Pi(\mathbf{r}) \left(-\frac{\hbar^2}{2\mu}\mathbf{D}^2 - eA_0\right) \Phi(\mathbf{r})$$
$$+ \frac{1}{2}\int_{\mathbf{r},\mathbf{r}'} [\Pi(\mathbf{r})\Phi(\mathbf{r}) - \bar{\rho}] V(\mathbf{r} - \mathbf{r}') [\Pi(\mathbf{r}')\Phi(\mathbf{r}') - \bar{\rho}] .$$
$$(6.28)$$

Exercise 6.20 *Check that the Hamiltonian above is a hermitian quantity.*

We now find J which gives rise to the effective vector potential $\mathbf{A}_{eff}$. The gradient of J was given by the equation:

$$-\frac{i\hbar c}{e}\nabla J = -\frac{i\hbar c}{e}\nabla\left[m\int d^2x' \rho(x')\log(z-z') - |z|^2\right] . \quad (6.29)$$

Define $\mathbf{a}(x) = \frac{m\hbar c}{e}\int d^2x' \rho(x')\nabla\theta(z-z')$. Therefore, $\nabla \times \mathbf{a} = \frac{m\hbar c}{e}2\pi\rho(x) = m\phi_0\rho(x)$. We know that

$$\nabla \log|\mathbf{r}-\mathbf{r}'| = -\hat{k}_3 \times \nabla\theta(\mathbf{r}-\mathbf{r}') \quad \text{(Cauchy Riemann conditions)}.$$
$$(6.30)$$

(Here $\hat{k}_3$ is the unit vector along the z-direction.) Using these, Eq.(6.29) becomes:

$$-\frac{i\hbar c}{e}\nabla J = \mathbf{a} + i(\hat{k}_3 \times \mathbf{a}) + \frac{i\hbar c}{2el^2}\cdot\mathbf{r} \quad (6.31)$$

Exercise 6.21 *Show that the last term in the above equation can be written as $i(\hat{k}_3 \times \mathbf{A})$.*

The effective potential $\mathbf{A}_{eff} = \mathbf{A} - \frac{i\hbar c}{e}\nabla J$, therefore becomes $\mathbf{A}_{eff} = (\mathbf{A}+\mathbf{a}) + i\hat{k}_3 \times (\mathbf{a}+\mathbf{A})$. To decouple the $\mathbf{a}$ field from the ρ field, we use Lagrange multipliers. This way we release the constraint on $\mathbf{a}$. The Lagrangian therefore becomes:

$$\begin{aligned}\mathcal{L} &= \left(\frac{e}{2m\phi_0}\int \epsilon^{\mu\nu\sigma}a_\mu\partial_\nu a_\sigma\right) + \int_\mathbf{r} \lambda(\mathbf{r})\partial_i a_i \\ &+ \int \Pi\left(-i\hbar\frac{\partial}{\partial t} + e(a_0+A_0)\right)\Phi + \frac{\hbar^2}{2\mu}\Pi D^2\Phi \\ &- \int_\mathbf{r}\int_{\mathbf{r}'}\delta\rho(\mathbf{r})V(\mathbf{r}-\mathbf{r}')\delta\rho(\mathbf{r}') ,\end{aligned} \quad (6.32)$$

where $a^\mu = (a^0, \mathbf{a}); \mu = 0, 1, 2 \ldots$.

The Euler-Lagrange equations of motion give:
(a) The a_0 equation:

$$-e\pi\phi = -\frac{2e}{2m\phi_0}\nabla\times\mathbf{A} \Rightarrow \nabla\times\mathbf{a} = m\phi_0\rho . \quad (6.33)$$

(b) The λ equation:
$$\partial_i a_i = 0 . \quad (6.34)$$

(c) The a_i equations:

$$\frac{e}{m\phi_0}(-\partial_j a_0 - \partial_0 a_j) - \partial_i\lambda = -\frac{e}{c}\left(\vec{J} - i\hat{k}_3\times\vec{J}\right) \quad (6.35)$$

where, $\vec{J} = \frac{\hbar}{2i\mu}(\Pi D\Phi - D^*\Pi\Phi)$.

(d) The Π equation:

$$\left[i\hbar\frac{\partial}{\partial t} + e(a_0+A_0)\right]\Phi(x) = -\frac{\hbar^2}{2\mu}D^2\Phi + \int_{x'}\delta\rho(x')V(x-x')\Phi(x) . \quad (6.36)$$

6.9. Chern-Simons theory

Now take m to be odd and equal to $1/\nu$. Look at the classical solutions when $A_0 = 0$. Let $\Phi(x) = \sqrt{\bar\rho} = \Pi(x)$ be the translationally invariant ground state, where, $\bar\rho = \nu B/\phi_0 = B/m\phi_0$. Also choose $\mathbf{a}(x) = -\mathbf{A}(x)$ and $a_0 = \lambda = 0$. This means that $\mathbf{A}_{eff} = 0$ and hence $\mathbf{D} = \nabla$. Since $\rho = \bar\rho$, $\delta\rho = 0$. This corresponds to condensation of particles into a particular state.

Exercise 6.22 *For the above situation of the uniform translationally invariant ground state show that the curl of the statistical field (Chern-Simons field) is equal to the external magnetic field i.e., show that $\nabla \times \mathbf{a} = \mathbf{B}$.*

Consider the zero momentum state

$$|G> = \left(\int_x \Pi(x)\right)^N |0> . \qquad (6.37)$$

The N particle wavefunction in this state is

$$\psi(x_1, x_2, \ldots x_N) = <0|\Psi(x_1)\Psi(x_2)\ldots\Psi(x_N)|G>$$
$$= <0|e^{J(x_1)}\Phi(x_1)e^{J(x_2)}\Phi(x_2)\ldots e^{J(x_N)}\Phi(x_N)|G> . \qquad (6.38)$$

The ground state can be written as

$$|G> = \int_{y_1} \Pi(y_1) \int_{y_2} \Pi(y_2) \ldots \int_{y_N} \Pi(y_N)|0> . \qquad (6.39)$$

N is the mean number of particles in the zero momentum state. The N-electron wavefunction can therefore be written as

$$\psi(x_1, x_2, \ldots, x_N)$$
$$= \int_{y_1} \ldots \int_{y_N} <0|e^{\sum_i J(x_i)}\Phi(x_1)\ldots\Phi(x_N)\Pi(y_1)\ldots\Pi(y_N)|0> \qquad (6.40)$$

where $J(x) = m\int_{x'} \rho(x')\log(z-z') - |z|^2$. The final expression for the wavefunction is

$$\psi(x_1, \ldots, x_N) = \int_{y_1} \ldots \int_{y_N} e^{\sum_i |z_i|^2} \Pi_{i<j}(z_i - z_j)^m$$
$$<0|\Phi(x_1)\ldots\Phi(x_N)\Pi(y_1)\ldots\Pi(y_N)|0>$$
$$= \psi_{\text{Laughlin}} . \qquad (6.41)$$

6.10 Vortices in the CS field and quasiholes

In the previous lecture, we arrived at a classical solution which satisfied the equations of motion of the Lagrangian of the generalized Chern-Simons theory. We note that this same solution can be thought of as a mean field approximation familiar from Landau theory in statistical mechanics. We set $\rho(x) = \bar{\rho}$ in the interaction and in the expression for the Chern-Simons gauge field, $\mathbf{a}$. This results in $\nabla \times \mathbf{a}(x) = m\phi_0 \rho(x) \to m\phi_0\bar{\rho}$. Setting $\nu = 1/m$, we then find that $\nabla \times \mathbf{a}(x) = -\nabla \times \mathbf{A}(x)$, hence $(\mathbf{a} + \mathbf{A}) = 0$ which is identical to the classical solution we had earlier.

We now proceed to construct quasihole wavefunctions. This will be done by constructing vortex solutions to the Chern-Simons Lagrangian and seeing that the wavefunctions thus obtained are just the Laughlin quasihole wavefunctions. We begin by considering a solution $\Phi = \sqrt{\bar{\rho}}e^{-i\theta}/r$ and $\Pi = \sqrt{\bar{\rho}}\, re^{-i\theta}$ and $\mathbf{a} + \mathbf{A} = (\hbar c/e)\hat{\theta}/r$. These forms for the Φ, Π are valid at large r and one has to look for numerical solutions to obtain the short range behavior.

With this choice of $(\mathbf{a} + \mathbf{A})$, we see that $\oint_S (\nabla \times \mathbf{a}) \cdot d\mathbf{S} = B \times$ (area) $+ hc/e$. But from using $\mathbf{a} = m\phi_0\rho(x)$, this is just $m\phi_0 \int_S \rho dS = m\phi_0 N$, N being the number of particles in the enclosed area. Now $B \times$ (area) $= m\phi_0 \bar{N}$, where $\bar{N}$ is the mean number of particles in the area for the uniform solution, which just follows from the definition of the filling fraction $m = 1/\nu$. Thus we find $\bar{N} - N = 1/m$, so that the particle density is depleted around the vortex solution by an amount $1/m$. This is just the particle depletion around the quasihole that we obtained using the plasma analogy on the Laughlin quasihole wavefunction.

> **Exercise 6.23** Check the consistency of the above solution. We define the vortex creation field as $\Pi_v = \int d^2 r \Pi(x) r e^{i\theta}$ which is consistent with the idea that we want to create a state with the classical solution given above, using the field operator $\Pi(x)$ which acts as a creation operator at point x. Using this definition, obtain the wavefunction corresponding to the vortex

6.11 Jain's theory of composite fermions

solution and check that this is precisely the Laughlin quasihole wavefunction.

6.11 Jain's theory of composite fermions

Recall the expression for the field Φ and the conjugate momentum Π:

$$\Phi = e^{-J}\psi, \quad \Pi = \Psi^\dagger e^J. \quad (6.42)$$

The expression for the current was $J = m\int_{x'} \rho(x)\log(z-z') - |z|^2$. Now take $m = 2k$, then the current becomes

$$J = m\int_{x'} \rho(x)\log(z-z') - 2k\nu|z|^2. \quad (6.43)$$

The Φ field represents particles called composite fermions. The effective vector potential becomes $\mathbf{A}_{eff} = (\mathbf{a} + \mathbf{A}) + i\hat{k}_3 \times (\mathbf{a} + 2k\nu\mathbf{A})$. Now choose $2k$ such that $\mathbf{a} + 2k\nu\mathbf{A} = 0$. Therefore, $\mathbf{A}_{eff} = (1 - 2k\nu)\mathbf{A}$ and $\mathbf{B}_{eff} = (1 - 2k\nu)\mathbf{B}$. Define $\nu_{eff} = \nu/(1 - 2k\nu)$. Suppose ν_{eff} is some integer p. At ν such that $p = \nu/(1 - 2k\nu)$, the composite fermions exhibit integer quantum Hall effect. Therefore one has plateaux at $\nu = p/(2kp \pm 1)$.

The fractional quantum hall effect in the original electron system can now be viewed as the integer quantum Hall effect in the composite fermion system. Consider the N electron wavefunction in the state ν -

$$\begin{aligned}
\psi_\nu^{elec} &= <0|\Psi(r_1)\ldots\Psi(r_N)|G_\nu> \\
&= <0|e^{J(r_1)}\Phi(r_1)\ldots e^{J(r_N)}\Phi(r_N)|G_\nu> \\
&= \Pi_{i<j}(z_i - z_j)^{2k} <0|e^{-\sum_i J(r_i)}\Phi(r_1)\ldots\Phi(r_N)|G_\nu> \\
&= \Pi_{i<j}(z_i - z_j)^{2k} e^{-2k\nu\sum_i |z_i|^2} <0|\Phi(r_1)\ldots\Phi(r_N)|G_\nu>.
\end{aligned} \quad (6.44)$$

We identify the last term with the composite fermionic Slater determinant.

Exercise 6.24 *The composite fermions live in a magnetic field* $\mathbf{B}_{eff} = \mathbf{B}(1 - 2k\nu)$ *which leads to an effective magnetic length* l_{eff}. *Show that the gaussian factor of the associated IQHE wavefunction of the composite fermions along with the gaussian factor explicitly appearing in the above fermionic wavefunction leads to* $e^{-|z|^2}$ *as expected.*

Bibliography

[] *General Quantum Hall Physics Reviews

[1] *Quantum Hall Effect*, ed. by R.E. Prange and S.M. Girvin, Springer, (New York), (1990).

[2] *The Quantum Hall Effect: A Perspective*, ed. by A.H. MacDonald, Klewer, (Boston), (1989).

[3] T. Chakraborty and P. Pietilainen *The Fractional Quantum Hall Effect*, Springer Verlag, (Berlin), (1988).

[4] A. Karlhede, S.A. Kivelson and S.L. Sondhi, *The Quantum Hall Effect – The Article*, Lectures at the 9th Jerusalem Winter School on Theoretical Physics, (1992) in *Correlated Electron Systems*, ed. by V.J. Emery, World Scientific, (1992).

[5] *Quantum Hall Effect*, ed. by M. Stone, World Scientific, (Singapore), (1992).

[] **Anyons and Chern-Simons reviews

[6] Frank Wilczek, *Fractional Statistics and Anyon Superconductivity*, World Scientific, (Singapore), (1990).

[7] Sumathi Rao, *An Anyon Primer*, (hep-th/9209066) in *Models and Techniques of Statistical Physics* ed. by S. M. Bhattacharjee, Narosa Publications, 1995.

[] *** Chern-Simons Field, Landau-Ginsberg theory and Composite Fermions (Original papers related to our lectures)

[8] S.M. Girvin and A.H. Macdonald, Phys. Rev. Lett. **58**, 1252 (1987).

[9] S.C. Zhang, T.H. Hansson and S. Kivelson, Phys.Rev.Lett. **62**, 82 (1989).

[10] N. Read, Phys. Rev. Lett. **62**, 86 (1989).

[11] R. Rajaraman and S.L. Sondhi, Int. J. Mod. Phys. B. **7**, vol. 10, 793 , (1996).

[12] J.K. Jain, Phys. Rev. Lett. **63**, 199 (1989); Phys.Rev.B **40**, 8079 (1989).

[13] A. Lopez and E. Fradkin, Phys. Rev. Lett. **69** , 2126 (1992); Phys. Rev. B **44**, 5246 (1991); ibid. **51**, 4347 (1995).

[14] R. Rajaraman, Phys. Rev. B **56**, 6788 (1997).

Chapter 7

Low-dimensional Quantum Spin Systems

Indrani Bose

Department of Physics
Bose Institute, 93/1, A.P.C. Road
Calcutta – 700009

Low-dimensional quantum spin systems are interacting many body systems for which several rigorous results are known. Powerful techniques like the Bethe Ansatz provide exact knowledge of the ground state energy and low-lying excitation spectrum. A large number of compounds exist which can effectively be described as low-dimensional spin systems. Thus many of the rigorous results are also of experimental relevance. In these lectures, an introduction is given to quantum spin systems and the rigorous results known for such systems. Some general theorems and exactly-solvable models are discussed. An elementary introduction to the Bethe Ansatz technique is given.

7.1 Introduction

In condensed matter physics, we study many body systems of different kinds. In a many body system, a large number of entities interact with each other and usually quantum effects are important. The entities may be fermions (electrons in a metal, He^3 atoms), bosons (He^4atoms) or localised spins on some lattice. In this set of lectures, we will discuss interacting spin systems in low dimensions. The spins are located at the different sites of a lattice. The most well-known model of interacting spins is the Heisenberg model with the Hamiltonian

$$H = \sum_{\langle ij \rangle} J_{ij} \mathbf{S}_i \cdot \mathbf{S}_j. \tag{7.1}$$

$\mathbf{S}_i$ is the spin operator located at the lattice site i. J_{ij} denotes the strength of the exchange interaction. The spin $|\mathbf{S}_i|$ can have a magnitude 1/2, 1, 3/2, 2,......, etc. The lattice, at the sites of which the spins are located, is d-dimensional. Examples are a linear chain (d = 1), the square lattice (d = 2) and the cubic lattice (d = 3). Ladders have structures interpolating between the chain and the square lattice. A ν-chain ladder consists of ν chains

7.1. Introduction

coupled by rungs. The spin dimensionality n denotes the number of components of a spin. The Ising, XY and Heisenberg models correspond to n =1, 2 and 3 respectively. The generalised Ising and XY Hamiltonians are:

$$H_{Ising} = \sum_{\langle ij \rangle} J_{ij} S_i^z S_j^z \qquad (7.2)$$

$$\text{and } H_{xy} = \sum_{\langle ij \rangle} J_{ij} \left[S_i^x S_j^x + S_i^y S_j^y \right]. \qquad (7.3)$$

For the anisotropic XY model, $J_{ij}^x \neq J_{ij}^y$. Usually, the sites i and j are nearest-neighbours (n.n.s) on the lattice and the J_{ij}'s have the same magnitude J for all the n.n. interactions. The Hamiltonian in Eq.(7.1) becomes

$$H = J \sum_{\langle ij \rangle} \mathbf{S}_i \cdot \mathbf{S}_j. \qquad (7.4)$$

There are, however, other possibilities. We mention some examples.
(a) Alternating chain system :

$$H = \sum_j \left[1 - (-1)^j \delta \right] \mathbf{S}_j \cdot \mathbf{S}_{j+1} \qquad (7.5)$$

where $0 \leq \delta \leq 1$.
(b) Further-neighbour interactions :
The well-known Majumdar-Ghosh chain [1] is described by the Hamiltonian

$$H_{MG} = J \sum_{i=1}^{N} \mathbf{S}_i \cdot \mathbf{S}_{i+1} + J/2 \sum_{i=1}^{N} \mathbf{S}_i \cdot \mathbf{S}_{i+2}. \qquad (7.6)$$

The Haldane-Shastry model [2] has a Hamiltonian of the form

$$H = J \sum_{\langle ij \rangle} \frac{1}{|i-j|^2} \mathbf{S}_i \cdot \mathbf{S}_j. \qquad (7.7)$$

(c) J_{ij} 's can be random in sign and magnitude [3].
(d) Quantum antiferromagnetic chain with incommensuration [4].
(e) Mixed-spin system:
A good example is a chain of alternating spins 1 and 1/2 .
(f) Fully anisotropic Heisenberg Hamiltonian :
The Hamiltonian in one dimension (1d) and for n.n. interaction is given by

$$H_{XYZ} = \sum_{i=1}^{N} \left[J_x S_i^x S_{i+1}^x + J_y S_i^y S_{i+1}^y + J_z S_i^z S_{i+1}^z \right] . \qquad (7.8)$$

The Ising and XY Hamiltonians are special cases of H_{XYZ} .

Once we know the Hamiltonian of a spin system, we want to determine the eigenvalues and the eigenfunctions of the system. This is, however, a difficult task as the number of eigenstates is huge . For a spin-1/2 system, the number is 2^N where N is the number of spins. We usually determine the ground state and the low-lying excited states in an approximate manner. Exact knowledge can be obtained only in a few cases. The low-lying excitation spectrum specifies the low-temperature thermodynamics and the response to weak external fields. The spectrum may have some universal features. In the case of electronic systems, there are two well-known universal classes: the Fermi liquids and the Luttinger liquids. We will learn about the universal features of spin excitation spectra as we go on. The spin excitation spectrum can be gapless or can have a gap. In the case of a gapless excitation spectrum, there is at least one momentum wave vector for which the excitation energy is zero. For a spectrum with gap, the lowest excitation is separated from the ground state by an energy gap Δ. The temperature dependence of thermodynamic quantities is determined by the nature of the excitation spectrum (with or without gap). Again, one can give examples from an electronic system. For a Fermi liquid, the electronic excitation spectrum is gapless and the electronic specific heat C_n has a linear temperature dependence. A conventional superconductor is characterised by an energy gap Δ in the electronic excitation spectrum and the

7.1. Introduction

electronic specific heat $C_n \sim \exp-(\Delta/k_B T)$ in the superconducting state.

The partition function Z of a thermodynamic system is

$$Z = \sum_i e^{-(E_i/k_B T)}, \tag{7.9}$$

where E_i's are the energy levels and the sum is over all the states of the system. The free energy is given by

$$F = -k_B T \ln Z. \tag{7.10}$$

The thermodynamic quantities of a magnetic system are:
Magnetic energy :

$$U_m = \langle H \rangle = -Z^{-1} \frac{\partial Z}{\partial(1/k_B T)} \tag{7.11}$$

Entropy S_H at constant field H :

$$S_H = -\left(\frac{\partial F}{\partial T}\right)_H \tag{7.12}$$

Magnetization M_T at constant temperature :

$$M_T = -\left(\frac{\partial F}{\partial H}\right)_T \tag{7.13}$$

Specific heat at constant field :

$$C_H = \left(\frac{\partial U}{\partial T}\right)_H = T\left(\frac{\partial S}{\partial T}\right)_H = -T\left(\frac{\partial^2 F}{\partial T^2}\right)_H \tag{7.14}$$

and isothermal susceptibility :

$$\chi_T = \left(\frac{\partial M}{\partial H}\right)_T = -\left(\frac{\partial^2 F}{\partial H^2}\right)_T. \tag{7.15}$$

The spontaneous magnetization M_S ($H = 0$) is known as the order parameter and has a non-zero value in the ordered phase of a magnetic system. Susceptibility, known as a response function of the system, is a measure of how the magnetization changes as the external field H is changed. The zero-field susceptibility is

$$\chi = \left(\frac{\partial M}{\partial H}\right)_{T, H=0}. \tag{7.16}$$

At low temperatures, only the lowest energy levels are excited and a knowledge of the low-lying excitation spectrum enables one to determine the low temperature thermodynamics.

Exchange interaction can give rise to magnetic order below a critical temperature. However, for some spin systems, there is no magnetic order even at $T = 0$. The three major types of magnetic order are ferromagnetism (FM), antiferromagnetism (AFM) and ferrimagnetism (Fi). For FM order, the n.n. spins have a tendency towards parallel alignment; for AFM order, the n.n. spins tend to be antiparallel; whereas for Fi order, the n.n. spins favour antiparallel alignment but the associated magnetic moments have different magnitudes. In all the three cases, the magnetic order survives upto a critical temperature T_c. The order parameter in the case of an AFM is the spontaneous sublattice magnetization. Another measure of the spontaneous order is in the two-spin correlation function. Long range order (LRO) exists in the magnetic system if

$$\lim_{R \to \infty} \langle \mathbf{S}(0) \cdot \mathbf{S}(\mathbf{R}) \rangle \neq 0 \tag{7.17}$$

where $\mathbf{R}$ denotes the spatial location of the spin. At $T = 0$, the expectation value is in the ground state and at $T \neq 0$, the expectation value is the usual thermodynamic average.

The dynamical properties of a magnetic system are governed by the time-dependent pair correlation functions or their Fourier transforms. An important time-dependent correlation function is

$$G(R, t) = \langle \mathbf{S}_R(t) \cdot \mathbf{S}_0(0) \rangle. \tag{7.18}$$

The dynamical correlation function is the quantity measured in inelastic neutron scattering experiments. The differential scattering cross-section in such an experiment is given by

$$\frac{d^2\sigma}{d\Omega d\omega} \propto S^{\mu\mu}(\mathbf{q}, \omega)$$

7.2. Ground and excited states

$$= \frac{1}{N} \sum_R e^{i\mathbf{q}\cdot\mathbf{R}} \int_{-\infty}^{+\infty} dt\, e^{i\omega t} \langle S_R^\mu(t) S_0^\mu(0) \rangle \qquad (7.19)$$

where $\mathbf{q}$ and ω are the momentum wave vector and energy of the spin excitation. The spin components S^μ (μ = x, y, z) are perpendicular to $\mathbf{q}$. For a particular $\mathbf{q}$, the peak in $S^{\mu\mu}(\mathbf{q},\omega)$ occurs at a value of ω which gives the excitation energy. At $T = 0$,

$$S^{\mu\mu}(\mathbf{q},\omega) = \sum_\lambda M_\lambda^\mu \delta(\omega + E_g - E_\lambda). \qquad (7.20)$$

E_g (E_λ) is the energy of the ground (excited) state and

$$M_\lambda^\mu = 2\pi \, |\, \langle G \,|\, S^\mu(\mathbf{q}) \,|\, \lambda \rangle \,|^2 \,. \qquad (7.21)$$

Quantities of experimental interest include the dynamical correlation function, various relaxation functions and associated lineshapes.

7.2 Ground and excited states of the FM and AFM Heisenberg Hamiltonian

Consider the isotropic Heisenberg Hamiltonian

$$H = J \sum_{\langle ij \rangle} \mathbf{S}_i \cdot \mathbf{S}_j \qquad (7.22)$$

where $\langle ij \rangle$ denotes a n.n. pair of spins. The sign of the exchange interaction determines the favourable alignment of the n.n. spins. $J > 0$ ($J < 0$) corresponds to AFM (FM) exchange interaction. To see how exchange interaction leads to magnetic order, treat the spins as classical vectors. Each n.n. spin pair has an interaction energy $JS^2 \cos\theta$ where θ is the angle between n.n. spin orientations. When $J < 0$, the lowest energy is achieved when $\theta = 0$, i.e., the interacting spins are parallel. The ground state has all the spins parallel and the ground state energy $E_g = -JNzS^2/2$ where z is the coordination number of the lattice. When $J > 0$, the lowest energy is achieved when $\theta = \pi$, i.e., the n.n. spins are antiparallel. The ground state is the Néel state in which n.n.

spins are antiparallel to each other. The ground state energy $E_g = -JNzS^2/2$.

Magnetism, however, is a purely quantum phenomemon and the Hamiltonian in Eq.(7.22) has to be treated quantum mechanically rather than classically. For simplicity, consider the case of a chain of spins of magnitude 1/2. Periodic boundary condition is assumed, $i.e., \mathbf{S}_{N+1} = \mathbf{S}_1$. So

$$H = J \sum_{i=1}^{N} \mathbf{S}_i \cdot \mathbf{S}_{i+1} = J \sum_{i=1}^{N} \left[S_i^z S_{i+1}^z + \frac{1}{2} \left(S_i^+ S_{i+1}^- + S_i^- S_{i+1}^+ \right) \right] \tag{7.23}$$

where

$$S_i^\pm = S_i^x \pm i S_i^y \tag{7.24}$$

are the raising and lowering operators. For spins of magnitude $+\frac{1}{2}$, S^z has two possible values $1/2$ and $-1/2$ ($\hbar = 1$). The corresponding states are denoted as $|\alpha\rangle$ (up-spin) and $|\beta\rangle$ (down-spin) respectively. The spin algebra is given by

$$\begin{aligned}
S^z |\alpha\rangle &= \frac{1}{2} |\alpha\rangle \\
S^z |\beta\rangle &= -\frac{1}{2} |\beta\rangle \\
S^+ |\alpha\rangle &= 0 \\
S^+ |\beta\rangle &= |\alpha\rangle \\
S^- |\alpha\rangle &= |\beta\rangle \\
S^- |\beta\rangle &= 0 .
\end{aligned} \tag{7.25}$$

It is easy to check that in the case of a FM, the classical ground state is still the quantum mechanical ground state with the same ground state energy. However, the classical AFM ground state (the Néel state) is not the quantum mechanical ground state. The determination of the exact AFM ground state is a tough many body problem and the solution can be obtained with the help of the Bethe Ansatz technique (Sec. 4).

For a FM, the low-lying excitation spectrum can be determined exactly. Excitations are created by deviating spins from their

7.2. Ground and excited states

ground state arrangement. Deviate a single spin in the FM ground state $|\alpha\alpha\alpha\alpha......\alpha\alpha\alpha\alpha\rangle$. The number of such states is N as the deviated spin can be at any one of the N locations (sites). The actual eigenfunction Ψ is a linear combination of these states:

$$\Psi = \sum_{m=1}^{N} a(m)\psi(m) \qquad (7.26)$$

where m denotes the location of the deviated spin and runs from 1 to N. The wave function $\psi(m)$ has a down spin β at the mth site.

$$\psi(m) = |\alpha\alpha\alpha\alpha.....\beta\alpha\alpha....\alpha\alpha\rangle \;. \qquad (7.27)$$

To solve the eigenvalue problem

$$H\Psi = E\Psi , \qquad (7.28)$$

one has to determine the unknown coefficients $a(m)$ using

$$H \sum_m a(m)\psi(m) = E \sum_m a(m)\psi(m) \;. \qquad (7.29)$$

Multiply by $\psi^*(l)$ on both sides to get

$$\begin{aligned}
\sum_m a(m) \langle \psi(l)| H |\psi(m)\rangle &= E \sum_m a(m) \langle \psi(l)| \psi(m)\rangle \\
&= E \sum_m a(m)\delta_{ml} \\
&= Ea(l). \qquad (7.30)
\end{aligned}$$

On the left hand side (lhs), only those terms survive for which $H|\psi(m)\rangle = |\psi(l)\rangle$. Consider the FM version of the Heisenberg Hamiltonian in Eq.(7.23), i.e., replace J by $-J$ with $J > 0$. The Hamiltonian H is a sum of two parts -

$$H = H_z + H_{XY}. \qquad (7.31)$$

We consider the effect of each piece of the Hamiltonian on a n.n. spin pair. The following rules hold true. H_z acting on a parallel (antiparallel) spin pair gives $-J/4(J/4)$ times the same spin pair as shown below -

$$-JS_i^z S_{i+1}^z |\alpha\alpha\rangle = -\frac{J}{4}|\alpha\alpha\rangle, \quad -JS_i^z S_{i+1}^z |\alpha\beta\rangle = \frac{J}{4}|\alpha\beta\rangle. \quad (7.32)$$

H_{XY} acting on a parallel spin pair gives zero and acting on an antiparallel spin pair interchanges the spins with the coefficient $-J/2$ as shown below -

$$-\frac{J}{2}(S_i^+ S_{i+1}^- + S_i^- S_{i+1}^+)|\alpha\alpha\rangle = 0$$

$$-\frac{J}{2}(S_i^+ S_{i+1}^- + S_i^- S_{i+1}^+)|\alpha\beta\rangle = -\frac{J}{2}|\beta\alpha\rangle$$

$$-\frac{J}{2}(S_i^+ S_{i+1}^- + S_i^- S_{i+1}^+)|\beta\alpha\rangle = -\frac{J}{2}|\alpha\beta\rangle. \quad (7.33)$$

In the wave function $\psi(m)$, there are $(N-2)$ parallel and 2 antiparallel pairs of spins. Applying the rules of spin algebra, we get from (30),

$$Ea(l) = \left[-\frac{J}{4}(N-2) + \frac{J}{4} \times 2\right]a(l) - \frac{J}{2}[a(l+1) + a(l-1)],$$

$$\epsilon a(l) = \frac{J}{2}[2a(l) - a(l+1) - a(l-1)] \quad (7.34)$$

where $\epsilon = E + JN/4$ is the energy of the excited state measured with respect to the ground state energy $E_g = -JN/4$. The solution for $a(l)$ in Eq.(7.33) is

$$a(l) = e^{ikl} \quad (7.35)$$

and we get

$$\epsilon = J(1 - \cos(k)). \quad (7.36)$$

From the periodic boundary condition (PBC),

$$a(l) = a(l+N), \quad (7.37)$$

which leads to

$$e^{ikN} = 1 \quad (7.38)$$

or, $k = 2\pi\lambda/N, \lambda = 0, 1, 2, ..., N - 1$. Thus, the excited state consists of a deviated down spin propagating along the chain. This excitation is called a spin wave or a magnon. For r down spins ($r = 2, 3, 4, ...$, etc.), , we have r magnon excitations. The r magnons can scatter against each other giving rise to a continuum of scattering states or they can form bound states. The exact bound state spectrum can be derived using the Bethe ansatz (BA) (Sec. 4). The case of the AFM Heisenberg Hamiltonian will be considered in this Section.

The BA technique is applicable to quantum many body systems only in 1d. However, not all 1d quantum many body problems can be solved by the BA. For such problems as well as for models in higher dimensions, approximate techniques are used to gain knowledge of the ground state and the excitation spectrum [5, 6].

7.3 Theorems and rigorous results for Heisenberg AFMs

7.3.1 Lieb-Mattis theorem

For general spin and for all dimensions and also for a bipartite lattice, the entire eigenvalue spectrum satisfies the inequality

$$E_0(S) \leq E_0(S+1)$$

where $E_0(S)$ is the minimum energy corresponding to total spin S. The weak inequality becomes a strict inequality for a FM exchange coupling between spins of the same sublattice. The theorem is valid for any range of exchange coupling and the proof does not require PBC. The ground state of the $S = 1/2$ Heisenberg AFM with an even number N of spins is a singlet according to the Lieb-Mattis theorem [7].

7.3.2 Marshall's sign rule

The rule [6, 8] specifies the structure of the ground state of a n.n. $S = 1/2$ Heisenberg Hamiltonian defined on a bipartite lattice. The rule can be generalised to spin S, n.n.n. FM interaction but not to n.n.n. AFM interaction. A bipartite lattice is a lattice which can be divided into two sublattices A and B such that the n.n. spins of a spin belonging to the A sublattice are located in the B sublattice and vice versa. Examples of such lattices are the linear chain, the square and the cubic lattices. According to the sign rule, the ground state ψ has the form

$$|\psi\rangle = \sum_\mu C_\mu |\mu\rangle \qquad (7.39)$$

where $|\mu\rangle$ is an Ising basis state. The coefficient C_μ has the form

$$C_\mu = (-)^{p_\mu} a_\mu \qquad (7.40)$$

with a_μ real and ≥ 0 and p_μ is the number of up-spins in the A sublattice.

Proof: We write the AFM Heisenberg Hamiltonian in the form

$$\begin{aligned} H &= 2J \sum_{\langle ij \rangle} \left(\mathbf{S}_i \cdot \mathbf{S}_j - \frac{1}{4} \right) \\ &= 2J \sum_{\langle ij \rangle} \left(S_i^z S_j^z - \frac{1}{4} \right) + J \sum_{\langle ij \rangle} \left[S_i^+ S_j^- + S_i^- S_j^+ \right] \\ &= H_z + H_{XY}. \end{aligned} \qquad (7.41)$$

Let m_μ be the z-component of the total spin and q_μ the number of antiparallel bonds in the spin state $|\mu\rangle$. H_z acting on a parallel spin pair gives zero and acting on an antiparallel spin pair gives $-J$. H_{XY} acting on a parallel spin pair gives zero and interchanges the spins in the case of an antiparallel spin pair. In the latter case, m_μ and q_μ are unchanged and $p_\mu \to p_\mu \pm 1$.

$$H |\psi\rangle = J \sum_\mu C_\mu \{ -q_\mu |\mu\rangle + \sum_\nu |\nu\rangle \} \qquad (7.42)$$

7.3. Theorems and rigorous results for antiferromagnets

where $|\nu\rangle$ differs from $|\mu\rangle$ by one interchange of antiparallel nearest neighbours. There are q_μ such states whose energy is given by

$$\begin{aligned} E_c &= \frac{\langle \psi | H | \psi \rangle}{\langle \psi | \psi \rangle} \\ &= \frac{J \sum_\mu \{-q_\mu C_\mu^* C_\mu + \sum_\nu C_\nu^* C_\mu\}}{\sum_\mu |C_\mu|^2} \\ &= \frac{J \sum_\mu \sum_\nu C_\mu \left[-C_\mu^* + C_\nu^*\right]}{\sum_\mu |C_\mu|^2}. \end{aligned} \qquad (7.43)$$

Let

$$C_\mu = (-1)^{p_\mu} a_\mu \qquad (7.44)$$

where a_μ is arbitrary. Since $p_\nu = p_\mu \pm 1$, from Eq.(7.44),

$$\begin{aligned} C_\mu C_\nu^* &= (-1)^{2p_\mu \pm 1} a_\mu a_\nu^* \\ &= -a_\mu a_\nu^*. \end{aligned} \qquad (7.45)$$

Also, $C_\mu C_\mu^* = a_\mu a_\mu^*$. Hence, from Eq.(7.41),

$$E_c = -J \frac{\sum_\mu \sum_\nu a_\mu \left[a_\mu^* + a_\nu^*\right]}{\sum_\mu a_\mu^* a_\mu}. \qquad (7.46)$$

Now, since μ, ν are dummy indices

$$\sum_\mu \sum_\nu a_\mu a_\nu^* = \frac{1}{2} \sum_\mu \sum_\nu \left[a_\mu a_\nu^* + a_\mu^* a_\nu\right]. \qquad (7.47)$$

If a_μ, a_ν are real and positive, $a_\mu \to |a_\mu|$, $a_\nu \to |a_\nu|$ and

$$a_\mu a_\nu = |a_\mu| |a_\nu|. \qquad (7.48)$$

If a_μ, a_ν are complex, the term in Eq.(7.47) is

$$\begin{aligned} \frac{1}{2}\{|a_\mu| e^{i\phi_\mu} |a_\nu| e^{-i\phi_\nu} &+ |a_\mu| e^{-i\phi_\mu} |a_\nu| e^{i\phi_\nu}\} \\ &= |a_\mu| |a_\nu| \cos(\phi_\mu - \phi_\nu) \\ &\leq |a_\mu| |a_\nu|. \end{aligned} \qquad (7.49)$$

The energy E_c in Eq.(7.41) can be reduced by choosing a_μ's to be real and ≥ 0 (proved).

7.3.3 Lieb, Schultz and Mattis theorem

A half-integer S spin chain described by a reasonably local Hamiltonian respecting translational and rotational symmetry either has gapless excitation spectrum or has degenerate ground states, corresponding to spontaneously broken translational symmetry[6, 9].

Proof: We consider the Heisenberg AFM Hamiltonian in Eq.(7.22). A rigorous proof of uniqueness (non-degeneracy) of the ground state $|\psi_0\rangle$ exists in this case. We wish to prove that there is a low energy excitation of $O(1/L)$, where L is the length of the chain. The proof consists of two steps :
(i) Construct a state $|\psi_1\rangle$ which has low energy, i.e.,

$$\langle\psi_1|(H-E_0)|\psi_0\rangle = O(1/L). \tag{7.50}$$

(ii) Show that $|\psi_1\rangle$ is orthogonal to the ground state $|\psi_0\rangle$. The state $|\psi_1\rangle$ is constructed by applying an unitary transformation on $|\psi_0\rangle$ which gives rise to a slowly varying rotation of the spins about the z-axis.

We construct

$$\begin{aligned} |\psi_1\rangle &= U|\psi_0\rangle \\ \text{with} \quad U &= exp\left[i\frac{2\pi}{L}\sum_n nS_n^z\right]. \end{aligned} \tag{7.51}$$

The operator U rotates the nth spin by the small angle $2\pi n/L$ about the z-axis. The spin coordinates are rotated by 2π about the z-axis between the first and last sites. Note that the spin identities

$$\begin{aligned} e^{i\theta S^z}S^+e^{-i\theta S^z} &= e^{i\theta}S^+, \\ e^{i\theta S^z}S^-e^{-i\theta S^z} &= e^{-i\theta}S^- \end{aligned} \tag{7.52}$$

are satisfied. Now,

$$\langle\psi_1|(H-E_0)|\psi_1\rangle = \langle\psi_0|U^{-1}(H-E_0)U|\psi_0\rangle. \tag{7.53}$$

7.3. Theorems and rigorous results for antiferromagnets

The Hamiltonian H in Eq.(7.23) consists of two parts H_z and H_{XY}. H_z commutes with U. Therefore,

$$\begin{aligned}
\langle\psi_0|U^{-1}HU|\psi_0\rangle &= \langle\psi_0|H_z|\psi_0\rangle + \langle\psi_0|U^{-1}H_{XY}U|\psi_0\rangle \\
&= \langle\psi_0|H|\psi_0\rangle - \langle\psi_0|H_{XY}|\psi_0\rangle \\
&\quad + \langle\psi_0|U^{-1}H_{XY}U|\psi_0\rangle \\
&= E_0 - \{\frac{1}{2}\langle S_i^+ S_{i+1}^-\rangle + h.c.\} \\
&\quad + \langle\psi_0|U^{-1}H_{XY}U|\psi_0\rangle.
\end{aligned} \quad (7.54)$$

In the last term, making use of the spin identities, we get

$$\begin{aligned}
U^{-1}S_i^+ S_{i+1}^- U &= U^{-1}S_i^+ U U^{-1} S_{i+1}^- U \\
&= e^{-i\frac{2\pi}{L}i} S_i^+ e^{i\frac{2\pi}{L}(i+1)} S_{i+1}^- \\
&= e^{i\frac{2\pi}{L}} S_i^+ S_{i+1}^-.
\end{aligned} \quad (7.55)$$

So, from Eqs.(7.53 - 7.55),

$$\langle\psi_1|(H-E_0)|\psi_1\rangle = \frac{1}{2}\left[e^{i\frac{2\pi}{L}} - 1\right]\sum_i \langle\{S_i^+ S_{i+1}^-\rangle + h.c.\}. \quad (7.56)$$

Since the ground state is unique, it must be isotropic implying

$$\langle S_i^+ S_{i+1}^-\rangle = \langle S_i^- S_{i+1}^+\rangle. \quad (7.57)$$

Thus,

$$\langle\psi_1|(H-E_0)|\psi_1\rangle = \left[\cos(\frac{2\pi}{L}) - 1\right]\sum_i \langle S_i^+ S_{i+1}^-\rangle. \quad (7.58)$$

Since $S_i^+ S_{i+1}^-$ is a bounded operator,

$$\langle\psi_1|(H-E_0)|\psi_1\rangle = O(1/L). \quad (7.59)$$

In the limit $L \to \infty$, the right hand side (rhs) of Eq.(7.59) $\to 0$.

Constructing a low-energy state, however, proves nothing. The state may become equal to the ground state as $L \to \infty$. To complete the proof, we have to show that this does not happen.

This can be done by showing that $|\psi_1\rangle$ are $|\psi_0\rangle$ are orthogonal to each other. We will now show that $|\psi_1\rangle$ has momentum π relative to the ground state $|\psi_0\rangle$ which, being unique, is also a momentum eigenstate. Thus the two states are orthogonal to each other. Let T be the translation operator which produces a translation by one site -

$$TS_j T^{-1} = S_{j+1}, \quad TS_N T^{-1} = S_1 \qquad (7.60)$$

and
$$\begin{aligned} TUT^{-1} &= exp\left[i\frac{2\pi}{L}(\sum_{n=1}^{L-1} nS^z_{n+1} + LS^z_1)\right] \\ &= U exp\left[-i\frac{2\pi}{L}\sum_{n=1}^{L} S^z_n\right] e^{i2\pi S^z_1}. \qquad (7.61) \end{aligned}$$

The ground state has total spin zero since it is unique and so the first exponential has the value 1. Thus,

$$\begin{aligned} T|\psi_1\rangle &= TU|\psi_0\rangle = TUT^{-1}T|\psi_0\rangle \\ &= e^{i2\pi S^z_1} UT|\psi_0\rangle. \qquad (7.62) \end{aligned}$$

The exponential is -1 since S^z_1 has eigenvalues $\pm\frac{1}{2}$. So,

$$\begin{aligned} T|\psi_1\rangle &= -UT|\psi_0\rangle = e^{i(\pi+k_0)}U|\psi_0\rangle \\ &= e^{i(\pi+k_0)}|\psi_1\rangle, \qquad (7.63) \end{aligned}$$

where k_0 is the momentum wave vector of the ground state. The state $|\psi_1\rangle$ has momentum wave vector π relative to the ground state $|\psi_0\rangle$ and is thus orthogonal to it. This completes the proof of the Lieb-Scuhltz-Mattis (LSM) theorem for the $S = 1/2$ Heisenberg AFM Hamiltonian in 1d. The proof extends immediately to arbitrary half-odd integer S but fails in the integer S case. For the latter case, the exponential in Eq.(7.62) is $+1$ and it cannot be proved that $|\psi_1\rangle$ and $|\psi_0\rangle$ are orthogonal to each other. If the translational symmetry is spontaneously broken in the ground state, the ground state develops degeneracy. (Eq.(7.59) is still true but $|\psi_1\rangle$ is another ground state as $L \to \infty$). The excitation spectrum has a gap in this case.

7.3. Theorems and rigorous results for antiferromagnets

In the presence of a magnetic field, the $S = 1/2$ HAFM chain has gapless excitation spectrum from zero field upto a saturation field, at which the ground state becomes the fully polarized ferromagnetic ground state. For integer S, Haldane made a conjecture that the excitation spectrum has a gap. This conjecture has been verified both theoretically and experimentally [10]. In the presence of a magnetic field, the gap persists up to a critical field, equal to the gap. The spectrum is gapless from the critical field to a saturation field. The LSM theorem can be extended to the case of an applied magnetic field [11]. In this case, a striking phenomenon analogous to the quantum Hall effect occurs. The phenomenon is that of magnetization plateaus. The content of the extended theorem is that translationally invariant spin chains in an applied field can have a gapped excitation spectrum, without breaking translational symmetry, only when the magnetization/site m obeys the relation

$$S - m = integer \qquad (7.64)$$

where S is the magnitude of the spin. The proof is an easy extension of that of the LSM theorem. The gapped phases correspond to magnetization plateaus in the m vs. H curve at the quantized values of m which satisfy Eq.(7.64). Whenever there is a gap in the spin excitation spectrum, it is obvious that the magnetization cannot change by changing external field. Fractional quantization can also occur, if accompanied by (explicit or spontaneous) breaking of the translational symmetry. In this case, the plateau condition is given by

$$n(S - m) = integer \qquad (7.65)$$

where n is the period of the ground state. Hida [12] has considered a $S = 1/2$ HAFM chain with period 3 exchange coupling. A plateau in the magnetization curve occurs at $m=1/6$ (1/3 of full magnetization). In this case, $n = 3, S = 1/2$ and $m = 1/6$ and the quantization condition in Eq.(7.65) is obeyed.

7.3.4 Mermin-Wagner theorem

There cannot be any AFM LRO at finite T in dimensions d = 1 and 2 [6, 13]. AFM LRO can, however, exist in the ground states of spin models in d = 2. LRO exists in the ground state of the 3d HAFM model for spin $S \geq 1/2$ [14]. At finite T, the LRO persists upto a critical temperature T_c. For square [15] and hexagonal [16] lattices, LRO exists in the ground state for $S \geq 1$. The above results are based on rigorous proofs. No such proof exists as yet for $S=1/2$, d = 2 (this case is of interest because the CuO_2 plane of the high-T_c cuprate systems is a $S = 1/2$ 2d AFM).

7.4 Possible ground states and excitation spectra

The Néel state is not the ground state of the quantum Heisenberg Hamiltonian. The Néel state, we know, has perfect AFM LRO. It is still possible that the true ground state of a quantum Hamiltonian has Néel-type order, *i.e.*, non-zero sublattice magnetization. The ground state is essentially the Néel state with quantum fluctuations mixed in. This picture does indeed describe the ground state of the HAFM in d $\geq$ 2. The ground state of the HAFM in 1d has no LRO and is an example of a spin liquid . Even in higher dimensions, one can have disordered ground states due to what is known as frustration. Frustration occurs when the interaction energies associated with all the spin pairs cannot be simultaneously minimized. This may occur due to lattice topology or due to the presence of further-neighbour interactions. The triangular [17], Kagomé [18] and pyrochlore lattices [19] are well-known examples of frustrated lattices. Another recent example is the pentagonal lattice [20]. In these lattices, the number of bonds in each elementary plaquette is odd. We illustrate frustration by considering Ising spins on a triangular plaquette. Each Ising spin, sitting at a vertex of the triangle, can be either up or down. AFM exchange interaction energy is minimized if the n.n. spin pairs are antiparallel. As Fig.(7.1) shows, not all the three n.n. spin pairs can be

7.4. Possible ground states and excitation spectra

simultaneously antiparallel. The parallel spin pair may be located on any one of the bonds of the triangle. In fact, the ground state of the Ising AFM on the triangular lattice is highly degenerate. The entropy/spin has a non-zero value in the ground state and there is no ordering of the spins. If there is a mixture of FM and AFM interactions, then a spin model defined on an unfrustrated lattice like the square lattice (each elementary plaquette, a square, has four bonds) can also exhibit frustration, if the number of n.n. AFM interactions in a plaquette is odd. Frustration also occurs if both n.n. as well as n.n.n. interactions are present. Frustration favours disordered ground states but does not rule out AFM LRO in the ground state. The Ising AFM on the triangular lattice has a disordered ground state but various calculations suggest that the HAFM Hamiltonian on the triangular lattice has AFM LRO [21].

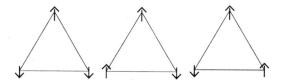

Figure 7.1: Ising spins on a triangular plaquette.

Various types of disordered ground states are possible: dimer, resonating valence bond (RVB), twisted, chiral and strip or collinear states. The lowest energy configuration of two spins interacting via the HAFM exchange interaction is a spin singlet with energy $-3J/4$. The singlet is often termed a valence bond (VB) or dimer (an object which occupies two sites). A pictorial representation of a VB between spins at sites i and j is a solid line joining the sites. The square lattice with N sites can be covered by $N/2$ VBs in various possible ways. Each possibility corresponds to a VB state. A RVB state is a coherent linear superposition of VB states (Fig.(7.2)). The 2d $S = 1/2$ AFM CaV_4O_9 has a lattice structure corresponding to a 1/5-depleted square lattice. The lattice consists of plaquettes of four spins connected by bonds. The ground state of the AFM is the Plaquette RVB (PRVB) state [22]. In this state, the spin configuration in each plaquette is a linear combination of two VB configurations. In one of these, the two singlet bonds are horizontal and in the other they are vertical. Bose and Ghosh [23] have constructed a spin model for which the PRVB state is the exact ground state. The triplet excitation is separated from the ground state by an energy gap and singlet excitations fall within the gap.

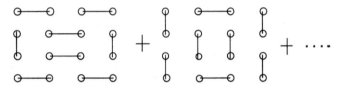

Figure 7.2: An example of a RVB state.

In a columnar dimer state, the dimers or VBs are arranged in columns. A description of the other disordered ground states can be obtained from Ref. [24]. The disordered ground states do not have conventional LRO but are not simple paramagnets. One can define suitable operators, the expectation values of which in the appropriate ground states are non-zero. The states, with different types of order, are described as quantum paramagnets. There is a huge literature on AFM spin models with disordered ground states. This is mainly because of the relevance of such studies for high temperature superconductivity. The common structural ingredient of the high-T_c cuprate systems is the CuO_2 plane. It looks like a square lattice with Cu^{2+} ions located at the lattice sites and oxygen ions sitting on the bonds in between. The Cu^{2+} ions carry spin-1/2 and the n.n. spins interact via the HAFM exchange interaction Hamiltonian. In the undoped state, the cuprates exhibit AFM LRO below the Néel temperature. On doping the systems with a few percent of holes, the AFM LRO is rapidly destroyed leaving behind a spin disordered state. It is important to understand the nature of the spin disordered state as this state serves as the reference state into which holes are introduced on dop-

7.4. Possible ground states and excitation spectra

ing. The holes are responsible for charge transport in the metallic state and form bound pairs in the superconducting state. We now discuss some spin models for which the ground state properties (energy and wave function as well as correlation functions) can be calculated exactly. The first model is described by the Majumdar-Ghosh Hamiltonian [1, 6]. This is defined in 1d for spins of magnitude 1/2. The Hamiltonian is AFM in nature and includes both n.n. as well as n.n.n. interactions. The strength of the latter is half that of the former.

$$H_{MG} = J \sum_{i=1}^{N} \mathbf{S}_i \cdot \mathbf{S}_{i+1} + J/2 \sum_{i=1}^{N} \mathbf{S}_i \cdot \mathbf{S}_{i+2} . \qquad (7.66)$$

The exact ground state of H_{MG} is doubly degenerate and the states are

$$\begin{aligned} \phi_1 &\equiv [12][34].......[N-1\,N] \\ \phi_2 &\equiv [23][45].......[N1] \end{aligned} \qquad (7.67)$$

where [lm] denotes a singlet spin configuration for spins located at the sites l and m. Also, PBC is assumed. One finds that translational symmetry is broken in the ground state. The proof that ϕ_1 and ϕ_2 are the exact ground states can be obtained by the method of 'divide and conquer'. One can verify that ϕ_1 and ϕ_2 are exact eigenstates of H_{MG} by applying the spin identity

$$\mathbf{S}_n \cdot (\mathbf{S}_l + \mathbf{S}_m)[lm] = 0. \qquad (7.68)$$

Let E_1 be the energy of ϕ_1 and ϕ_2. Let E_g be the exact ground state energy. Then

$$E_g \leq E_1 . \qquad (7.69)$$

One divides the Hamiltonian H into sub-Hamiltonians, H_i's, such that $H = \sum_i H_i$. H_i can be exactly diagonalised and let E_{i0} be the ground state energy. Let ψ_g be the exact ground state wave function. By variational theory,

$$E_g = \langle \psi_g | H | \psi_g \rangle = \sum_i \langle \psi_g | H_i | \psi_g \rangle \geq \sum_i E_{i0} . \qquad (7.70)$$

From Eqs.(7.69) and (7.70), one gets

$$\sum_i E_{i0} \leq E_g \leq E_1 . \qquad (7.71)$$

If one can show that $\sum_i E_{i0} = E_1$, then E_1 is the exact ground state energy. For the MG-chain, the sub-Hamiltonian H_i is

$$H_i = \frac{J}{2}(\mathbf{S}_i \cdot \mathbf{S}_{i+1} + \mathbf{S}_{i+1} \cdot \mathbf{S}_{i+2} + \mathbf{S}_{i+2} \cdot \mathbf{S}_i) . \qquad (7.72)$$

There are N such sub-Hamiltonians. One can easily verify that $E_{i0} = -3J/8$ and $E_1 = -3JN/8$ ($-3J/4$ is the energy of a singlet and there are $N/2$ VBs in ϕ_1 and ϕ_2). From Eq.(7.71), one finds that the lower and upper bounds of E_g are equal and hence ϕ_1 and ϕ_2 are the exact ground states with energy $E_1 = -3JN/8$. There is no LRO in the two-spin correlation function in the ground state:

$$K^2(i,j) = \langle S_i^z S_j^z \rangle = \frac{1}{4}\delta_{ij} - \frac{1}{8}\delta_{|i-j|,1} . \qquad (7.73)$$

The four-spin correlation function has off-diagonal LRO -

$$\begin{aligned} K^4(ij,lm) &= \langle S_i^x S_j^x S_l^y S_m^y \rangle \\ &= K^2(ij)K^2(lm) + \frac{1}{64}\delta_{|i-j|,1}\delta_{|l-m|,1} \\ &\times exp(i\pi(\frac{i+j}{2} - \frac{l+m}{2})) . \end{aligned} \qquad (7.74)$$

Let T be the translation operator for unit displacement. Then

$$T\phi_1 = \phi_2, \quad T\phi_2 = \phi_1 . \qquad (7.75)$$

The states

$$\phi^+ = \frac{1}{\sqrt{2}}(\phi_1 + \phi_2), \quad \phi^- = \frac{1}{\sqrt{2}}(\phi_1 - \phi_2) \qquad (7.76)$$

correspond to momentum wave vectors $k = 0$ and $k = \pi$. The excitation spectrum is not exactly known. Shastry and Sutherland [25] have derived the excitation spectrum in the basis of 'defect' states. A defect state has the wave function

7.4. Possible ground states and excitation spectra

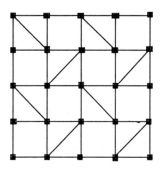

Figure 7.3: The Shastry-Sutherland model.

$$\psi(p, m) = \quad ....[2p - 3, 2p - 2]\alpha_{2p-1}[2p, 2p + 1]$$
$$....[2m - 2, 2m - 1]\alpha_{2m}[2m + 1, 2m + 2] \quad (7.77)$$

where the defects (α_{2p-1} and α_{2m}) separate two ground states. The two defects are up-spins and the total spin of the state is 1. Similarly, the defect spins can be in a singlet spin configuration so that the total spin of the state is zero. Because of PBC, the defects occur in pairs. A variational state can be constructed by taking a linear combination of the defect states. The excitation spectrum consists of a continuum with a lower edge at $J(5/2 - 2 \mid \cos k \mid)$. A bound state of the two defects can occur in a restricted region of momentum wave vectors. The MG chain has been studied for general values αJ of the n.n.n. interaction [26]. The ground state is known exactly only at the MG point $\alpha = 1/2$. The excitation spectrum is gapless for $0 < \alpha < \alpha_{cr}$ ($\simeq 0.2411$). Generalizations of the MG model to two dimensions exist [27,28]. The Shastry-Sutherland model [27] is defined on a square lattice and includes diagonal interactions as shown in Fig.(7.3). The n.n. and diagonal exchange interactions are of strength J_1 and J_2 respectively. For J_1/J_2 below a critical value ~ 0.7, the exact ground state consists of singlets along the diagonals. At the critical point, the ground state changes from the gapful disordered state to the AFM ordered gapless state. The compound $SrCu_2(BO_3)_2$ is well-described by the Shastry-Sutherland model [29]. Bose and Mitra [28] have constructed a J_1-J_2-J_3-J_4-J_5 spin-1/2 model on the square lattice. J_1, J_2, J_3, J_4 and J_5 are the strengths of the n.n., diagonal, n.n.n., knight's-move-distance-away and further-neighbour-diagonal exchange interactions (Fig.(7.4)). The four columnar dimer states (Fig.(7.5)) have been found to be the exact eigenstates of the spin Hamiltonian for the ratio of interaction strengths

$$J_1 : J_2 : J_3 : J_4 : J_5 = 1 : 1 : 1/2 : 1/2 : 1/4 \qquad (7.78)$$

It has not been possible as yet to prove that the four columnar dimer states are also the ground states. Using the method of 'divide and conquer', one can only prove that a single dimer state is the exact ground state with the dimer bonds of strength $7J$. The strengths of the other exchange interactions are as specified in Eq.(7.78).

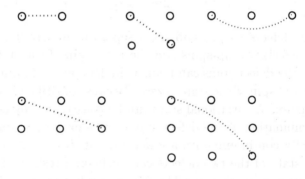

Figure 7.4: Five types of interactions in the $J_1 - J_2 - J_3 - J_4 - J_5$ model.

In Sec. 2, we have discussed the LSM theorem, the proof of which fails for integer spin chains. Haldane [30,10] in 1983 made the conjecture, based on a mapping of the HAFM Hamiltonian, in the long wavelength limit, onto the nonlinear σ model, that integer-spin HAFM chains have a gap in the excitation spectrum. The conjecture has now been verified both theoretically and experimentally [31]. In 1987, Affleck, Kennedy, Lieb and Tasaki

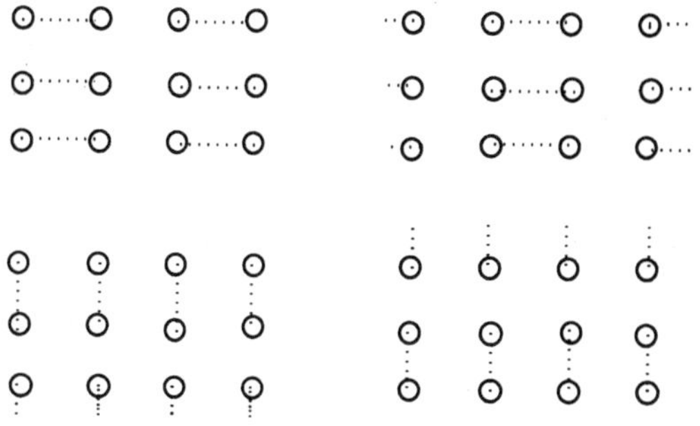

7.4. Possible ground states and excitation spectra

Figure 7.5: Four columnar dimer states.

(AKLT) constructed a spin-1 model in 1d for which the ground state could be determined exactly [32]. Consider a 1d lattice, each site of which is occupied by a spin-1. Each such spin can be considered to be a symmetric combination of two spin-1/2's. Thus, one can write down

$$\psi_{++} = |++\rangle, \quad S^z = +1$$
$$\psi_{--} = |--\rangle, \quad S^z = -1$$
$$\psi_{+-} = \frac{1}{\sqrt{2}}(|+-\rangle + |-+\rangle), \quad S^z = 0$$
$$\text{and } \psi_{-+} = \psi_{+-}, \quad (7.79)$$

where '+' ('−') denotes an up (down) spin.

AKLT constructed a valence bond solid (VBS) state in the following manner. In this state, each spin-1/2 component of a spin-1 forms a singlet (valence bond) with a spin-1/2 at a neighbouring site. Let $\epsilon^{\alpha\beta}$ ($\alpha, \beta = +$ or $-$) be the antisymmetric tensor:

$$\epsilon^{++} = \epsilon^{--} = 0, \quad \epsilon^{+-} = -\epsilon^{-+} = 1. \quad (7.80)$$

A singlet spin configuration can be expressed as $\epsilon^{\alpha\beta}|\alpha\beta\rangle/\sqrt{2}$, summation over repeated indices being implied. The VBS wave function (with PBC) can be written as

$$|\psi_{VBS}\rangle = 2^{-\frac{N}{2}} \psi_{\alpha_1\beta_1} \epsilon^{\beta_1\alpha_2} \psi_{\alpha_2\beta_2} \epsilon^{\beta_2\alpha_3} \ldots$$
$$\psi_{\alpha_i\beta_i} \epsilon^{\beta_i\alpha_{i+1}} \ldots \psi_{\alpha_N\beta_N} \epsilon^{\beta_N\alpha_1}. \quad (7.81)$$

$|\psi_{VBS}\rangle$ is a linear superposition of all configurations in which each $S^z = +1$ is followed by a $S^z = -1$ with an arbitrary number of $S^z = 0$ spins in between and vice versa. If one leaves out the zeroes, one gets a Néel-type of order. One can define a non-local string operator

$$\sigma_{ij}^\alpha = -S_i^\alpha exp(i\pi \sum_{l=i+1}^{j-1} S_l^\alpha) S_j^\alpha, \quad (\alpha = x, y, z) \tag{7.82}$$

and the order parameter

$$O_{string}^\alpha = \lim_{|i-j| \to \infty} \left\langle \sigma_{ij}^\alpha \right\rangle. \tag{7.83}$$

The VBS state has no conventional LRO but is characterised by a non-zero value 4/9 of O_{string}^α. After constructing the VBS state, AKLT determined the Hamiltonian for which the VBS state is the exact ground state. The Hamiltonian is

$$H_{AKLT} = \sum_i P_2(\mathbf{S}_i + \mathbf{S}_{i+1}) \tag{7.84}$$

where P_2 is the projection operator onto spin 2 for a pair of n.n. spins. The presence of a VB between each neighbouring pair implies that the total spin of each pair cannot be 2 (after two of the $S=1/2$ variables form a singlet, the remaining $S=1/2$'s could form either a triplet or a singlet). Thus, H_{AKLT} acting on $|\psi_{VBS}\rangle$ gives zero. Since H_{AKLT} is a sum over projection operators, the lowest possible eigenvalue is zero. Hence, $|\psi_{VBS}\rangle$ is the ground state of H_{AKLT} with eigenvalue zero. The AKLT ground state (the VBS state) is spin-disordered and the two-spin correlation function has an exponential decay. The total spin of two spin-1's is 2, 1 or 0. The projection operator onto spin j for a pair of n.n. spins has the general form

$$P_j(\mathbf{S}_i + \mathbf{S}_{i+1}) = \prod_{l \neq j} \frac{[l(l+1) - \mathbf{S}^2]}{[l(l+1) - j(j+1)]} \tag{7.85}$$

7.4. Possible ground states and excitation spectra

where $\mathbf{S} = \mathbf{S}_i + \mathbf{S}_{i+1}$. For the AKLT model, $j = 2$ and $l = 1, 0$. From Eqs.(7.84) and (7.85),

$$H_{AKLT} = \sum_i \left[\frac{1}{2}(\mathbf{S}_i \cdot \mathbf{S}_{i+1}) + \frac{1}{6}(\mathbf{S}_i \cdot \mathbf{S}_{i+1})^2 + \frac{1}{3}\right]. \quad (7.86)$$

The method of construction of the AKLT Hamiltonian can be extended to higher spins and to dimensions d >1. The MG Hamiltonian (apart from a numerical prefactor and a constant term) can be written as

$$H = \sum_i P_{3/2}(\mathbf{S}_i + \mathbf{S}_{i+1} + \mathbf{S}_{i+2}). \quad (7.87)$$

The $S = 1$ HAFM and the AKLT chains are in the same Haldane phase, characterised by a gap in the excitation spectrum. The physical picture provided by the VBS ground state of the AKLT Hamiltonian holds true for real systems [33]. The excitation spectrum of H_{AKLT} cannot be determined exactly. Arovas et al [34] have proposed a trial wave function

$$|k\rangle = N^{-\frac{1}{2}} \sum_{j=1}^{N} e^{ikj} S_j^\mu |\psi_{VBS}\rangle, \quad \mu = z, +, - \quad (7.88)$$

and obtained

$$\epsilon(k) = \frac{\langle k| H_{VBS} |k\rangle}{<k|k>} = \frac{25 + 15\cos(k)}{27}. \quad (7.89)$$

There exists a gap in the excitation spectrum $\Delta = 10/27$ at $k = \pi$. Another equivalent way of creating excitations is to replace a singlet bond by a triplet spin configuration [35].

We end this section with a brief mention of spin ladders. The subject of spin ladders is a rapidly growing area of research [36-38]. Ref. [38] gives an exhaustive overview of real systems with ladder-like structure and the results of various experiments on ladder compounds. A spin ladder consists of n chains ($n = 2, 3, 4, \ldots$, etc.,) coupled by rungs. The simplest spin ladder (Fig.(7.6)) corresponds to $n = 2$. Each site of the ladder is occupied by a

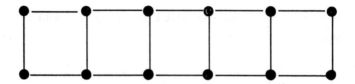

Figure 7.6: A two-chain spin ladder.

spin-1/2. Let J and J_R be the strengths of the intrachain n.n. and rung exchange interactions. When $J_R = 0$, the ladder decouples into two HAFM chains for which the excitation spectrum is known to be gapless. A gap opens up in the excitation spectrum even for an infinitesimal value of J_R and persists for all values of J_R. When $J_R \gg J$, the ground state predominantly consists of singlets along the rungs of the ladder. The ground state is spin-disordered. A triplet excitation is created by replacing a rung singlet by a triplet and letting it propagate. The n-chain spin ladder exhibits the 'odd-even' effect. The excitation spectrum is gapless (with gap) when n is odd (even). When n is odd, the spin-spin correlation function has a power-law decay as in the case of the HAFM $S = 1/2$ chain. For n even, the ground state correlation function has an exponential decay. For n sufficiently large, the square lattice limit is reached with AFM LRO in the ground state and a gapless excitation spectrum. Bose and Gayen [39] have constructed a two-chain spin ladder model which includes diagonal exchange interactions of strength J. When $J_R \gg 2J$, the exact ground state consists of singlets along the rungs. Later, Xian [39] has shown that the same exact ground state is obtained for $J_R/J > J_R/J_c \simeq 1.40148$. Ghosh and Bose [40] have generalised the two-chain ladder model to an n-chain model for which the 'odd-even' effect can be exactly demonstrated.

The spin ladder can be doped with holes. Two holes can form a bound state as in the case of high-T_c cuprate systems. The binding of two holes in a t-J ladder model has been shown through exact, analytic calculations [41]. A ladder compound has been discovered which shows superconductivity (maximum $T_c \sim 12K$)

on doping with holes and under pressure. In the superconducting state, the holes form bound pairs. Again, the analogy with the cuprate systems is strong. For detailed information on undoped and doped spin ladders, see the Refs. [36,38].

7.5 The Bethe Ansatz

The Bethe Ansatz (BA) was formulated by Bethe in 1931 [42] and describes a wave function with a particular kind of structure. Bethe considered a well-known model in magnetism, the spin-1/2 Heisenberg linear chain in which only n.n. spins interact. The wave function of the interacting many body system has the BA form. For ferromagnetic (FM) interactions, Bethe ,starting with the BA wave function derived the energy dispersion relations of spin wave (magnon) bound states exactly. Hulth/'en [43] used the BA method to derive an exact expression for the ground state energy of the antiferromagnetic (AFM) spin-1/2 Hamiltonian. In later years, it was realized that the method has a wider applicability and is not confined to magnetic spin chains. In fact, many of the exact solutions of interacting many body systems are but variations or generalizations of Bethe's method. Examples include the Fermi and Bose gas models in which particles on a line interact through delta function potentials [44], the Hubbard model in 1d [45], 1d plasma which crystallizes as a Wigner solid [46], the Lai-Sutherland model [47] which includes the Hubbard model and a dilute magnetic model as special cases, the Kondo model in 1d [48], the single impurity Anderson model in 1d [49], the supersymmetric t-J model ($J = 2t$) [50], etc. In the case of quantum models, the BA method is applicable only to 1d models. The BA method has also been applied to derive exact results for classical lattice statistical models in 2d. In the following we show how the BA method works by considering the spin-1/2 linear chain Heisenberg model mentioned in the beginning. The Hamiltonian for the linear chain is given by

$$H = -J \sum_{i=1}^{N} \mathbf{S}_i \cdot \mathbf{S}_{i+1}, \qquad (7.90)$$

describing the interaction of spins located at n.n. lattice sites. The total number of spins is given by N. The magnitude of the spin at each site is given by $1/2$. Periodic boundary condition (PBC) is assumed so that $N + 1 \equiv 1$. We consider the interaction to be ferromagnetic so that the exchange constant $J > 0$. The Hamiltonian describes magnetic insulators in which magnetic moments are localized on well-separated atoms. The Hamiltonian in Eq.(7.90) can be written as

$$H = -J \sum_{i=1}^{N} \left[S_i^z S_{i+1}^z + \frac{1}{2}(S_i^+ S_{i+1}^- + S_i^- S_{i+1}^+) \right]. \qquad (7.91)$$

There are two conserved quantities:

$$\begin{aligned} S^z &= \sum_{i=1}^{N} S_i^z, \quad [S^z, H] = 0 \\ \text{and } \mathbf{S} &= \sum_{i=1}^{N} \mathbf{S}_i, \quad [\mathbf{S}^2, H] = 0. \end{aligned} \qquad (7.92)$$

Since S^z is a conserved quantity, there is no mixing of subspaces corresponding to the different values of S^z, thus making it easier to solve the eigenvalue problem. Define the spin deviation operator

$$n = \frac{N}{2} + S^z. \qquad (7.93)$$

The FM ground state ψ_g corresponds to $n = 0$, i.e., $S^z = -N/2$ with all the spins pointing downwards:

$$\psi_g = \beta(1)\beta(2)\ldots\ldots\beta(N). \qquad (7.94)$$

The ground state energy is $E_g = -JN/4$.

7.5. The Bethe Ansatz

For $n = r$, we have r up-spins at the sites $m_1, m_2, \ldots m_r$. The eigenfunction ψ of H in the $n = r$ subspace is a linear combination of the ^NC_r functions $\psi(m_1, m_2, \ldots, m_r)$:

$$\psi = \sum_{\{m\}} a(m_1, m_2, \ldots, m_r) \psi(m_1, m_2, \ldots, m_r),$$
$$\{m\} = (m_1, m_2, \ldots, m_r) . \qquad (7.95)$$

The summation is over all the m_i's running over 1 to N subject to the condition

$$m_1 < m_2 < \ldots < m_r \qquad (7.96)$$

to avoid overcounting of states. We now consider the operation of the Hamiltonian H given in (92) on ψ. The rules of the spin algebra are:

(a) if two neighbouring spins are parallel, the term $S^+S^- + S^-S^+$ in H gives no contribution. The S^zS^z term on the other hand gives $-J/4$ times the same state $\psi(\{m\})$,

(b) for an antiparallel n.n. spin pair, the term $S^+S^- + S^-S^+$ interchanges the spins in $\psi(\{m\})$ with a multiplicative constant $-J/2$. The S^zS^z term gives a contribution $J/4$ multiplied by the same state $\psi(\{m\})$.

The eigenvalue equation is given by

$$H\psi = E\psi = E \sum_{\{m\}} a(\{m\})\psi(\{m\}). \qquad (7.97)$$

Take the scalar product with a particular $\psi^*(m_1, m_2, \ldots, m_r)$ and use orthogonality properties to get

$$Ea(m_1, m_2, \ldots, m_r) = -\frac{J}{2} \sum_{\{m'\}} a(\{m'\}) + \frac{J}{4} N' a(\{m\})$$
$$- \frac{J}{4}(N - N')a(\{m\}). \qquad (7.98)$$

N' is the total number of antiparallel spin pairs so that $(N - N')$ is the number of parallel spin pairs. The sum is over all N' distributions $(m'_1, m'_2, \ldots, m'_r)$ which arise from a n.n. exchange of antiparallel spins in $(m_1, m_2, ..., m_r)$. Eq.(7.98) can further be written as

$$(E + JN)a(m_1, m_2, ..., m_r)$$
$$= -\frac{J}{2} \sum_{\{m'\}} [a(m'_1, m'_2, ..., m'_r) - a(m_1, m_2, ..., m_r)] \quad (7.99)$$

or

$$2\epsilon a(m_1, m_2, ..., m_r)$$
$$= -\sum_{\{m'\}} [a(m'_1, m'_2, ..., m'_r) - a(m_1, m_2, ..., m_r)], \quad (7.100)$$

where

$$\epsilon = \frac{1}{J}(E + \frac{N}{4}J) \quad (7.101)$$

is the energy, in units of J, of the excited states with respect to that of the ground state energy E_g. Next, one considers the PBC from which one gets

$$a(m_1, ..., m_i, ..., m_r) = a(m_1, ..., m_r, m_i + N). \quad (7.102)$$

The spin waves of a FM system are obtained for $n = 1$ for which the equations are

$$\psi(m) = \sum_m a(m)\psi(m) \quad (7.103)$$

and $\quad 2\epsilon a(m) = -[a(m+1) + a(m-1) - 2a(m)]. \quad (7.104)$

The solution of Eq.(7.104) is given by

$$a(m) = e^{ikm}$$

and

$$\epsilon = 1 - \cos k . \quad (7.105)$$

From PBC, the allowed values of k are

$$k = \frac{2\pi\lambda}{N}, \quad \lambda = 0, 1, ..., N-1. \quad (7.106)$$

7.5. The Bethe Ansatz

There are N states. Consider now the case $n = 2$, i.e., there are two spin deviations. We have to take into account two cases. If the two spin deviations are not neighbours, one gets

$$\begin{aligned}2\epsilon a(m_1, m_2) &= -a(m_1 + 1, m_2) - a(m_1 - 1, m_2) - a(m_1, m_2 + 1) \\ &\quad - a(m_1, m_2 - 1) + 4a(m_1, m_2).\end{aligned} \quad (7.107)$$

If the spin deviations are neighbours, one gets

$$2\epsilon a(m_1, m_1+1) = -a(m_1-1, m_1+1) - a(m_1, m_1+2) + 2a(m_1, m_1+1) \quad (7.108)$$

and Eq.(7.107) is satisfied by the ansatz

$$a(m_1, m_2) = C_1 e^{i(k_1 m_1 + k_2 m_2)} + C_2 e^{i(k_2 m_1 + k_1 m_2)} . \quad (7.109)$$

Then

$$\epsilon = 1 - \cos k_1 + 1 - \cos k_2 \quad (7.110)$$

where C_1, C_2, k_1 and k_2 are to be determined. Eq.(7.108) can be satisfied by Eq.(7.109), if the coefficients C_1 and C_2 are chosen to make

$$a(m_1, m_1) + a(m_1 + 1, m_1 + 1) - 2a(m_1, m_1 + 1) = 0 . \quad (7.111)$$

For spin-1/2 particles, the amplitudes $a(m, m)$ have no physical significance but are defined by Eq.(7.111). Put Eq.(7.109) into Eq.(7.111) to get

$$\frac{C_1}{C_2} = \frac{\sin\frac{1}{2}(k_1 - k_2) + i\{\cos\frac{1}{2}(k_1 + k_2) - \cos\frac{1}{2}(k_1 - k_2)\}}{\sin\frac{1}{2}(k_1 - k_2) - i\{\cos\frac{1}{2}(k_1 + k_2) - \cos\frac{1}{2}(k_1 - k_2)\}} . \quad (7.112)$$

Let $C_1 = e^{i\frac{\phi}{2}}$ and $C_2 = e^{-i\frac{\phi}{2}}$. Then

$$2 \cot \frac{\phi}{2} = \cot \frac{k_1}{2} - \cot \frac{k_2}{2} . \quad (7.113)$$

Also, PBC gives

$$a(m_1, m_2) = a(m_2, m_1 + N) . \quad (7.114)$$

This is in accordance with the ordering of m_i's given in Eq.(7.96). From the PBC, one gets

$$Nk_1 - \phi = 2\pi\lambda_1,$$
$$Nk_2 + \phi = 2\pi\lambda_2$$
$$\text{and} \quad \lambda_1, \lambda_2 = 0, 1, 2, \ldots, (N-1). \quad (7.115)$$

We have three equations in Eqs.(7.113) and (7.115) for the three quantities ϕ_1, k_1, k_2. The sum of k_1 and k_2 is a constant of motion by translational symmetry.

$$k = k_1 + k_2 = \frac{2\pi}{N}(\lambda_1 + \lambda_2). \quad (7.116)$$

Since λ_1 and λ_2 can be interchanged without affecting the solution, we choose $\lambda_1 \leq \lambda_2$. For a given λ_1 and λ_2, the solutions are completely determined by Eqs.(7.113) and (7.115).

The following three cases are to be considered for determining the dependence of the wave numbers k_1 and k_2 on λ_1 and λ_2:
Case I : $\lambda_2 \geq \lambda_1 + 2$. Case II : $\lambda_1 = \lambda_2 = \frac{\lambda}{2}$, λ even.
Case III : $\lambda_1 = \lambda_2 - 1 = \frac{(\lambda-1)}{2}$, λ odd.
In Case I, for a given value of λ_2, λ_1 can take on the values $\lambda_1 = 0, 1, 2, \ldots, \lambda_2 - 2$. The total number of solutions is given by

$$\sum_{\lambda_2=2}^{N-1} (\lambda_2 - 1) = {}^{N-1}C_2. \quad (7.117)$$

In this case, all the wave numbers are real and for specified values of λ_1 and λ_2, k_1, k_2 and ϕ are determined uniquely from Eqs.(7.113) and (7.115). The energy eigenvalue ϵ_I is given by Eq.(7.110) with k_1, k_2 real. In Case II, from Eq.(7.115),

$$Nk_1 - \phi = \pi\lambda$$
$$\text{and} \quad Nk_2 + \phi = \pi\lambda. \quad (7.118)$$

We look for solutions of the form

$$k_1 = u + iv,$$
$$k_2 = u - iv,$$
$$\text{and} \quad \phi = \psi + i\chi. \quad (7.119)$$

7.5. The Bethe Ansatz

From Eqs.(7.118) and (7.119), we find that $\psi = 0$ and $\chi = Nv$. If v is non-zero, then for $N \to \infty$, χ also $\to \infty$. So

$$\cot \frac{\phi}{2} = \frac{\sin \psi - i \sinh \chi}{\cosh \chi - \cos \psi}$$
$$= -i(1 + 2e^{-\chi}).$$

To a first approximation,

$$2 \cot \frac{\phi}{2} = \cot \frac{k_1}{2} - \cot \frac{k_2}{2}$$
$$\Rightarrow -2i = \frac{\sin u - i \sinh v}{\cosh v - \cos u} - \frac{\sin u + i \sinh v}{\cosh v - \cos u}$$
$$\text{or,} \quad \sinh v = \cosh v - \cos u$$
$$\text{or,} \quad e^{-v} = \cos u . \qquad (7.120)$$

Hence,

$$\epsilon = 2 - \cos(u + iv) - \cos(u - iv)$$
$$= 2 - \cos u (\cos u + \frac{1}{\cos u})$$
$$= \frac{1}{2}(1 - \cos 2u) = \frac{1}{2}(1 - \cos k). \qquad (7.121)$$

From Eq.(7.120), $\cos u \geq 0$, so $-\pi/2 \leq u \leq \pi/2$ and k can be chosen to be in the range $-\pi \leq k \leq \pi$. Let us specify the energy obtained in Eq.(7.121), for complex values of k_1 and k_2, as ϵ_{II}. Let us now compare ϵ_I and ϵ_{II}. From Eq.(7.110),

$$\epsilon_I = 1 - \cos k_1 + 1 - \cos k_2$$
$$= 2 - 2 \cos \frac{1}{2}(k_1 - k_2) \cos \frac{1}{2}(k_1 + k_2)$$
$$= 2 - 2 \cos \frac{1}{2}(k_1 - k_2) \cos \frac{k}{2} . \qquad (7.122)$$

The value of k lies in the range $-\pi \leq k \leq \pi$. For each value of k, $k_1 - k_2$ lies in the range $0 \leq (k_1 - k_2) \leq 2\pi$. Thus if one plots ϵ_I against k, one gets a continuum of scattering states (Fig.(7.7)).

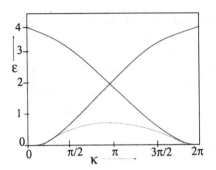

Figure 7.7: The continuum of scattering states (solid lines) and the bound state (dotted line) of two magnons.

Now, the minimum value of ϵ_I is

$$\epsilon_I^{min} = 2 - 2\cos\frac{k}{2} . \qquad (7.123)$$

From Eq.(7.121),

$$\epsilon_{II} = 1 - \cos^2\frac{k}{2} . \qquad (7.124)$$

Thus,

$$\frac{\epsilon_{II}}{\epsilon_I^{min}} = \frac{1}{2}(1 + \cos\frac{k}{2}) \leq 1 . \qquad (7.125)$$

We find that $\epsilon_{II} < \epsilon_I^{min}$ for $k \neq 0$. For $k = 0$, the bound state is degenerate with an eigenstate belonging to the continuum of scattering states and having the energy eigenvalue zero. Thus there are $N-1$ bound state solutions.

We now consider the wave functions in Cases I and II. For Case I, k_1, k_2 and ϕ are real and

$$a(m_1, m_2) = e^{i(k_1 m_1 + k_2 m_2 + \phi/2)} + e^{i(k_2 m_1 + k_1 m_2 - \phi/2)} . \qquad (7.126)$$

The state describes the scattering of spin waves. The 'phase shift' ϕ results from the mutual interaction between the spin waves. Since spin operators at two different sites commute, the excitations behave like bosons and the amplitude $a(m_1, m_2)$ is properly

7.5. The Bethe Ansatz

symmetrized to reflect the bosonic nature of the spin waves. In Case II, from Eq.(7.119) and with $\psi = 0$ and $\chi = Nv$, we get

$$\begin{aligned}
a(m_1, m_2) &= e^{i\phi/2} e^{i(k_1 m_1 + k_2 m_2)} + e^{-i\phi/2} e^{i(k_2 m_1 + k_1 m_2)} \\
&= e^{ik/2(m_1+m_2)} \left[e^{-Nv/2 - v(m_1 - m_2)} + e^{Nv/2 + v(m_1 - m_2)} \right] \\
&= 2 e^{i\frac{k}{2}(m_1+m_2)} \cosh v \left[\frac{N}{2} - (m_2 - m_1) \right].
\end{aligned} \quad (7.127)$$

If we normalise these states, we see that the two reversed spins tend to be localised at n.n. positions; $\mid a(m_1, m_2) \mid$ is a maximum for $m_2 = m_1 + 1$ and has an exponential decay for $m_2 > m_1 + 1$. The width of the bound state is given by v and u is the velocity of the centre of mass of the bound complex in the chain. The Case II solutions represent the bound states of two spin waves.

Finally, we consider Case III, i.e., $\lambda_1 = (\lambda - 1)/2$ and $\lambda_2 = (\lambda + 1)/2$ (λ odd). From Eq.(7.115), we get

$$\begin{aligned}
Nk_1 - \phi &= \pi(\lambda - 1) \\
\text{and } Nk_2 + \phi &= \pi(\lambda + 1).
\end{aligned} \quad (7.128)$$

A solution of Eq.(7.128) is $k_1 = k_2$ and $\phi = \pi$. Then from Eq.(7.126),

$$a(m_1, m_2) = e^{ik/2(m_1+m_2)} \left[exp(i\pi/2) + exp(-i\pi/2 \right] = 0. \quad (7.129)$$

For complex k_1, k_2 and ϕ, $\psi = \pi$ and $\chi = Nv$ from Eq.(7.119). For $N \to \infty$, we obtain the same solutions as we find in Case II.

For the general case $n = r$, Bethe proposed a form for the amplitude $a(\{m\})$ (the celebrated BA) which is a generalisation of the amplitude in the $n = 2$ case given by

$$a(m_1, m_2, ..., m_r) = \sum_P exp \left[i \sum_{j=1}^{r} k_{Pj} m_j + \frac{1}{2} i \sum_{j<l}^{1,r} \phi_{Pj,Pl} \right]. \quad (7.130)$$

The sum over P denotes a sum over all the $r!$ permutations of the integers $1, 2, ..., r$. Pj is the image of j under the permutation P. Each term in Eq.(7.130) has r plane waves scattering against one another. For each pair of plane waves, there is a phase shift $\phi_{j,l}$. The sum over permutations is in accordance with the bosonic nature of the waves, the spin waves, propagating along the chain. For r spin deviations several possibilities occur: none of the spin deviations occur at n.n. sites, only two of the spin deviations occur at n.n. sites, more than a single pair of spin deviations occur at n.n. sites, three spin deviations occur at n.n. sites and so on. However, only the first two possibilities need to be considered to solve the eigenvalue problem; the other possibilities do not give any new information [51]. For $r = 3$, let us now write down the amplitude $a(m_1, m_2, m_3)$ in the BA form given by Eq.(7.130). Since $r = 3$, the sum contains $3! = 6$ terms given by

$$\begin{aligned}
a(m_1, m_2, m_3) &= e^{i(k_1 m_1 + k_2 m_2 + k_3 m_3) + i/2(\phi_{12} + \phi_{13} + \phi_{23})} \\
&+ e^{i(k_1 m_1 + k_3 m_2 + k_2 m_3) + i/2(\phi_{12} + \phi_{13} + \phi_{32})} \\
&+ e^{i(k_2 m_1 + k_1 m_2 + k_3 m_3) + i/2(\phi_{21} + \phi_{13} + \phi_{23})} \\
&+ e^{i(k_3 m_1 + k_1 m_2 + k_2 m_3) + i/2(\phi_{12} + \phi_{31} + \phi_{32})} \\
&+ e^{i(k_2 m_1 + k_3 m_2 + k_1 m_3) + i/2(\phi_{21} + \phi_{31} + \phi_{23})} \\
&+ e^{i(k_3 m_1 + k_2 m_2 + k_1 m_3) + i/2(\phi_{21} + \phi_{31} + \phi_{32})}.
\end{aligned}$$
(7.131)

For the general case $n = r$, i.e., r up-spins, one can proceed as in the $n = 2$ case and obtain the following equations:

$$\epsilon = \sum_{i=1}^{r}(1 - \cos k_i) , \qquad (7.132)$$

$$2 \cot \frac{1}{2}\phi_{j.l} = \cot \frac{k_j}{2} - \cot \frac{k_l}{2}, -\pi \leq \phi_{j.l} \leq \pi \qquad (7.133)$$

$$\text{and } Nk_i = 2\pi \lambda_i + \sum_j \phi_{ij} . \qquad (7.134)$$

The Eqs. (7.133) are $r(r-1)/2$ in number. Since $\phi_{jl} = -\phi_{lj}$, there are only $r(r-1)/2$ distinct ϕ's. The Eqs. (7.134) are r in number. These r equations along with the $r(r-1)/2$ equations of the type

7.5. The Bethe Ansatz

in (7.133) give $r(r+1)/2$ equations in as many unknowns. Thus the equations can be expected to have solutions. Bethe showed that they also give the correct number of solutions.

We now consider a particular type of solution, in which the r deviated spins form a bound state and move together. The following simplified analysis has been given by Ovchinnikov [52]. One assumes that $r \ll N$ and that only the phases $\phi_{12}, \phi_{23}, \ldots, \phi_{r-1,r}$ are large. One may put $\text{Im}\phi_{l-1,l} > 0$ without any loss of generality. Then from Eq.(7.134),

$$\begin{aligned} \text{Im}\phi_{12} &= N\text{Im}k_1 = Nx_1, \\ \text{Im}(\phi_{23} - \phi_{12}) &= N\text{Im}k_2 = Nx_2 \\ \text{and } \text{Im}\phi_{r-1,r} &= Nx_r. \end{aligned} \quad (7.135)$$

One has $\sum_{i=1}^{r} x_i \geq 0$ for all r. Substituting in Eq.(7.133), one gets to an accuracy e^{-N} the following equation -

$$2\cot\frac{\phi_{12}}{2} = 2i\frac{e^{i(\text{Re}\phi_{12}+i\text{Im}\phi_{12})} + e^{-i(\text{Re}\phi_{12}+i\text{Im}\phi_{12})}}{e^{i(\text{Re}\phi_{12}+i\text{Im}\phi_{12})} - e^{-i(\text{Re}\phi_{12}+i\text{Im}\phi_{12})}} = -2i. \quad (7.136)$$

Thus,

$$\begin{aligned} -2i &= \cot\frac{k_1}{2} - \cot\frac{k_2}{2}, \\ -2i &= \cot\frac{k_2}{2} - \cot\frac{k_3}{2}, \quad \text{etc.}, \end{aligned} \quad (7.137)$$

So,

$$\cot\frac{k_l}{2} = 2i + \cot\frac{k_{l-1}}{2}. \quad (7.138)$$

The solution is given by

$$\cot\frac{k_l}{2} = 2il + C. \quad (7.139)$$

To determine C, introduce the total momentum of the deviated up-spins

$$k = \sum_{i=1}^{r} k_i = \frac{2\pi}{N}\sum_{i=1}^{r} \lambda_i$$

The wave function ψ is multiplied by e^{ik} due to a shift $m_i \to m_i+1$ and the energy levels are characterised by k. One gets

$$C = \frac{2ri - i(e^{ik} - 1)}{e^{ik} - 1},$$

$$e^{ikl} = \frac{r + l(e^{ik} - 1)}{r + (l-1)(e^{ik} - 1)}, \quad (7.140)$$

$$\text{and } \epsilon = \sum_{l=1}^{r}(1 - \cos k_l)$$

$$= \frac{1}{r}(1 - \cos k).$$

The results can be generalised to the Hamiltonian with longitudinal anisotropy given by

$$H(J,\sigma) = -J \sum_{i=1}^{N} \left[S_i^z S_{i+1}^z + \sigma(S_i^x S_{i+1}^x + S_i^y S_{i+1}^y) \right]. \quad (7.141)$$

The anisotropy parameter σ lies between 0 and 1. The multimagnon bound states were first detected in the quasi-one-dimensional magnetic system $CoCl_2.2H_2O$ at pumped helium temperature in high magnetic fields by far infrared spectroscopy [53]. Later improvements [54] made use of infrared HCN/DCN lasers, the high intensity of which made possible observation of even 14 magnon bound states.

We now describe the calculation of the ground state energy of the HAFM Hamiltonian using the BA. The sign of the exchange integral changes from $-J$ to $J (J > 0)$ and $\epsilon = (E - JN/4)/J$. The BA equations are still given by Eqs.(7.132)-(135); only there is an overall 'minus' sign on the rhs. of Eq.(7.132). In the AFM ground state, $N/2$ spins are up ($r = N/2$) and $N/2$ spins down. The numbers λ_i's are ordered as

$$0 < \lambda_1 \leq \lambda_2 \leq \lambda_3 \leq \ldots\ldots \quad (7.142)$$

Again, $\lambda_{j+1} \geq \lambda_j + 2$, for real k_i's (Case I). There is a unique choice of the λ_i's as

$$\lambda_1 = 1, \lambda_2 = 3, \lambda_3 = 5, \ldots, \lambda_{\frac{N}{2}} = N - 1. \quad (7.143)$$

7.5. The Bethe Ansatz

The ground state has total spin $S = 0$ and is non-degenerate. The total ground state wave vector is

$$P = \sum_{i=1}^{N/2} k_i = \frac{2\pi}{N}(1 + 3 + \ldots\ldots + (N-1)) = \frac{\pi}{2}N. \quad (7.144)$$

If $N = 4m$, m positive, the wave vector is 0 (mod 2π); if $N = 4m + 2$, it is π (mod 2π). The ground state energy is calculated by a passage to the continuum limit :

$$\lambda_j = \frac{2j-1}{N} \to x, \quad (7.145)$$

$$\frac{1}{N}\sum_{l=1}^{N/2} \phi_{jl} \to \frac{1}{2}\int_0^1 \phi(x,y)dy. \quad (7.146)$$

Eqs.(7.132) - (7.134) become

$$2\cot\frac{1}{2}\phi(x,y) = \cot\frac{1}{2}k(x) - \cot\frac{1}{2}k(y), \quad (7.147)$$

$$k(x) = 2\pi x + \frac{1}{2}\int_0^1 \phi(x,y)dy \quad (7.148)$$

$$\text{and } \epsilon = -\sum_{j=1}^{N/2}(1-\cos k_j)$$

$$= -\frac{N}{2}\int_0^1 (1-\cos k(x))dx. \quad (7.149)$$

We note that when $x = y$, ϕ jumps from π to $-\pi$ and

$$\frac{\partial \phi(x,y)}{\partial x} = -2\pi\delta(x-y) + 2\frac{d}{dx}(\cot^{-1}\frac{B}{A}), \quad (7.150)$$

where $A = 2$ and $B = \cot\frac{k(x)}{2} - \cot\frac{k(y)}{2}$. Thus from Eq.(7.148), we get

$$\frac{dk}{dx} = \pi + \mathrm{cosec}^2(\frac{k}{2})\frac{dk}{dx}\int_0^1 \frac{dy}{4 + \{\cot\frac{1}{2}k(x) - \cot\frac{1}{2}k(y)\}^2}. \quad (7.151)$$

Put
$$g = \cot \frac{1}{2}k \qquad (7.152)$$

and
$$\rho_0(g) = -\frac{dx}{dy}. \qquad (7.153)$$

Then from Eq.(7.151), we get the integral equation derived by Hulthén given by

$$\frac{2}{\pi(1+g^2)} = \rho_0(g) + \frac{2}{\pi}\int_{-\infty}^{+\infty} \frac{\rho_0(g')dg'}{(4+(g-g')^2)}. \qquad (7.154)$$

The integral equation can be solved by the method of iterations,

$$\rho_0(g) = \frac{2}{\pi(1+g^2)} - \lambda \int_{-\infty}^{+\infty} \frac{\rho_0(g')dg'}{(4+(g-g')^2)}, \qquad (7.155)$$

(where, at the end of the calculation λ is put equal to $2/\pi$). On iteration, Eq.(7.155) becomes

$$\rho_0(g) = \frac{2}{\pi(1+g^2)} - \lambda \int_{-\infty}^{+\infty} \frac{dg_1}{(g_1-g)^2+4}$$
$$\times \left[\frac{2}{\pi(1+g_1^2)} - \lambda \int_{-\infty}^{+\infty} \frac{\rho_0(g_2)dg_2}{(4+(g_2-g_1)^2)}\right]. (7.156)$$

One can continue to iterate Eq.(7.156) to generate higher order terms in λ.

$$\rho_0(g) = \frac{2}{\pi(1+g^2)} - \lambda\frac{2}{\pi}\int_{-\infty}^{+\infty} \frac{dg_1}{(4+(g_1-g)^2)(g_1^2+1)}$$
$$+ \lambda^2 \frac{2}{\pi}\int_{-\infty}^{+\infty} \frac{dg_1 dg_2}{(4+(g_1-g)^2)(4+(g_2-g_1)^2)(g_2^2+1)}$$
$$- \lambda^3 \int \dotsc\dotsc\dotsc \qquad (7.157)$$

One can calculate the integrals in Eq.(7.157) by the method of residues. For example,

$$\frac{2}{\pi}\int_{-\infty}^{+\infty} \frac{dg_1}{(4+(g_1-g)^2)(g_1^2+1)} = \frac{3}{g^2+9}. \qquad (7.158)$$

7.5. The Bethe Ansatz

Putting $\lambda = 2/\pi$, we finally obtain the solution of the integral Eq.(7.155) as

$$\rho_0(g) = \frac{2}{\pi}\left[\frac{1}{g^2+1} - \frac{3}{g^2+3^2} + \frac{5}{g^2+5^2} - \ldots\ldots\right]. \quad (7.159)$$

The series in Eq.(7.159) is uniformly convergent and one gets,

$$\rho_0(g) = \frac{1}{2}\text{sech}\frac{\pi g}{2}. \quad (7.160)$$

Thus,

$$\epsilon = -N\int_{-\infty}^{+\infty}\frac{\rho_0(g)}{1+g^2}dg = -N\ln 2. \quad (7.161)$$

In absolute measure, the ground state energy E_g is

$$E_g = \frac{NJ}{4} - JN\ln 2. \quad (7.162)$$

The low-lying excitation spectrum has been calculated by des Cloizeaux and Pearson (dCP) [55] by making appropriate changes in the distribution of $\lambda'_i s$ in the ground state. The spectrum is given by

$$\epsilon = \frac{\pi}{2}|\sin k|, -\pi \leq k \leq \pi \quad (7.163)$$

for spin 1 states. The wave vector k is measured with respect to that of the ground state. A more rigorous calculation of the low-lying excitation spectrum has been given by Faddeev and Takhtajan [56]. There are $S = 1$ as well as $S = 0$ states. We give a qualitative description of the excitation spectrum; for details Ref. [56] should be consulted. The energy of the low-lying excited states can be written as $E(k_1, k_2) = \epsilon(k_1) + \epsilon(k_2)$ with $\epsilon(k_i) = \pi \sin k_i/2$ and total momentum $k = k_1 + k_2$. At a fixed total momentum k, one gets a continuum of scattering states (Fig.(7.8)). The lower boundary of the continuum is given by the dCP spectrum (one of the $k'_i s = 0$). The upper boundary is obtained for $k_1 = k_2 = k/2$ and

$$\epsilon_k^U = \pi|\sin\frac{k}{2}|. \quad (7.164)$$

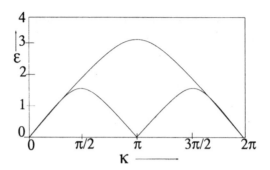

Figure 7.8: Continuum of scattering states for the $S = 1/2$ Heisenberg AFM chain.

The energy-momentum relations suggest that the low-lying spectrum is actually a combination of two elementary excitations known as spinons. The energies and the momenta of the spinons just add up, showing that they do not interact. A spinon is a $S = 1/2$ object, so on combination they give rise to both $S = 1$ and $S = 0$ states.

Figure 7.9: A two-spinon configuration in an AFM chain.

In the Heisenberg model, the spinons are only noninteracting in the thermodynamic limit $N \to \infty$. For an even number N of sites, the total spin is always an integer, so that the spins are always excited in pairs. The spinons can be visualised as kinks in the AFM order parameter (Fig.(7.9)). Due to the exchange interaction, the individual spinons get delocalized into plane wave states. Inelastic neutron scattering study of the linear chain $S = 1/2$ HAFM compound $KCuF_3$ has confirmed the existence of unbound spinon pair excitations [57]. In the case of integer spin chains, the

7.5. The Bethe Ansatz

spinons are bound and the excitation spectrum consists of spin-wave-like modes exhibiting the Haldane gap. The BA technique described in this Section is the one originally proposed by Bethe. There is an algebraic version of the BA which is in wide use and which gives the same final results as the earlier technique. For an introduction to the algebraic BA method, see the Refs. [58,59].

Bibliography

[1] C.K. Majumdar and D.K. Ghosh, J.Math. Phys. **10**, 1388 (1969); see also Ref. [6].

[2] F.D.M. Haldane, Phys. Rev. Lett. **60**, 635 (1988); B. S. Shastry, Phys. Rev. Lett. **60**, 639 (1988).

[3] E. Westerberg, A. Furusaki, M. Sigrist and P.A. Lee, Phys. Rev. Lett. **75**, 4302. (1995).

[4] S.M. Bhattacharjee and S.Mukherji, J.Phys. A: Math. Gen. **31**, L695 (1998).

[5] E. Manousakis, Rev. Mod. Phys. **63**, 1 (1991).

[6] *Interacting Electrons and Quantum Magnetism* by A. Auerbach (Springer-Verlag, New York 1994).

[7] E. Lieb and D.C. Mattis, J. Math. Phys. **3**, 749 (1962).

[8] W. Marshall, Proc. R. Soc. London Ser. **A 232**, 48 (1955).

[9] E. Lieb, T.D. Schultz and D.C. Mattis, Ann. Phys. **16**, 407 (1961).

[10] I. Affleck, J. Phys.: Condens. Matter **1**, 3047 (1989).

[11] M. Oshikawa, M.Yamanaka and I. Affleck, Phys. Rev. Lett. **78**, 1984 (1997).

[12] K. Hida, J.Phys. Soc. Jpn. **63**, 2359 (1994).

[13] N. D. Mermin and H. Wagner, Phys. Rev. Lett. **22**, 1133 (1966).

[14] F. Dyson, E.H. Lieb and B. Simon, J. Stat. Phys. **18**, 335 (1978); T. Kennedy, E.H. Lieb and B.S. Shastry, J. Stat. Phys. **53**, 1019 (1988).

[15] E.J. Neves and J.F. Perez, Phys. Lett. **114** A, 331 (1986).

[16] I. Affleck, T. Kennedy, E.H. Lieb and H. Tasaki, Commun. Math. Phys. **115**, 447 (1988).

[17] Th. Jolicoeur and J.C. Le Guillou, Phys. Rev. B **40**, 2727 (1989).

[18] C. Zeng and V. Elser, Phys. Rev. B **51**, 8318 (1995).

[19] B. Canals and C. Lacroix, Phys. Rev. Lett. **80**, 2933 (1998).

[20] U. Bhaumik and I. Bose, Phys. Rev. B **58**, 73 (1998).

[21] B. Bernu, P. Lecheminant, C. Lhuillier and L. Pierre, Phys. Rev. B **50**, 10048 (1994).

[22] K. Ueda, H. Kontani, M. Sigrist and P.A. Lee, Phys. Rev. Lett. **76**, 1932 (1996).

[23] I. Bose and A. Ghosh, Phys. Rev. B **56**, 3149 (1997).

[24] E. Dagotto, Int. J. Mod. Phys. B **5**, 907 (1991); I. Bose, Physica A 186, 298 (1992).

[25] B.S. Shastry and B. Sutherland, Phys. Rev. Lett. **47**, 964 (1981).

[26] S. Rao and D. Sen, Nucl. Phys. B **424**, 547 (1994); S. Rao and D. Sen, J. Phys. : Condens. Matter **9**, 1837 (1997); K. Nomura and K. Okamoto, J. Phys. A: Math. Gen. **27**, 5773 (1994).

[27] B.S. Shastry and B. Sutherland, Physica B **108**, 1069 (1981).

[28] I. Bose and P. Mitra, Phys. Rev. B **44**, 443 (1991); see also U. Bhaumik and I. Bose, Phys. Rev. B **52**, 12489 (1995).

[29] S. Miyahara and K. Ueda, Phys. Rev. Lett. **82**, 3701 (1999).

[30] F.D.M. Haldane, Phys. Rev. Lett. **50**, 1153 (1983); Phys. Lett. A **93**, 464 (1983).

[31] J.P. Renard et al, Europhys. Lett. **3**, 945 (1987); M. Date and K. Kindo, Phys. Rev. Lett. **65**, 1659 (1990).

[32] I. Affleck, T. Kennedy, E.H. Lieb and H. Tasaki, Phys. Rev. Lett. **59**, 799 (1987).

[33] M. Hagiwara et al, Phys. Rev. Lett. **65**, 3181 (1990).

[34] D.P. Arovas, A. Auerbach and F. D. M. Haldane, Phys. Rev. Lett. **60**, 531 (1988).

[35] G. Fáth and J. Sólyom, J. Phys.: Condens. Matter **5**, 8983 (1993).

[36] E. Dagotto and T.M. Rice, Science **271**, 618 (1996).

[37] T.M. Rice, Z.Phys.B 103, 165 (1997).

[38] E. Dagotto, Rep. Prog. Phys. **62**, 1525 (1999), cond-mat/9908250.

[39] I. Bose and S. Gayen, Phys. Rev. B **48**, 10653 (1993); Y.Xian, Phys. Rev. B **52**, 12485 (1995).

[40] A. Ghosh and I. Bose, Phys. Rev. B **55**, 3613 (1997).

[41] I. Bose and S. Gayen, J.Phys.: Condens. Matter **11**, 6427 (1999).

[42] H. Bethe, Z. Physik **71**, 205 (1931).

[43] L. Hulthén, Arkiv. Mat. Astron. Fys. **26**A, 11 (1938).

[44] E. Lieb and W. Liniger, Phys. Rev. **130**, 1605 (1963); E. Lieb, Phys. Rev. **130**, 1616 (1963); M.Gaudin, Phys. Lett. **24**A, 55 (1967); C.N. Yang, Phys. Rev. Lett. **19**, 1312 (1967); C.N. Yang and C.P. Yang, J.Math.Phys. **10**, 1115 (1969).

[45] E. Lieb and F.Y. Wu, Phys. Rev. Lett. **20**, 1445 (1968).

[46] B. Sutherland, Phys. Rev. Lett. **34**, 1083 (1975); ibid **35**, 185 (1975).

[47] B. Sutherland, Phys. Rev. B **12**, 3795 (1975).

[48] N. Andrei, Phys. Rev. Lett. **45**, 379 (1980).

[49] P.B. Wiegmann, J.Phys. C **14**, 1463 (1981); V.M. Filyov, A.M. Tsvelick and P.G. Wiegmann, Phys. Lett. **81** A, 175 (1981).

[50] P.A. Bares and G. Blatter, Phys. Rev. Lett. **64**, 2567 (1990).

[51] I. Bose in *Models and Techniques in Statistical Physics* ed. by S.M. Bhattacharjee (Narosa, New Delhi India 1997), p.144.

[52] A. Ovchinnikov, Zh. Eksperim. i Teor. Fiz. **56**, 1354 (1969).

[53] J.B. Torrance and M. Tinkham, Phys. Rev. **187**, 587 (1969); ibid **187**, 59 (1969).

[54] D.F. Nicoli and M. Tinkham, Phys. Rev. B **9**, 3126 (1974).

[55] J. des Cloizeaux and J.J. Pearson, Phys. Rev. **128**, 2131 (1962).

[56] L.D. Faddeev and L.A. Takhtajan, Phys. Lett. **85**A, 375 (1981); see also C.K. Majumdar in *Exactly Solvable Problems in Condensed Matter and Relativistic Field Theory* ed. by B.S. Shastry, S.S. Jha and V. Singh (Springer-Verlag, Berlin Heidelberg 1985).

[57] D.A. Tennant, T.G. Perring, R.A. Cowley and S.E. Nagler, Phys. Rev. Lett. **70**, 4003 (1993).

[58] L.A. Takhtajan in *Exactly Solvable Problems in Condensed Matter and Relativistic Field Theory* ed. by B.S. Shastry, S.S. Jha and V.Singh (Springer-Verlag, Berlin Heidelberg 1985).

[59] I. Bose in *Models and Techniques of Statistical Physics* ed. by S.M. Bhattacharjee (Narosa, New Delhi India 1997), p.156.